Peptide Transport and Hydrolysis

The Ciba Foundation for the promotion of international cooperation in medical and chemical research is a scientific and educational charity established by CIBA Limited – now CIBA-GEIGY Limited – of Basle. The Foundation operates independently in London under English trust law.

Ciba Foundation Symposia are published in collaboration with Elsevier Scientific Publishing Company / Excerpta Medica / North-Holland Publishing Company in Amsterdam

Elsevier / Excerpta Medica / North-Holland, P.O.Box 211, Amsterdam

Peptide Transport and Hydrolysis

Ciba Foundation Symposium 50 (new series)

1977

Elsevier · Excerpta Medica · North-Holland

Amsterdam · Oxford · New York

ISBN 0-444-15272-5

Published in August 1977 by Elsevier/Excerpta Medica/North-Holland, P.O. Box 211, Amsterdam and Elsevier North-Holland Inc., 52 Vanderbilt Avenue, New York, N.Y. 10017.

Suggested series entry for library catalogues: Ciba Foundation Symposia.
Suggested publisher's entry for library catalogues: Elsevier/Excerpta Medica/North-Holland

Ciba Foundation Symposium 50 (new series)

395 pages, 105 figures, 42 tables

Library of Congress Cataloging in Publication Data

Symposium on Peptide Transport and Hydrolysis, Ciba Foundation, 1976.
Peptide transport and hydrolysis.

(Ciba Foudation symposium; 50 (new ser.))
Includes bibliographies and index.

1. Peptides–Congresses. 2. Peptidase–Congresses. 3. Biological transport–Congresses. 4. Intestinal absorption–Congresses. I. Title. II. Series: Ciba Foundation. Symposium; 50 (new ser.) [DNLM: 1. Peptides–Metabolism–Congresses. 2. Hydrolysis–Congresses. W3 C161F v. 50 1976 / QU60 S992p 1976] QP552.P4S93 1976 574.1'9256 77-8378
ISBN 0-444-15272-5

Printed in The Netherlands by Van Gorcum, Assen

Contents

Participants

Symposium on Peptide Transport and Hydrolysis, held at the Ciba Foundation, London 9th–11th November, 1976

D. S. PARSONS (*Chairman*) Department of Biochemistry, University of Oxford, South Parks Road, Oxford OX1 3QU, UK

S. A. ADIBI Gastroenterology and Nutrition Unit, University of Pittsburgh School of Medicine, Montefiore Hospital, 3459 Fifth Avenue, Pittsburgh, Pa. 15213, USA

J. G. ALLEN Department of Biochemistry, Roche Products Limited, PO Box 8, 40 Broadwater Road, Welwyn Garden City, Hertfordshire AL7 3AY, UK

*S. J. BAKER Department of Biochemistry, Wellcome Research Unit, Christian Medical College Hospital, Vellore-632004, Tamil Nadu, India

D. BURSTON Department of Experimental Chemical Pathology, The Vincent Square Laboratories of Westminster Hospital, 124 Vauxhall Bridge Road, London SW1V 2RH, UK

P. DE LAEY 46 Chaussée de Bruxelles, 1490 Court-St-Etienne, Belgium

T-M. ENARI Biotechnical Laboratory, Technical Research Centre, Bulevardi 29-31, PO Box 192, SF-00121 Helsinki 12, Finland

P. F. FOTTRELL Department of Biochemistry, University College, Galway, Ireland

M. L. G. GARDNER Department of Biochemistry, University of Edinburgh Medical School, Teviot Place, Edinburgh EH8 9AG, UK

* *Present address:* Department of Medical Research, Rangoon, Burma

W. A. HEMMINGS ARC Immunology Group, Department of Zoology, The Brambell Laboratories, University College of North Wales, Bangor, Gwynedd LL57 2UW, UK

L. JOSEFSSON Department of Biochemistry C, University of Copenhagen, Panum Instituttet, 3 Blegdamsvej, DK-2200 Copenhagen N, Denmark

A. J. KENNY Department of Biochemistry, University of Leeds, 9 Hyde Park Terrace, Leeds LS2 9LS, UK

Y. S. KIM Gastrointestinal Research Laboratory, University of California Service, Veterans Administration Hospital, 4150 Clement Street, San Francisco, California 94121, USA

J. K. LLOYD Department of Child Health, St George's Hospital, Blackshaw Road, London SW17 0QT, UK

S. MAROUX CNRS, Centre de Biochimie et de Biologie Moléculaire, 31 Chemin Joseph-Aiguier, 13274 Marseilles Cedex 2, France

D. M. MATTHEWS Department of Experimental Chemical Pathology, The Vincent Square Laboratories of Westminster Hospital, 124 Vauxhall Bridge Road, London SW1V 2RH, UK

A. MEISTER Department of Biochemistry, Cornell University Medical College, 1300 York Avenue, New York, NY 10021, USA

O. MELLANDER Department of Medical Biochemistry, Faculty of Medicine, University of Göteborg, Fack S-400 33 Göteborg 33, Sweden

M. D. MILNE Medical Unit, Page Street Wing, Westminster Hospital, 17 Page Street, London SW1P 2AP

J. A. NICHOLSON Gastrointestinal Research Laboratory, University of California Service, Veterans Administration Hospital, 4150 Clement Street, San Francisco, California 94121, USA; (temporarily at: Cell Biology Unit, Royal Postgraduate Medical School, DuCane Road, London W12 0HS, UK)

O. NORÉN Department of Biochemistry C, University of Copenhagen, Panum Instituttet, 3 Blegdamsvej, DK-2200 Copenhagen N, Denmark

W. NÜTZENADEL Klinikum der Universität Heidelberg, Kinderklinik, Im Neuenheimer Feld 150, 6900 Heidelberg, West Germany

J. W. PAYNE Department of Botany, Science Laboratories, University of Durham, South Road, Durham DH1 3LE, UK

*A. N. RADHAKRISHNAN Department of Biochemistry, Wellcome Research Unit, Christian Medical College Hospital, Vellore-632004, Tamil Nadu, India

J. B. RHODES Department of Internal Medicine, Division of Gastroenterology, University of Kansas Medical Center, Rainbow Boulevard at 39th, Kansas City, Kansas 66103, USA

A. RUBINO Department of Paediatrics, Institute of Child Health, Università di Napoli, 2ª Facoltà di Medicina, Via S. Pansini 5, 80131 Naples, Italy

G. SEMENZA Laboratorium für Biochemie, Eidgenössische Technische Hochschule, Universitätstrasse 16, CH-8006 Zürich, Switzerland

D. B. A. SILK Liver Unit, King's College Hospital and Medical School, Denmark Hill, London SE5 9RS, UK

D. H. SMYTH MRC Unit for Metabolic Studies in Psychiatry, University Department of Psychiatry, Middlewood Hospital, PO Box 134, Sheffield S6 1TP, UK

T. SOPANEN Department of Biochemistry, University of Helsinki, Unioninkatu 35, 00170 Helsinki 17, Finland; (temporarily at: Department of Experimental Chemical Pathology, The Vincent Square Laboratories of Westminster Hospital, 124 Vauxhall Bridge Road, London SW1V 2RH, UK)

†A. M. UGOLEV Pavlov Institute of Physiology, USSR Academy of Sciences, Nab. Makarova 6, Leningrad 199164, USSR

Editors: KATHERINE ELLIOTT (*Organizer*) and MAEVE O'CONNOR

* *Present address:* School of Life Sciences, University of Hyderabad, Hyderabad-500001, India.

†Contributed *in absentia*

Chairman's opening remarks

D. S. PARSONS

Department of Biochemistry, University of Oxford

A major aim of the Ciba Foundation is to advance the study of chemistry, medicine and surgery and to promote international cooperation in medical, chemical and pharmaceutical research. The present symposium has been organized in accordance with this aim and those participating are drawn from widely differing areas of biological science. A previous symposium on Peptide Transport (Ciba Foundation 1972) brought together microbiologists and gastroenterologists and there is little doubt that subsequent work on the intestinal mucosa in this field has benefited from an appreciation of the work on bacteria. However, for meetings such as this to succeed, it is essential that everyone speaks the same language. It is a fact that groups of people sharing a common interest develop a private language or dialect for their own purposes of communication. To an outsider, the words used may have a meaning that can be quite different from the meaning in communication within the group. One example is the word 'biophysics'. To one group of scientists this could mean the study of the molecular structure of substances of biological importance by the use of the techniques of, say, X-ray diffraction. Yet, to another group, it could properly mean the study of electrical phenomena in excitable tissues. Also, different groups may give to similar objects different names. I have noticed that surgeons talk of the 'bowel' (of which they take 'specimens') while physicians and also a few physiologists, speak of the 'gut' (of which physicians take 'biopsies'). Other physiologists and also biochemists usually refer to the 'intestine' (of which they take 'samples'). It follows that one important function of these symposia is to facilitate the exchange of scientific concepts by clarifying linguistic inconsistencies and obscurities.

The title of the present symposium is Peptide Transport and Hydrolysis. The word *peptide* seems to have been used in English later than what can be called its derivatives, *dipeptide*, *tripeptide* and *polypeptide*. The earliest use of the term *peptide* appears to have been in 1906 in the *Journal of the Chemical Society*, in an abstract of a paper published in that year by Abderhalden & Terbuchi (1906*a*, *b*). It is interesting that the word is used without comment, as if it were already known and part of the vocabulary of certain groups. The term *polypeptide* was first printed in English in 1903, again in the *Journal of the Chemical Society* in an abstract of an article by Emil Fischer (1903*a*, *b*) who had originated the word. Note that before using *polypeptide*, he had previously coined *di*, *tri* and *tetrapeptide* (Fischer 1902).

The word *protein*, originally considered to be *the* essential constituent of animal and vegetable bodies, was first used in English by Watts in his translation of Gmelin's *Handbook of Chemistry* (1872). By the beginning of this century the nomenclature of nitrogenous macromolecules had become so confused (words like proteid, albuminose, proteoses being used in different senses in different parts of the world), that in 1906 a committee of the Physiological Society was appointed to 'consider and report on the question of proteid nomenclature.' After consultation with the Chemical Society, a Report prepared by Drs Halliburton and Hopkins (Physiological Society 1906) recommended, *inter alia*, that use of the word *proteid* be abolished and that the word *protein* be adopted as the general name of the whole group of substances under consideration.

In contrast, the use of the word *transport* as a verb (meaning to convey across; to carry, convey or remove from one place to another), goes back—according to the Oxford English Dictionary—to 1483 when it was used by Caxton in *The Golden Legend*. Incidentally, the first recorded use of the word *hydrolysis*, in the sense of a decomposition of a substance and which involved the fixation of the elements of water, seems to be that of Armstrong (1880).

The subject, peptide transport and hydrolysis, can be considered to be a branch of the science of nutrition, for peptides are nutrients not only of animals but also of plants and microorganisms. We must also remember that peptide transport systems may be involved in the uptake by cells of substances other than those derived from protein digestion. The folates represent one example and, in the field of therapeutics, some antibiotics are peptide in nature.

During our deliberations we shall be considering peptide transport across the plasma membranes of animal and vegetable cells, including bacteria, and also into and across epithelia—notably those of the intestinal tract. The energetic aspects of the membrane transport of peptides are of great importance. Clearly if the peptide is moved from the environment, through the membrane, to be deposited in the cell interior at a higher concentration (or electrochemical potential if the peptide carries a significant charge) than outside, work will have been done to sustain the transport and a source of power must be available.

On the other hand, if the peptide undergoes some form of decomposition during the phase of membrane transport and the products of the decomposition become distributed asymmetrically as on either side of the membrane, say largely on the inside, then the energetics of the transport process will be quite different. For example, is the free energy of hydrolysis available for the forward propulsion of the products of decomposition? Accordingly, when considering the uptake of peptides by epithelia such as the intestine and kidney, the localization of the site of hydrolysis, if any, has important energetic implications.

As far as my own subject, intestinal absorption, is concerned, the older physiologists were well aware of the role of the intestinal epithelium in digestion of protein as well as carbohydrate, as becomes clear if the article on the 'Chemistry of the digestive processes', written by Benjamin Moore (1898) in Schafer's *Textbook of Physiology* is consulted. It is also evident that A. E. Garrod (1909, 1923) recognized that certain aspects of the excretion of N in the urine of cystinurics raised questions as to the form in which N was absorbed from the intestinal tract. It seems to have been the fact that free amino acids were found in the intestinal lumen during the course of protein digestion and that there was an associated increase in the plasma amino-N that gave rise to the dogma that all protein is absorbed into the epithelium in the form of the constituent amino acids. If there are a number of alternative, non-exclusive hypotheses that could account for a phenomenon, then dreadful mistakes can be made if evidence in support of one hypothesis is taken as sufficient to exclude the others, on the grounds, for example, of economy of hypotheses. The concept of 'fail-safe' systems is not confined to engineering.

I am grateful to Mr A. M. Hughes, Science Editor of the Oxford English Dictionary Supplement, for telling me about the early history of the words *peptide* and *polypeptide*.

References

ABDERHALDEN, E. & TERBUCHI, Y. (1906*a*) Ueber den Abbau einiger Aminosäuren und Peptide im Organismus des Hundes. *Z. Physiol. Chem. 47*, 159-172

ABDERHALDEN, E. & TERBUCHI, Y. (1906*b*) Fate of certain amino acids and peptides in the animal body. *J. Chem. Soc. 90*, Pt. 2, 293 (Abstr.)

ARMSTRONG, H. E. (1880) *Introduction to the Study of Organic Chemistry*, 2nd edn, p. 190*n*, Longmans, Green, London

CIBA FOUNDATION (1972) *Peptide Transport in Bacteria and Mammalian Gut (Ciba Found. Symp. 4)*, Associated Scientific Publishers, Amsterdam

FISCHER, E. (1902) Ueber die Hydrolyse der Proteinstoffe. *Chem. Zeitung. 26*, 939-940

FISCHER, E. (1903*a*) Synthese von Derivaten der Polypeptide. *Sber. Preuss. Akad. Wiss. (Berlin)*, Pt. 1, 387-400

FISCHER, E. (1903*b*) Synthesis of derivatives of polypeptides. *J. Chem. Soc. 84*, Pt. 1, 466 (Abstr.)
GARROD, A. E. (1909) *Inborn Errors of Metabolism*, p. 80, Oxford University Press/Hodder & Stoughton, London
GARROD, A. E. (1923) *Inborn Errors of Metabolism*, 2nd edn, pp. 132-133, Oxford University Press/Hodder & Stoughton, London
GMELIN, L. (1872) *Handbook of Chemistry* (Watts, H., trans.), vol. 18, p. 232, Works of the Cavendish Society, London
MOORE, B. (1898) Chemistry of the digestive processes, in *Textbook of Physiology* (Schafer, E.A., ed.), vol. 1, pp. 312-474, Young J. Pentland, Edinburgh
PHYSIOLOGICAL SOCIETY (1906) *Proteid Nomenclature 25*, xvii-xx

Introduction

D. M. MATTHEWS

Department of Experimental Chemical Pathology,
Vincent Square Laboratories of Westminster Hospital, London

Abstract Carrier-mediated membrane transport of small peptides is now realized to be a process of wide biological distribution, occurring not only in the small intestine and elsewhere in the animal body but also in bacteria, yeast, the mould *Neurospora crassa*, and probably in higher plants during the germination of seeds. The important features of peptide transport are outlined, and possible relationships between peptide transport and hydrolysis are discussed. Peptide transport is a stereochemically specific active process, and is independent of the transport of free amino acids. It is frequently, though not always, more rapid than the transport of amino acids. In the intestine, it is probably limited to dipeptides and tripeptides, but certain other animal cells and bacteria can take up larger peptides of seven or more amino acid residues. The ability to take up small peptides on a large scale is nutritionally important in some microorganisms, and might be of nutritional importance to the intact animal and to animal cells in culture. In the absorptive cells of the small intestine, and in *Escherichia coli*, peptide transport into the cells is followed by intracellular hydrolysis; transport and hydrolysis are quite distinct processes. Whether hydrolysis and amino acid transport can be coupled processes, or whether peptide transport and hydrolysis are different aspects of the same process, remains to be seen. This question is one of those where a close integration of studies of peptide transport with those of peptide hydrolysis should be particularly helpful.

> *'For the physiology of the late nineteenth century, peptones were important because they diffused through membranes...'*
>
> (FRUTON 1972)

When a Ciba Foundation Symposium on *Peptide Transport in Bacteria and Mammalian Gut* was held in 1971 (Ciba Foundation 1972), the concept of carrier-mediated membrane transport of small peptides by mechanisms independent of those for free amino acids was already quite familiar to investigators of transport mechanisms in bacteria (Payne & Gilvarg 1971), but the existence of such transport in mammalian small intestine was not yet widely recognized. In the last five years, it has become clear that carrier-mediated membrane transport of small peptides is a process of very wide biological distribution. It occurs not only in the intestine of many species (see Matthews 1975*a,b;* Adibi 1976; Matthews & Adibi 1976) but elsewhere in the animal body, for example in kidney, liver, skeletal muscle, brain and erythrocytes; indeed, it is possible that the ability to take up small peptides on a substantial scale is a property of many types of animal cell, including malignant cells (see Matthews & Payne 1975*a,b;* Adibi, this volume; Burston *et al.*, this volume). This might explain some, though not all, of the effects of peptides on the growth of animal cells in tissue culture (Matthews & Payne 1975*a*). Peptide transport also occurs in bacteria, in yeast (see Payne 1975*a*, 1976), in the mould *Neurospora crassa* (Wolfinbarger & Marzluf 1974) and in the scutellum of germinating barley (Burston *et al.*, this volume). Thus there is a need for workers on peptide transport in different organs and in different kinds of organism to come together to compare results and ideas. In the last few years, investigations of the hydrolysis of small peptides in the intestinal mucosa and in microorganisms have yielded much new information (Lindberg *et al.*, 1975; Payne 1975*b*; Peters 1975; Kim 1977), yet investigators of peptide hydrolases have worked in comparative isolation, and need to discuss their findings with each other and with investigators of peptide transport. These are the reasons for holding the present symposium on *Peptide Transport and Hydrolysis.*

In this introductory paper, I intend to outline some of the important features of membrane transport of small peptides, and to discuss possible relationships between peptide transport and hydrolysis. The account of peptide transport will deal almost exclusively with cellular uptake of peptides of no more than a few amino acid residues, which appears to be the result of a carrier-mediated active process. It should be recognized, however, that membrane transport of a peptide may involve one or more of several mechanisms, depending on the molecular size and structural features of the peptide (Matthews 1975*a,b*). Very large peptides may well utilize the mechanisms for membrane transport of whole proteins. Lipid-soluble peptides may be expected to diffuse through membrane lipid. Small α-linked peptides of the type resulting from protein digestion will utilize the high-capacity carrier-mediated transport mechanism already referred to, undergoing transmembrane transport on a massive scale. Small water-soluble peptides whose structural features make them unsuitable for carrier-mediated transport may possibly be transported, on a limited scale, by diffusion through aqueous areas in the membrane. Unfortunately, we still know virtually nothing about the mechanisms of membrane transport of peptides which are apparently unsuitable for carrier-mediated transport owing to their size or structural features. To take an example: what is the mechanism of intestinal absorption of a tripeptide such as thyroliberin (thyrotropin-releasing factor), pyroGlu-His-Pro-NH_2? This peptide is known to be absorbed from human intestine and, unlike most small peptides, enters the blood intact, though only on a small scale. It is actively secreted by the renal tubules, but owing to the substitution of the terminal amino and carboxyl groups it appears to have little or no affinity for carrier-mediated uptake by the absorptive cells (see Matthews 1975*a,b*; Matthews & Payne 1975*b*). Does it in fact have enough affinity for the carrier-mediated uptake mechanism to enter the absorptive cells by this means, or is its absorption the result of diffusion through 'aqueous pores' in the cell membranes, or via an intercellular route? If it does enter the absorptive cells by a carrier-mediated mechanism, by what mechanism does it leave them? At the moment, these questions cannot be answered. There is a multitude of biologically active peptides, including hormones, kinins, toxins, antibiotics and peptides, which influence the central nervous system (see Matthews & Payne 1975*b*). In many cases such peptides are known to be capable of crossing biological membranes, but very little is known of how they do this. At least one biologically active peptide, the antibiotic cephalexin, appears to share the carrier-mediated transport mechanism used by small peptides of dietary origin (see Matthews 1975*b*). Investigation of the mechanisms of membrane transport of peptides of medium size and of unusual structure, especially biologically active peptides, could hardly fail to be rewarding.

CARRIER-MEDIATED TRANSPORT OF SMALL PEPTIDES

Carrier-mediated transport of small peptides has been most extensively studied in bacteria, especially *Escherichia coli* (see Payne 1975*a*, 1976), and in mammalian small intestine (see Matthews 1975*a,b*; Adibi 1976; Matthews & Adibi 1976). There are many resemblances between peptide transport by the intestinal absorptive cells and by bacteria, but it is now apparent that there are also certain interesting differences.

Among the resemblances are these. In both intestine and bacteria, peptide transport is an active process, as shown by its energy-dependence and the ability of the cells to concentrate peptides which are exceptionally resistant to hydrolysis, such as glycylsarcosine. In bacteria, the ability to concentrate intact peptides may also be shown in circumstances in which intracellular peptidase activity is reduced by experimental manipulation, or a peptidase is lacking owing to mutation. In the intestine, peptide transport is Na^+-dependent and may well be coupled to that of Na^+, as with amino acids. In bacteria, the question of whether peptide transport is coupled to that of an ion such as H^+ or Na^+ has not yet been investigated. In both intestine and bacteria, the carriers responsible for transport of peptides and those for transport of free amino acids appear to be completely independent. The main evidence for this comes from the results of experiments on competition for transport, which have shown that peptides compete with one another for transport, but not with free amino acids, and from the study of genetic transport defects. In bacterial mutants, the ability to take up peptides or to take up free amino acids may be lost independently. In man, the ability of the absorptive cells to take up certain groups of free amino acids may be lost, as in Hartnup disorder (neutral amino acids) or cystinuria (dibasic amino acids and cystine), while the ability to take up peptides, including those containing or made up of the affected amino acids, is retained. No genetic defect of intestinal peptide uptake has yet been described, but there is little doubt that when it is, it will be found that the ability to take up amino acids is unimpaired. There are several resemblances between the molecular structural requirements for peptide uptake in the intestine and in *E. coli*. In the intestine, substitution of the terminal α-NH_2 or α-COOH groups reduces or abolishes affinity for transport. The same is true of the dipeptide uptake system of *E. coli*, though not of the oligopeptide transport system (see below) which requires a free α-NH_2-terminal group but has no requirement for the α-COOH-terminal group. In both intestine and bacteria, peptide transport is stereochemically specific, preferring peptides made up of L-amino acid residues and, in both, the peptide bond may be methylated (as in glycylsarcosine and glycylsarcosylsarcosine) without abolishing affinity for peptide transport. In both, it seems that α-linked peptides are preferred substrates, and γ-linked peptides have little or no affinity for transport. The influence of the amino acid side chains, in particular the influence of the charge on these side chains, on peptide transport has not been adequately investigated in either intestine or bacteria—though it is probable that, as with amino acids, large lipophilic side chains increase affinity for transport. In *E. coli*, a wide variety of peptides apparently compete for transport, and the peptide transport systems have been described as indifferent to the nature of the side chains—but the existence of subsidiary transport systems, determined by the nature of these chains, has not been excluded. In the small intestine, it has been shown that very many dipeptides and tripeptides containing or made up of various combinations of neutral, basic and acidic amino acid residues appear to share the same transport system, but here again the existence of subsidiary systems has not been ruled out (see Burston *et al.*, this volume). There are theoretical reasons for supposing that the number of peptide transport systems in any biological membrane is likely to be fairly limited, and that if there are in fact multiple systems, each will favour a particular class of peptide rather than some individual peptide. There might conceivably be a special transport system for each of the 20 amino acids occurring in proteins—but hardly a special system for each of the 400 dipeptides or the 8000 tripeptides which can be formed from these amino acids.

The main differences which have so far been observed between peptide transport by the small intestine and peptide transport by *E. coli* concern the size of peptide that can be transported and the existence, in the bacterium, of distinct dipeptide and oligopeptide transport systems. In *E. coli*, the ability to take up peptides appears to be limited by molecular volume, probably owing to a 'sieving' effect of the pores in the peripheral cell wall which surrounds the plasma membrane, but it extends to peptides of up to six or seven amino acid residues. In the intestine, on the other hand, the ability to take up peptides on a large scale seems to be limited to much smaller peptides than this. Indeed, an investigation using the series glycylsarcosine, glycylsarcosylsarcosine and glycylsarcosylsarcosylsarcosine (Addison *et al.* 1975*a,b*; Burston *et al.*, this volume) indicated that in hamster jejunum the dipeptides and tripeptides were taken up by an active mechanism, but that the tetrapeptide was not taken up and had no measurable affinity for the uptake mechanism used by the dipeptide and tripeptide. The conclusion that only dipeptides and tripeptides are likely to enter the absorptive cells fits well with the observation by Kim *et al.* (1974) that the peptide hydrolases of the cytosol, in contrast to those of the brush border, hydrolyse dipeptides and tripeptides but not tetrapeptides or higher peptides. In *E. coli*, there is an uptake system specific for dipeptides, and also an oligopeptide uptake system requiring a free terminal α-NH_2 group but with no requirement for a terminal α-COOH group; dipeptides can utilize both their own system and the oligopeptide system, but oligopeptides cannot utilize the dipeptide system. In the intestine, however, there is no evidence for the existence of distinct systems of this type. Neither the dipeptide glycylsarcosine nor the tripeptide glycylsarcosylsarcosine can utilize a transport system unavailable to the other (Sleisenger *et al.* 1976; Burston *et al.*, this volume).

Peptide transport on a substantial scale is not necessarily limited to very small peptides in all animal cells. Grahl-Nielsen *et al.* (1974) have reported that cultures of cell line RPM1 No. 2402, originating in a carcinoma of the small intestine of the hamster, can take up a series of homopeptides of Lys from di-Lys to deca-Lys, hydrolysing them intracellularly. Another difference between these malignant cells and normal absorptive cells was that the rate of uptake (as judged by growth response) was inversely proportional to peptide size.

SPECIAL FEATURES OF PEPTIDE TRANSPORT

Since peptides are transported intact, competition for transport between the constituent amino acids is naturally avoided. This is an obvious but important feature of peptide transport. The rapidity with which peptides may be transported, in comparison to the transport rates of the equivalent free amino acids, is a particularly striking feature of peptide transport in both small intestine and bacteria. Peptides may be taken up over 10 times more rapidly than their constituent amino acids in the free form (though they are not always transported more rapidly than free amino acids and may, in some circumstances, be transported more slowly). The rapidity of peptide transport is not accounted for by the absence of competition for transport between the constituent amino acids, for these amino acids may be transported in peptide form not only more rapidly than when present in the equivalent mixture, but more rapidly than when present alone. The kinetic characteristics of transport of peptides and transport of their constituent amino acids are different, and it has been suggested that in the intestine peptide transport is characterized by higher V_{max} values than amino acid transport (Adibi & Soleimanpour 1974). In the intestine, the relative rapidity of peptide transport in relation to that of the constituent amino acids seems, in most cases, to become progressively greater as the concentration is raised. Amino acids which are particularly slowly absorbed from a mixture of free amino acids, such as glycine, lysine and the dicarboxylic amino acids, may be taken up much more rapidly in peptide form. It has been suggested that in animals this feature of intestinal absorption might be of nutritional importance, leading to more effective protein synthesis than that occurring on diets in which the protein component is replaced by a mixture of free amino acids (Matthews & Payne 1975*a*). In several species of bacterium, including *E. coli*, it is established that the ability to take up peptides is of major nutritional importance. This is why 'peptones' figure so prominently in the composition of many bacteriological media and why, for some years, there was thought to be a vitamin-like peptide growth factor, 'strepogenin' (Payne 1975*a*, 1976). The coexistence of amino acid and peptide transport systems in both intestine and microorganisms means that their total uptake capacity for protein digestion products is much greater than their uptake capacity for free amino acids would suggest.

Certain theoretical suggestions concerning peptide transport should be mentioned, even though there is as yet little relevant experimental evidence. Peptide transport is likely to be economical in its requirement for metabolic energy, since theoretically it should require only about the same energy expenditure for a molecule containing several amino acids to be transported in peptide form as for a molecule of one amino acid to be transported in the free form. With peptides which undergo rapid intracellular hydrolysis, as most probably do, a concentration gradient may be maintained between the exterior and the interior of the cell, peptides entering down this gradient (Payne & Gilvarg 1971; Smyth 1972). This would enable peptide entry to take place by facilitated diffusion, even in the absence of active transport. It has also been suggested that the free energy released by hydrolysis of the peptide bond, which is fairly large, might be utilized in driving transport (Parsons 1972; Payne 1976). A further suggestion is that exit of amino acids from a cell, resulting from intracellular hydrolysis of peptides which that cell has taken up, might enable it to accumulate free amino acids by a process of counter-transport. Stimulation of uptake of [^{14}C]glycine by glycylglycine has been observed in streptococci (Brock & Wooley 1964), and more recently glycylleucine and leucylglycine have been reported to stimulate uptake of tritiated glycine and leucine by isolated intestinal vesicles of the brush border membrane (Sigrist-Nelson 1975).

RELATIONSHIPS BETWEEN PEPTIDE TRANSPORT AND HYDROLYSIS

There is a close spatial relationship between the transport and hydrolysis of small peptides. In *E. coli*, peptide transport into the cells is followed by intracellular hydrolysis; extracellular, periplasmic and plasma membrane-bound peptidase activity is absent, as in many other bacterial species. Peptide transport into the absorptive cells, followed by intracellular hydrolysis, also takes place in the small intestine, but here the situation is complicated by a second process: hydrolysis of peptides in the brush border and intestinal lumen, followed by uptake of free amino acids by the usual carrier systems. The existence of the latter process, the importance of which varies from one peptide to another, partly obscures the characteristic features of peptide transport. Thus, during the absorption of a peptide which is readily hydrolysed by the brush border peptidases, competition for transport between the constituent amino acids is not completely avoided, and intestinal absorption of amino acids A and B from a mixed peptide AB is usually unequal, not equal as it would be if uptake of intact peptide were the only transport mechanism involved. Furthermore, absorption of amino acids from such a peptide may be reduced in the presence of another free amino acid, owing to competition for uptake between this amino acid and the amino acids released from the peptide in the brush border—even though the process of peptide uptake is unaffected. The experimental use of peptides which are exceptionally resistant to hydrolysis (such as carnosine (β-alanylhistidine), glycylsarcosine and glycylsarcosylsarcosine), so that their absorption is almost exclusively the result of peptide

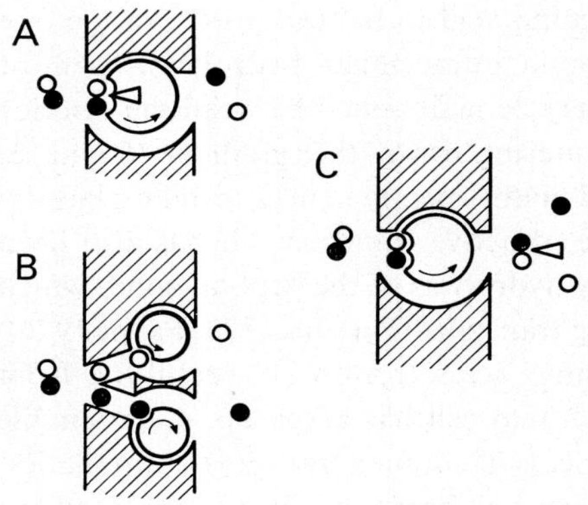

FIG. 1. Three schemes of peptide absorption and hydrolysis. A hydrolase is represented by a wedge and a carrier by a rotating circle. The intestinal lumen is on the left and the interior of cell on the right of each diagram. A: intramembrane hydrolysis: a hydrolase acts also as an amino acid carrier, hydrolysis of dimer and transport of monomers being different aspects of same process. B: intramembrane hydrolysis: peptides are hydrolysed and the resultant amino acids located on amino acid carriers inaccessible to free amino acids in the intestinal lumen. C: intracellular hydrolysis: peptide transport is followed by hydrolysis deep to the transport mechanism, in the interior of the cell. Redrawn from Matthews (1975*b*).

transport, has enabled the special features of peptide transport, including its independence of transport of free amino acids, to be demonstrated particularly clearly.

The use of peptides resistant to hydrolysis, which are accumulated intact by the intestinal wall *in vitro*, has also shown unequivocally that intestinal peptide transport and peptide hydrolysis may be dissociated. Similar evidence has enabled peptide transport and hydrolysis to be dissociated in bacteria. In addition, studies of bacterial mutants have shown that the ability to hydrolyse a peptide may be lost without loss of transport ability, or peptide transport ability may be lost without loss of hydrolytic ability. The fact that in both small intestine and bacteria transmural transport of peptides and hydrolysis of peptides can be shown to be distinct processes is of considerable theoretical importance. The suggestion has often been made that transport and hydrolysis might be linked in some way, as is apparently the case in the intestinal absorption of disaccharides (Ramaswamy *et al.* 1974). Some possible schemes of transport and hydrolysis are shown in Fig. 1. In Scheme A (Rubino *et al.* 1971) the hydrolase also acts as an amino acid carrier, taking up peptides on the outer (luminal) side of the plasma membrane and releasing amino acids on the inner side. In Scheme B (Ugolev 1972; Matthews 1972) peptides are hydrolysed on the luminal side of the membrane and the resulting amino acids are located on amino acid carriers inaccessible to free amino acids. In Scheme C (Smyth 1972; Matthews 1972) a distinct peptide transport step is followed by intracellular hydrolysis. The pros and cons of these schemes have been discussed at length elsewhere (Matthews 1975*a*,*b*). All that need be said here is that Scheme C is the only scheme which will account without difficulty for all the phenomena associated with peptide transport in the intestine and in bacteria and has some positive evidence in its favour. Whether transport/hydrolysis (A) or 'coupled transport' (B) do in fact occur remains to be seen. In further investigations of this question, a closer integration of studies of peptide transport with those of peptide hydrolysis should be particularly helpful.

References

ADDISON, J. M., BURSTON, D., PAYNE, J. W., WILKINSON, S. & MATTHEWS, D. M. (1975*a*) Evidence for active transport of tripeptides by hamster jejunum *in vitro*. *Clin. Sci. Mol. Med. 49*, 305-312

ADDISON, J. M., BURSTON, D., DALRYMPLE, J. A., MATTHEWS, D. M., PAYNE, J. W., SLEISENGER, M. H. & WILKINSON, S. (1975*b*) A common mechanism for transport of di- and tripeptides by hamster jejunum *in vitro*. *Clin. Sci. Mol. Med. 49*, 313-322

ADIBI, S. A. (1976) Intestinal phase of protein assimilation in man. *Am. J. Clin. Nutr. 29*, 205-215

ADIBI, S. A. (1977) Clearance of dipeptides from plasma: role of kidney and intestine, in this volume, pp. 265-280

ADIBI, S. A. & SOLEIMANPOUR, M. R. (1974) Functional characterization of dipeptide transport system in human jejunum. *J. Clin. Invest. 53*, 1368-1374

BROCK, T. D. & WOOLEY, S. O. (1964) Glycylglycine uptake in streptococci and a possible role of peptides in amino acid transport. *Arch. Biochem. Biophys. 105*, 51-57

BURSTON, D., MARRS, T. C., SLEISENGER, M. H., SOPANEN, T. & MATTHEWS, D. M. (1977) Mechanisms of peptide transport, in this volume, pp. 79-98

CIBA FOUNDATION (1972) *Peptide Transport in Bacteria and Mammalian Gut (Ciba Found. Symp. 4)*, Associated Scientific Publishers, Amsterdam

FRUTON, J. S. (1972) *Molecules and Life*, pp. 115, Wiley-Interscience, New York

GRAHL-NIELSEN, O., ØDEGAARD, P. & TRITSCH, G. L. (1974) Oligopeptides as sources of indispensable amino acids for mammalian cells in culture. *In Vitro 9*, 414-420

KIM, Y. S. (1977) Intestinal mucosal peptidases. *Physiol. Rev.* in press.

KIM, Y. S., KIM, Y. W. & SLEISENGER, M. H. (1974) Specificities of peptide hydrolases in brush border and cytosol fractions of rat small intestine. *Biochim. Biophys. Acta 370*, 283-296

LINDBERG, T., NORÉN, O. & SJÖSTRÖM, H. (1975) Peptidases of the intestinal mucosa, in *Peptide Transport in Protein Nutrition* (Matthews, D. M. & Payne, J. W., eds.), pp. 204-242, North-Holland, Amsterdam

MATTHEWS, D. M. (1972) Rates of peptide uptake by small intestine, in *Peptide Transport in Bacteria and Mammalian Gut (Ciba Found. Symp. 4)*, pp. 71-88, Associated Scientific Publishers, Amsterdam

MATTHEWS, D. M. (1975*a*) Absorption of peptides by mammalian intestine, in *Peptide Transport in Protein Nutrition* (Matthews, D. M. & Payne, J. W., eds.), pp. 61-146, North-Holland, Amsterdam

MATTHEWS, D. M. (1975*b*) Intestinal absorption of peptides. *Physiol. Rev. 55*, 537-608

MATTHEWS, D. M. & ADIBI, S. A. (1976) Peptide absorption. *Gastroenterology 71*, 151-161

MATTHEWS, D. M. & PAYNE, J. W. (1975*a*) Peptides in the nutrition of microorganisms and peptides in relation to animal nutrition, in *Peptide Transport in Protein Nutrition* (Matthews, D. M. & Payne, J. W., eds.), pp. 1-60, North-Holland, Amsterdam

MATTHEWS, D. M. & PAYNE, J. W. (1975*b*) Occurrence and biological activities of peptides, in *Peptide Transport in Protein Nutrition* (Matthews, D. M. & Payne, J. W., eds.), pp. 392-463, North-Holland, Amsterdam

PARSONS, D. S. (1972) Summary, in *Transport Across the Intestine* (Burland, W. L. & Samuel, P. D., eds.), pp. 253-275, Churchill Livingstone, Edinburgh

PAYNE, J. W. (1975*a*) Transport of peptides in microorganisms, in *Peptide Transport in Protein Nutrition* (Matthews, D. M. & Payne, J. W., eds.), pp. 283-364, North-Holland, Amsterdam

PAYNE, J. W. (1975*b*) Microbial peptidohydrolases, in *Peptide Transport in Protein Nutrition* (Matthews, D. M. & Payne, J. W., eds.), pp. 365-391, North-Holland, Amsterdam

PAYNE, J. W. (1976) Peptides and microorganisms, in *Advances in Microbial Physiology*, vol. 13 (Rose, A. H. & Tempest, D. W., eds.), pp. 55-113, Academic Press, London

PAYNE, J. W. & GILVARG, C. (1971) Peptide transport, in *Advances in Enzymology*, vol. 35 (Meister, A., ed.), pp. 187-244, Wiley, New York

PETERS, T. J. (1975) The subcellular localisation of intestinal peptide hydrolases, in *Peptide Transport in Protein Nutrition* (Matthews, D. M. & Payne, J. W., eds.), pp. 243-267, North-Holland, Amsterdam

RAMASWAMY, K., MALATHI, P., CASPARY, W. F. & CRANE, R. K. (1974) Studies on the transport of glucose and disaccharides by hamster small intestine *in vitro*. II. Characteristics of the disaccharidase-related transport system. *Biochim. Biophys. Acta 345*, 39-48

RUBINO, A., FIELD, M. & SHWACHMAN, H. (1971) Intestinal transport of amino acid residues of dipeptides. I. Influx of the glycine residue of glycyl-L-proline across mucosal border. *J. Biol. Chem. 246*, 3542-3548

SIGRIST-NELSON, K. (1975) Dipeptide transport in isolated intestinal brush border membrane. *Biochim. Biophys. Acta 394*, 220-226

SLEISENGER, M. H., BURSTON, D., DALRYMPLE, J. A., WILKINSON, S. & MATTHEWS, D. M. (1976) Evidence for a single common carrier for uptake of a dipeptide and a tripeptide by hamster jejunum *in vitro*. *Gastroenterology 71*, 76-81

SMYTH, D. H. (1972) Peptide transport by mammalian gut, in *Peptide Transport in Bacteria and Mammalian Gut (Ciba Found. Symp. 4)*, pp. 59-66, Associated Scientific Publishers, Amsterdam

UGOLEV, A. M. (1972) Membrane digestion and peptide transport, in *Peptide Transport in Bacteria and Mammalian Gut (Ciba Found. Symp. 4)*, pp. 123-137, Associated Scientific Publishers, Amsterdam

WOLFINBARGER, L. & MARZLUF, G. A. (1974) Peptide utilization by amino acid auxotrophs of *Neurospora crassa*. *J. Bacteriol. 119*, 371-378

Amino acid and peptide absorption in man

D. B. A. SILK*

Medical Unit, Department of Gastroenterology, St Bartholomew's Hospital, London, and Gastrointestinal Research Unit, Veterans Administration Hospital, Department of Medicine, University of California, San Francisco

Abstract An intestinal perfusion technique has been used in man to study absorption of glycine and L-alanine from the dipeptide glycyl-L-alanine and the tripeptide L-alanylglycylglycine. Glycine was absorbed faster from both peptides than from their equivalent free amino acid mixtures, and in the dipeptide experiments the differential in absorption rate of glycine and L-alanine observed when the equivalent mixtures of free amino acids were perfused was not observed during the perfusions with glycyl-L-alanine. These findings suggest that peptide-bound and free amino acids are absorbed from the normal human intestine by separate transport processes.

Absorption of individual amino acids from two partial enzymic hydrolysates of casein which consisted predominantly of small peptides was then studied. Absorption of α-amino nitrogen was greater from the casein hydrolysates than from their two equivalent mixtures of amino acids. There was considerable variation in absorption of individual amino acids during perfusion of the two free amino acid mixtures, but this was much reduced when the enzymic hydrolysate solutions were perfused. This was so because a number of individual amino acid residues that were poorly absorbed from the free amino acid mixtures were absorbed significantly faster from the enzymic hydrolysates. These results suggested that there could be an advantage in using mixtures of peptides rather than free amino acids as the nitrogen source in elemental diets. To test this hypothesis, plasma α-amino nitrogen increments and extent as well as site of absorption of amino acid residues were compared after two liquid test meals containing a protein hydrolysate or its equivalent amino acid mixture had been administered to normal human subjects. The finding of less secretion of fluid into the intestinal lumen, and higher concentrations of α-amino nitrogen in plasma after the protein hydrolysate meal, and the fact that after the two meals amino acid residues were absorbed to the same extent and at the same location, suggest that it would be worth considering the use of peptide mixtures as the nitrogen source in elemental diets.

---

* *Present address:* Liver Unit, King's College Hospital, London

The first evidence suggesting that peptides and free amino acids might be absorbed by separate transport processes in the human small intestine was derived from the observations made in man that higher concentrations of glycine occurred in plasma after oral administration of glycine peptides than after equivalent equimolar quantities of free glycine (Craft *et al.* 1968). Subsequent experiments in patients with the two inherited disorders of amino acid transport, Hartnup disease and cystinuria, confirmed the existence of separate transport processes for the absorption of peptides and amino acids in these two conditions. Group-specific defects in intestinal transport were demonstrated for neutral amino acids in Hartnup disease, and dibasic amino acids in cystinuria. Despite these defects, however, the 'affected' amino acids were absorbed normally or almost normally when they were presented to the mucosa in the form of oligopeptides (Asatoor *et al.* 1970*a,b*, 1971; Hellier *et al.* 1972; Silk *et al.* 1975*a*).

After the realization that separate transport processes for free and peptide-bound amino acids could exist in the human intestine, a number of workers began investigating the intestinal absorption of peptides in man. This paper reports a series of intestinal perfusion experiments designed to investigate several aspects of peptide absorption. The results led us to believe that there could be advantages in using peptides rather than free amino acids as the nitrogen source in elemental diets. This paper therefore also describes some preliminary results of a study performed to investigate absorption of a partial enzymic hydrolysate of protein given orally to normal subjects.

Finally, during this work, it has become clear that the brush border, as well as the cytoplasmic fraction of intestinal mucosa, contains enzymes capable of hydrolysing a wide range of dipeptides (Peters 1970, 1973; Donlon & Fottrell 1972; Kim *et al.* 1972; Fujita *et al.* 1972; Wojnarowska & Gray 1975). Experiments are reported here that were designed to investigate the role of these two groups of mucosal enzymes during the *in vivo* absorption of two dipeptides. The results suggest that whereas some dipeptides are predominantly absorbed intact and hydrolysed by cytoplasmic peptidases, significant proportions of others may be hydrolysed at the luminal surface of the cell by brush border peptidases and absorbed in the form of free amino acids. The potential implications of these findings in the interpretation of other peptide absorption experiments will be discussed.

INTESTINAL PERFUSION EXPERIMENTS WITH INDIVIDUAL PEPTIDES

Normal human subjects were intubated with a double lumen perfusion tube incorporating a proximal occlusive balloon (Sladen & Dawson 1970). For

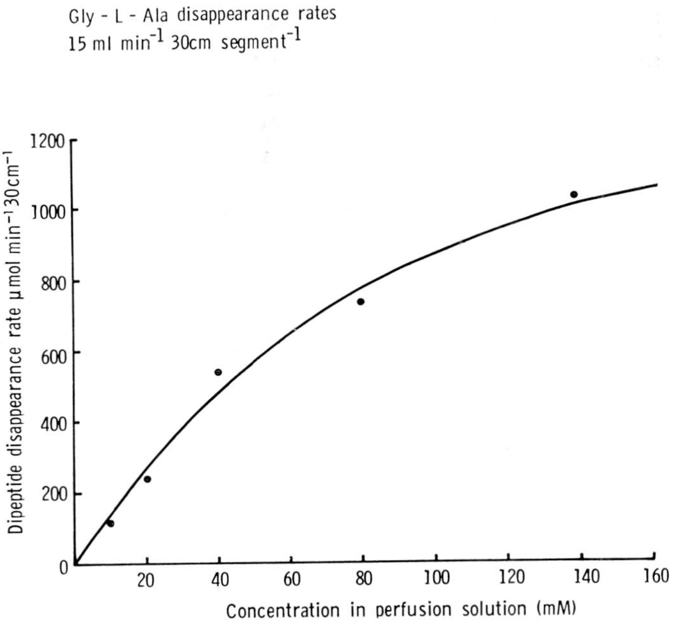

FIG. 1. Dipeptide disappearance rates during perfusion of 30-cm segments of human jejunum with test solutions containing glycyl-L-alanine. Each point is the mean of studies on at least four individual volunteers.

jejunal studies the tube was allowed to pass through the intestine until the infusion orifice was positioned distal to the duodeno-jejunal junction. Test solutions containing amino acid or peptide were rendered isotonic with plasma by the addition of sodium chloride, and they contained the non-absorbable marker, polyethylene glycol. The pH of the solutions was adjusted to pH 7.0 immediately before perfusion. Fuller details of the perfusion procedure and methods of analysis have been described (Silk *et al.* 1973*a*, 1974*a*).

Initially, jejunal absorption of amino acid residues was measured during perfusion of test solutions containing either the dipeptide glycyl-L-alanine (Gly-Ala) or equivalent equimolar concentrations of its constituent amino acids, glycine and L-alanine. The results (Fig. 1) showed that the rates of Gly-Ala disappearance as a function of concentration followed the kinetics of a saturable transport system. A limiting absorption rate was approached in the range of Gly-Ala concentrations used. Furthermore, Lineweaver–Burk plots of the reciprocals of absorption rates versus the reciprocals of geometric mean intraluminal concentrations showed a straight-line relationship (Fig. 2). At all

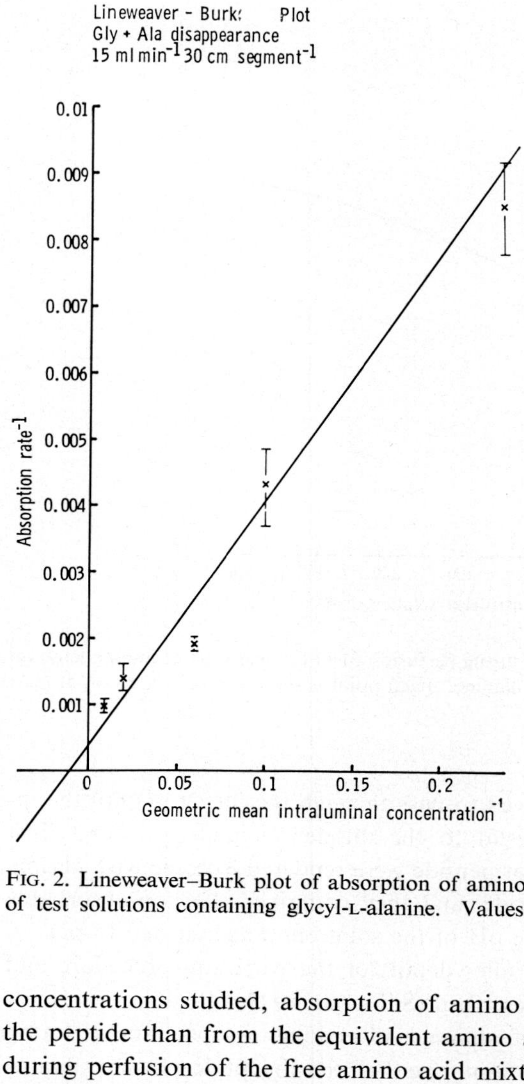

FIG. 2. Lineweaver–Burk plot of absorption of amino acid residues during jejunal perfusion of test solutions containing glycyl-L-alanine. Values are mean $\pm$ S.E.M., $n = 4$ or more.

concentrations studied, absorption of amino acid residues was greater from the peptide than from the equivalent amino acid solutions (Fig. 3). Finally, during perfusion of the free amino acid mixtures, L-alanine was consistently absorbed faster than glycine. This differential in absorption rates of glycine and L-alanine when presented to the mucosa in the free form was not apparent during perfusion of the dipeptide. The more rapid absorption of glycine from the dipeptide than from the equivalent free amino acid solution, and the lack of differential in absorption rates of glycine and L-alanine from the dipeptide solution, were also observed during ileal perfusion experiments (Silk *et al.* 1974*b*).

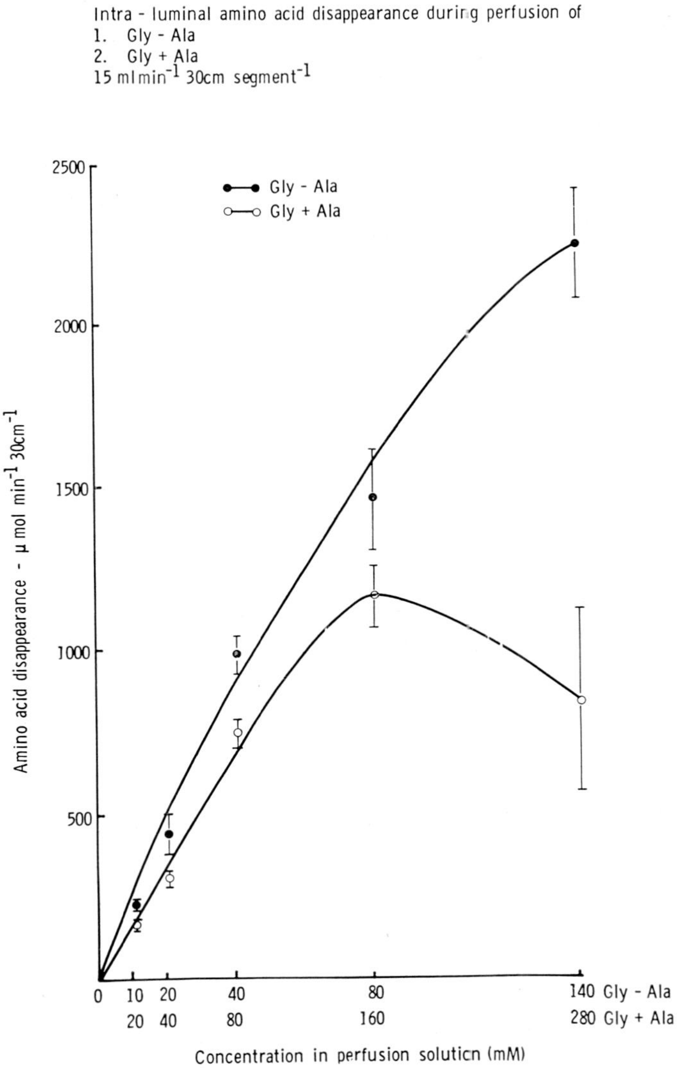

FIG. 3. Comparisons of intraluminal disappearance rates of amino acids during perfusion of 30-cm segments of human jejunum with test solutions containing glycyl-L-alanine (closed circles) or equivalent equimolar concentration of glycine + L-alanine (open circles). Values are mean ± S.E.M., $n = 4$ or more.

These studies were extended to include investigation of tripeptide absorption. Solutions containing either 10 mM-L-alanylglycylglycine or 10 mM-L-alanine +

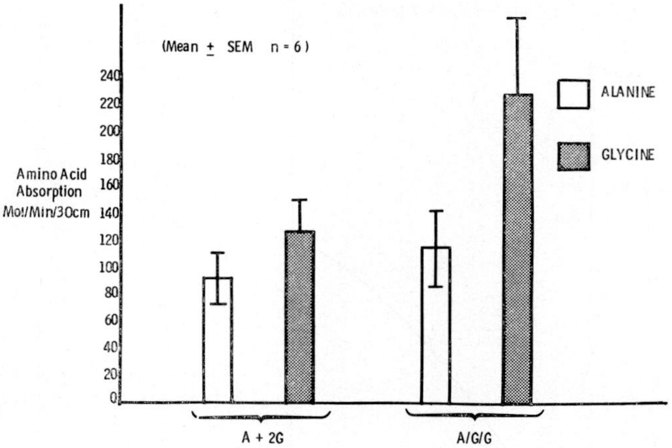

FIG. 4. Comparison of absorption rates of amino residues during perfusion of 30-cm segments of human jejunum with test solutions containing (1) 10mM-L-alanine + 20 mM-glycine (A + 2G) and (2) 10 mM-L-alanylglycylglycine (A/G/G).

20 mM free glycine were perfused through the jejunum of another group of six human subjects. As shown in Fig. 4, both glycine ($P < 0.001$) and L-alanine ($P < 0.05$) were absorbed significantly faster during perfusion of the tripeptide than the equivalent free amino acid solution. Similar results were obtained when absorption of the tripeptide glycylglycylglycine was studied. Thus glycine was absorbed faster during perfusion of 3.3 mM tripeptide than during perfusion of 10 mM free glycine (Silk *et al.* 1974*a*).

The finding of more rapid absorption of amino acid residues from peptide than from equivalent equimolar amino acid solutions cannot be explained if complete luminal or brush border hydrolysis of peptides precedes uptake of liberated free amino acids. If this were the case, at the best, absorption rates of amino acid residues from the two solutions would be the same. In addition, if complete luminal or surface dipeptide hydrolysis was a prerequisite for absorption, the normal differential between absorption of the constituent free amino acids would have been expected. This was clearly not the case. These findings therefore suggest that the dipeptides and tripeptides studied are absorbed from the lumen of the normal human small intestine by transport processes that are separate from those used for absorption of their constituent free amino acids. Absorption of two other dipeptides (glycylglycine and glycyl-L-leucine: Adibi 1971; Adibi & Soleimanpour 1974) and two tripeptides (glycylglycylglycine and L-leucyl-L-leucyl-L-leucine: Adibi *et al.* 1975) have been extensively studied in man. Very similar results to those described above were obtained. When considered as a whole, therefore, all these experiments with individual peptides provide strong evidence that, as in Hartnup disease and cystinuria, separate transport systems exist for the absorption of peptides and free amino acids in the normal human small intestine.

INTESTINAL PERFUSION EXPERIMENTS WITH PARTIAL ENZYMIC HYDROLYSATES OF WHOLE PROTEIN

The finding of more rapid absorption of amino acid residues from individual peptides than from their equivalent free amino acid mixtures prompted us to investigate the absorption of peptide-bound amino acids from two heterogeneous mixtures of oligopeptides. For the first study, a trypsin hydrolysate of casein was prepared *in vitro* (Silk *et al.* 1973*b*). About 66% of the amino acid residues in this preparation existed in the form of small peptides (chain length two to three residues) and 33% in the free form. Isotonic test solutions containing either the tryptic hydrolysate of casein (total α-NH_2 nitrogen, 40 mM), or a free amino acid mixture simulating exactly its pattern and molar concentration were perfused through the jejunum of six normal subjects. Fig. 5 shows that the extent of absorption of amino acid residues from the free amino acid solution varied considerably. Thus, whereas 73% of methionine was absorbed, only 26% of aspartic acid was absorbed. This variation was not nearly so marked when the tryptic hydrolysate of casein was perfused. The reason for this is that six amino acids (phenylalanine, lysine, glutamic acid, alanine, histidine and aspartic acid) that were relatively poorly absorbed from the amino acid mixture were absorbed to a greater extent from the tryptic hydrolysate of casein. Finally total absorption of α-NH_2 nitrogen was greater during perfusion of the tryptic hydrolysate of casein than its equivalent free amino acid mixture ($P < 0.05$).

For the second study, a peptide mixture was prepared by incubating casein *in vitro* with papain and hog kidney peptidase (Silk *et al.* 1975*b*). In this preparation, about 50% of the amino acid residues existed in the form of peptides (chain length two to three amino acid residues) and 50% in the free form. Test solutions containing either the casein hydrolysate (total α-NH_2 nitrogen, 64 mM) or a free amino acid mixture simulating exactly its composition and molar concentration were perfused through the jejunum of another six subjects. Again the extent of absorption of amino acid residues from the free amino acid solution varied greatly (Fig 6). As before, however, there was less variation in absorption from the casein hydrolysate. In this experiment, seven amino acids (phenylalanine, alanine, tyrosine, serine, aspartic acid, threonine, and histidine) that were relatively poorly absorbed from the amino acid mixture

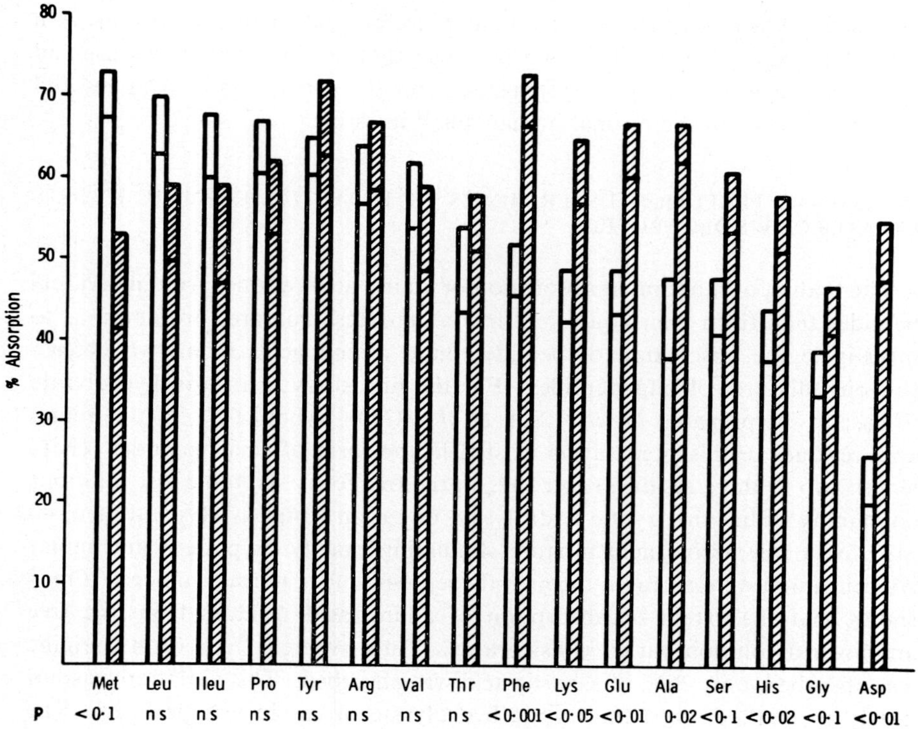

FIG. 5. Absorption of amino acids from an amino acid mixture simulating casein and a tryptic hydrolysate of casein. The total height of each column represents the mean value ($n = 6$), and the transverse line across each column shows one standard error. The significance of the difference between absorption from the amino acid mixture and from the tryptic hydrolysate is given below each pair of columns. Open columns, amino acid mixture; shaded columns, tryptic hydrolysate; n.s., not significant.

were absorbed to a greater extent from the casein hydrolysate. Total absorption of α-NH_2 nitrogen was also greater during perfusion of the casein hydrolysate than its equivalent free amino acid mixture ($P < 0.05$).

Currently, free amino acids are used as the nitrogen source in a number of 'elemental diets' available for oral alimentation of patients. Several features of the above experiments suggest that free amino acids may not necessarily be the optimal form in which dietary nitrogen should be administered to these patients. Thus, per unit time and unit length of bowel studied, absorption of several amino acid residues and total absorption of α-NH_2 nitrogen was greater when the peptide mixtures rather than the free amino acid mixtures were presented to the mucosa for absorption. Furthermore, a consistent finding in

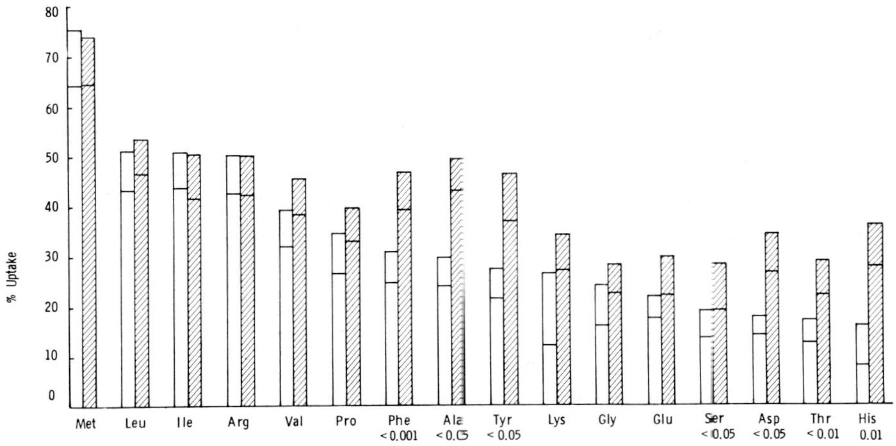

FIG. 6. Amounts of amino acids absorbed from a perfusion solution containing either an amino acid mixture simulating casein or an enzymic hydrolysate of casein containing oligopeptides and free amino acids, expressed as a percentage of the amounts in the perfusate. The total height of each column represents the mean value for six adult subjects with the standard error represented by a transverse line. Where the difference between absorption from the amino acid mixture and the enzymic hydrolysate was significant, the level of significance (*P*) is indicated. Open columns, amino acid mixture; shaded columns, enzymic hydrolysate.

the two experiments was of a more 'even' uptake of amino acid residues from the protein hydrolysates than from their equivalent free amino acid mixtures (Figs. 5 and 6). Present evidence suggests that this phenomenon could give rise to enhanced protein synthesis (Gitler 1964; Matthews 1971), which could be important in certain clinical situations when patients are in a protein catabolic state. Finally, additional support for the use of peptide mixtures rather than free amino acids as the nitrogen source in elemental diets is gained from recent studies which show that even in patients with abnormalities of the intestinal mucosa such as untreated adult coeliac disease, dermatitis herpetiformis or tropical sprue, absorption of amino acid residues is faster when they are presented to the mucosa as peptides than in the free form (Silk *et al.* 1974*c*; Adibi *et al.* 1974; Hellier *et al.* 1976).

ORAL FEEDING EXPERIMENTS

In the light of the results of the perfusion experiments just described, an experiment was designed to compare the effects of oral administration to normal human subjects of liquid test meals containing oligopeptides and free amino acids.

For the peptide meal, a pancreatic digest of fish protein was prepared *in vitro* (K. Berger, K. Conley & G. A. Spiller, Syntex Research, Palo Alto, California). Analysis has shown that 82% of the constituent amino acid residues in this digest exist in the form of oligopeptides. Of these peptide-bound residues, 90% were found to be constituents of peptides with a chain length of four amino acid units or less. An amino acid mixture simulating exactly the amino acid pattern and molar concentration of the pancreatic digest of fish protein was also prepared. Isonitrogenous liquid test meals containing polyethylene glycol (4 g/l) were prepared by dissolving either 50 g pancreatic digest of fish protein or 54.4 g amino acid mixture in distilled water to a final volume of 600 ml. These meals were administered orally on separate occasions to six normal subjects who had been intubated with a triple lumen tube with sampling orifices at 120, 160, and 200 cm from the mouth of each subject. Plasma amino acid and α-NH_2 nitrogen increments, and the extent and site of intestinal absorption of amino acid residues, were compared after ingestion of the two meals.

At 30 min and 1 h after ingestion of the peptide meal plasma increments of α-NH_2 nitrogen were larger than after the amino acid meal ($P < 0.05$; Table 1). There was no significant difference in plasma increments at later times. Indeed the mean value for α-NH_2 increments at three hours was higher after the ingestion of the amino acid meal.

As the two meals were isonitrogenous, the peptide meal had a lower osmolality than the free amino acid meal. As expected, therefore, in intestinal contents sampled from the proximal (120 cm) and middle orifices (160 cm) polyethylene glycol was more diluted after the amino acid meal ($P < 0.05$, $P < 0.01$ for proximal and mid-intestinal contents respectively) than after the peptide meal.

TABLE 1

Plasma α-amino nitrogen increments (mmol/l) in normal subjects at four different times after oral administration of isonitrogenous liquid test meals containing 50 g pancreatic digest of fish protein (peptide meal) or 54.4 g free amino acids (amino acid meal) (values are mean ± S.E., $n = 6$)

	30 min	*1 hour*	*2 hours*	*3 hours*
Amino acid meal	1.42 ± 0.14	1.72 ± 0.20	1.18 ± 0.35	1.02 ± 0.25
Peptide meal	2.15 ± 0.28	2.23 ± 0.41	1.44 ± 0.30	0.84 ± 0.12
P	< 0.05	< 0.05	N.S.	N.S.

N.S.: not significant

TABLE 2

Mean percentage absorption of the 18 free and peptide-bound amino acids at three different intestinal locations. Values are means of results from 4 subjects and were calculated from the relationship of amount of each amino acid to concentration of polyethylene glycol in test meals and the same relationship in intestinal contents sampled at 2 h

	Site of collection (cm from mouth)		
	120	*160*	*200*
Amino acid meal	53.9	85.4	95.5
Peptide meal	60.6	84.9	86.0

Table 2 shows that there was no significant difference in the extent of absorption of amino acid residues two hours after ingestion of the two meals at any of the three intestinal locations. Of interest, however, was the finding that at two hours absorption was virtually complete at 160 cm from the mouth of the control subjects.

The higher increments in plasma α-NH_2 nitrogen after ingestion of the peptide meal (Table 1) could be due to one of two factors. Faster absorption of peptide-bound than of free amino acids could explain the results. However, an alternative explanation has to be considered. As mentioned above, the osmolality of the peptide meal was nearly half that of the amino acid meal. It is possible, therefore, that gastric emptying of the peptide meal was faster than for the amino acid meal. If so, total absorption from the peptide meal would be expected to be greater at early time periods than from the amino acid meal. Studies of gastric emptying are being done now to answer this question.

Although the explanation for the higher plasma increments of α-NH_2 nitrogen after the peptide meal is not clear at present, the net result could be interpreted as favouring the use of peptides rather than free amino acids as a nitrogen source in elemental diets. Furthermore, the finding that polyethylene glycol was diluted less in the proximal gut after the peptide meal indicates that less fluid was secreted in this region of the bowel in response to the peptide than to the amino acid meal. This could be an advantage clinically, because diarrhoea often occurs when patients with the short bowel syndrome, inflammatory bowel disease or malabsorption are treated with ‘elemental diets’.

Finally peptide-bound and free amino acid residues were absorbed at the same site in the intestine (Table 2). The similar or higher increments in plasma α–NH_2 nitrogen observed after ingestion of the peptide meal than after the amino acid meal, and the lower intestinal secretion of fluid in response to the peptide meal, suggest that consideration could be given in the future to the use of oligopeptides as the nitrogen source in elemental diets.

THE ROLE OF BRUSH BORDER PEPTIDASES IN THE HYDROLYSIS OF LUMINAL PEPTIDES

One interesting aspect of the human perfusion experiments with the partial enzymic hydrolysates of whole protein (Figs. 5 and 6) was that the extent of absorption of many of the amino acid residues from the peptide solutions was the same as from the equivalent free amino acid solutions. In some respects this was an unexpected finding since the previous perfusion experiments with individual peptides had frequently shown faster absorption of peptide-bound than free amino acid residues. At the time the most likely explanation for the similar absorption rates seemed to be either that the amino acids in question were present in high concentrations in the free form in the hydrolysate solutions, or that they were indeed present as peptides but their concentrations were so low that differences in absorption rates could not be detected.

As already mentioned, however, it has now become clear that the brush border fraction of the intestinal mucosa cell contains significant peptidase activity (Peters 1970, 1973; Donlon & Fottrell 1972; Kim *et al.* 1972; Fujita *et al.* 1972; Wojnarowska & Gray 1975). An additional explanation for the lack of faster absorption of many amino acid residues during perfusion of the protein hydrolysates compared to perfusion of the equivalent free amino acid mixtures (Figs. 5 and 6) could be that significant components of the peptide-bound amino acid residues in the protein hydrolysates are in fact hydrolysed at the surface of the cell and absorbed in the form of free amino acids, not peptides.

To investigate this possibility further, an experiment was designed to evaluate the role of brush border peptidases during absorption of two marker dipeptides *in vivo*. The rat was used as an experimental model. To quantitate *in vivo* dipeptide disappearance rates, two dipeptides, glycyl-L-phenylalanine (Gly-Phe) and L-phenylalanyl-glycine (Phe-Gly), at a concentration of 10 mM were perfused through isolated segments of rat jejunum *in vivo*. Full details of the experimental technique have been published (Lane *et al.* 1975). At the end of the perfusions the pH of the intestinal contents was measured, and the perfused segments were removed. The mucosa was scraped from the segments, weighed, and homogenized in isotonic saline (4 °C; 10 ml/g mucosal scrapings). The activity of brush border peptidases in the mucosal homogenates obtained from the perfused segments was measured at the pH of the luminal contents, using substrate concentrations of 2 mM and 10 mM (Nicholson & Kim 1975). The lower substrate concentration was included in the assays because, although 10 mM concentrations of dipeptide were infused, the unstirred water layer could effectively lower the substrate concentration in contact with the externally orientated brush border enzymes (Wilson & Dietschy 1974). In this way it was

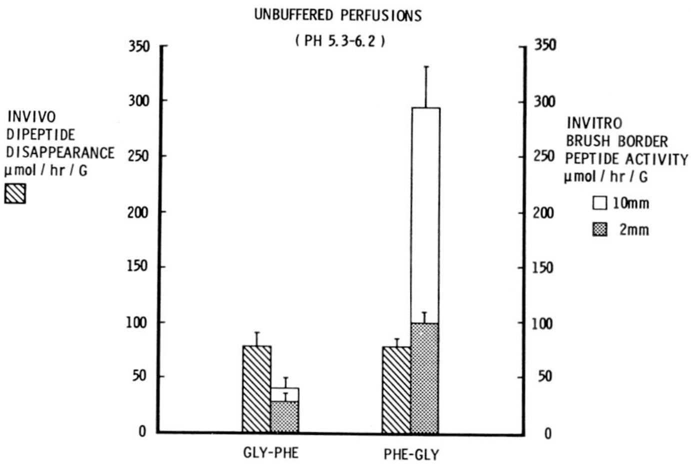

FIG. 7. Dipeptide disappearance rates during *in vivo* perfusion of rat jejunal segments with unbuffered test solutions containing 10 mM concentrations of glycyl-L-phenylalanine (Gly-Phe) and L-phenylalanylglycine (Phe-Gly) and rates of *in vitro* hydrolysis of the two dipeptides by brush border peptidase enzymes in the perfused segments. *In vitro* assays were done at the pH of the luminal contents, at substrate concentrations of 10 mM and 2 mM. Values are mean ± S.E., $n = 6$.

possible to compare *in vivo* luminal disappearance rates of dipeptide with the capacity of brush border peptidases in the perfused segments to hydrolyse luminal dipeptide.

The results are shown in Fig. 7. Luminal disappearance rates of Gly-Phe exceeded brush border hydrolysis rates of the dipeptide even when 10 mM substrate concentrations were used in the enzyme assay. It follows therefore that in the rat jejunum Gly-Phe is predominantly absorbed intact and hydrolysed in the intracellular compartment of the cell by cytoplasmic peptidases. In contrast, however, the capacity of brush border peptidases in the jejunal segments to hydrolyse Phe-Gly exceeded the luminal disappearance rates of the dipeptide. This was the case even when the substrate concentration used for the *in vitro* enzyme assays was one-fifth the concentration used for the *in vivo* perfusion studies. It is possible therefore that during *in vivo* absorption, significant components of Phe-Gly are hydrolysed at the luminal surface of the cell and absorbed in the form of free amino acids rather than in the peptide-bound form.

To date at least three brush border peptidases have been purified to homogeneity. They have been shown to hydrolyse a wide variety of other dipeptide substrates (Wojnarowska & Gray 1975; Kim & Brophy 1976). On the one hand, therefore, the results of the studies with Gly-Phe support the generally held concept that peptides are transported intact into the mucosal cell and hydrolysed by cytoplasmic peptidases. On the other hand, however, the results of the studies with Phe-Gly raise the possibility that the cytoplasmic compartment of the cell is not necessarily the major site of hydrolysis of all peptides.

If in fact the group of brush border peptidases plays a greater role in the hydrolysis of luminal peptides than previously realized, peptide absorption experiments that do not show greater absorption of peptide-bound than free amino acid residue could be explained on the basis that significant components of these peptides are hydrolysed at the luminal surface of the cell and absorbed in the free form.

ACKNOWLEDGEMENTS

I am grateful to the North East London Regional Hospital Board, the Board of Governors of St Bartholomew's Hospital, and the Medical Research Council for their financial support. It is a pleasure to acknowledge the help received from the following during the course of this work: Dr M. L. Clark, Dr A. M. Dawson, Professor D. M. Matthews, Dr Y. S. Kim, Dr D. Perrett, Dr J. A. Nicholson, Dr P. D. Fairclough, Dr M. Beigler, and Dr G. A. Spiller.

References

ADIBI, S. A. (1971) Intestinal transport of dipeptides in man: relative importance of hydrolysis and intact absorption. *J. Clin. Invest. 50*, 2266-2275

ADIBI, S. A. & SOLEIMANPOUR, M. R. (1974) Functional characterization of dipeptide transport system in human jejunum. *J. Clin. Invest. 51*, 1368-1374

ADIBI, S. A., FOGEL, M. R. & AGRAWAL, R. M. (1974) Comparison of free amino acid and dipeptide absorption in the jejunum of sprue patients. *Gastroenterology 67*, 586-591

ADIBI, S. A., MORSE, E. L., MASILAMANI, S. S. & AMIN, P. F. (1975) Triglycine absorption in human intestine: evidence for a common carrier for dipeptide and tripeptide transport. *J. Clin. Invest. 56*, 1355-1363

ASATOOR, A. M., BANDOH, J. K., LANT, A. F., MILNE, M. D. & NAVAB, F. (1970*a*) Intestinal absorption of carnosine and its constituent amino acids in man. *Gut 11*, 250-254

ASATOOR, A. M., CHENG, B., EDWARDS, K. D. G., LANT, A. F., MATTHEWS, D. M., MILNE, M. D., NAVAB, F. & RICHARDS, A. J. (1970*b*) Intestinal absorption of two dipeptides in Hartnup disease. *Gut 11*, 380-389

ASATOOR, A. M., CROUCHMAN, M. R., HARRISON, A. R., LIGHT, F. W., LOUGHRIDGE, L. W., MILNE, M. D. & RICHARDS, A. J. (1971) Intestinal absorption of oligopeptides in cystinuria. *Clin. Sci. (Oxf.), 41*, 23-33

CRAFT, I. L., GEDDES, D., HYDE, C. W., WISE, I. J. & MATTHEWS, D. M. (1968) Absorption and malabsorption of glycine and glycine peptides in man. *Gut 9*, 425-437

DONLON, J. & FOTTRELL, P. F. (1972) Studies on substrate specificities and subcellular location of multiple forms of peptide hydrolases in guinea-pig intestinal mucosa. *Comp. Biochem. Physiol. 41B*, 181-193

FUJITA, M., PARSONS, D. S. & WOJNAROWSKA, F. (1972) Oligopeptidases of brush border membranes of rat small intestinal mucosal cells. *J. Physiol. (Lond.) 227*, 377-394

GITLER, C. (1964) Protein digestion and absorption in nonruminants, in *Mammalian Protein*

Metabolism (Munro, H. N. & Allison, J. B., eds.), vol. 1, pp. 35-69, Academic Press, New York

HELLIER, M. D., HOLDSWORTH, C. D., PERRETT, D. & THIRUMALAI, C. (1972) Intestinal dipeptide transport in normal and cystinuric subjects. *Clin. Sci. (Oxf.) 43*, 659-668

HELLIER, M. D., RADHAKRISHNAN, A. N., GANAPATHY, V., MATHAN, V. I. & BAKER, S. J. (1976) Intestinal perfusion studies in tropical sprue. *Gut 17*, 511-516

KIM, Y. S. & BROPHY, E. J. (1976) Rat intestinal brush border membrane peptidases. 1. Solubilization, purification and physico-chemical properties of two different forms of the enzyme. *J. Biol. Chem. 251*, 3199-3205.

KIM, Y. S., BIRTWHISTLE, W. & KIM, Y. W. (1972) Peptide hydrolases in the brush border and soluble fractions of small intestinal mucosa of rat and man. *J. Clin. Invest. 51*, 1419-1430

LANE, A. E., SILK, D. B. A. & CLARK, M. L. (1975) Absorption of two proline containing peptides by rat small intestine *in vivo*. *J. Physiol. (Lond.) 248*, 143-149

MATTHEWS, D. M. (1971) Protein absorption. *J Clin. Pathol. (Lond.) 24, Suppl. 5*, 29-40

NICHOLSON, J. A. & KIM, Y. S. (1975) A one step L-amino acid oxidase assay for intestinal peptide hydrolase activity. *Anal. Biochem. 63*, 110-117

PETERS, T. J. (1970) The subcellular localisation of di- and tripeptide hydrolase activity in guinea-pig small intestine. *Biochem. J. 120*, 195-203

PETERS, T. J. (1973) The hydrolysis of glycine oligopeptides by guinea-pig intestinal mucosa and by isolated brush borders. *Clin. Sci. Mol. Med. 45*, 803-816

SILK, D. B. A., PERRETT, D. & CLARK, M. L. (1973*a*) Intestinal transport of two dipeptides containing the same neutral amino acids in man. *Clin. Sci. Mol. Med. 45*, 291-299

SILK, D. B. A., MARRS, T. C., CLEGG, K. M., ADDISON, J. M., BURSTON, D., CLARK, M. L. & MATTHEWS, D. M. (1973*b*) Absorption of amino acids from an amino acid mixture simulating casein and a tryptic hydrolysate of casein in man. *Clin. Sci. Mol. Med. 45*, 715-719

SILK, D. B. A., PERRETT, D., WEBB, J. P. W. & CLARK, M. L. (1974*a*) Absorption of two tripeptides by the human small intestine: a study using a perfusion technique. *Clin. Sci. Mol. Med. 46*, 393-402

SILK, D B. A., WEBB, J. P. W., LANE, A. E., CLARK, M. L. & DAWSON, A. M. (1974*b*) Functional differentiation of human jejunum and ileum: a comparison of the handling of glucose, peptides, and amino acids. *Gut 15*, 444-449

SILK, D. B. A., KUMAR, P. J., PERRETT, D., CLARK, M. L. & DAWSON, A. M. (1974*c*) Amino acid and peptide absorption in patients with coeliac disease and dermatitis herpetiformis. *Gut 1*, 1-8

SILK, D. B. A., PERRETT, D. & CLARK, M. L. (1975*a*) Jejunal and ileal absorption of dibasic amino acids and an arginine-containing dipeptide in cystinuria. *Gastroenterology 68*, 1426-1432

SILK, D. B. A., MARRS, T. C., CLEGG, K. M., ADDISON, J. M., BURSTON, D., CLARK, M. L. & MATTHEWS, D. M. (1975*b*) Absorption of an amino acid mixture simulating casein and enzymic hydrolysate of casein prepared for oral administration of patients. *Br. J. Nutr. 33*, 95-100

SLADEN, G. E. & DAWSON, A. M. (1970) Further studies on the perfusion method for measuring intestinal absorption in man. The effects of a proximal occlusive balloon and a mixing segment. *Gut 11*, 947-954

WILSON, F. A. & DIETSCHY, J. M. (1974) The intestinal unstirred layer: its surface area and effect on active transport kinetics. *Biochim. Biophys. Acta 363*, 112-126

WOJNAROWSKA, F. & GRAY, G. M. (1975) Intestinal surface peptide hydrolases. Identification and characterization of three enzymes from rat brush border. *Biochim. Biophys. Acta 403*, 147-160

Discussion of the two preceding papers

Parsons: Dr Silk has indicated that two quite separate systems may be involved in peptide transport and hydrolysis in the human intestinal mucosa, and also in that of the rat. Professor Matthews, were the three models you described in Fig. 1 (p. 12) mutually exclusive? You plumped for model C and gave A and B rather short shrift.

Matthews: I have maintained for a number of years that peptide absorption is a combination of two processess, brush border hydrolysis with uptake of amino acids, as described by Dr Silk, and peptide transport into the cell, with intracellular hydrolysis. I believe that brush border hydrolysis releases amino acids in free solution, and they behave just like free amino acids introduced into the human experimentally. I don't think there is 'coupled transport' of amino acids released by hydrolysis.

Parsons: It should be possible to test that by presenting the gut with amino acids which are released by peptide hydrolysis and at the same time presenting it with the parent peptide. If the amino acid concentration is high enough, the uptake system for free amino acids would be saturated and any additional amino-N uptake must then have occurred from peptide uptake (see, for example, Cheeseman & Parsons 1976). If there is no additional amino-N uptake then there is simply superficial hydrolysis and an associated amino acid scavenging system.

Dr Silk, you presented us with a slide with 16 comparisons on it. If you had had 20 comparisons, on the grounds of simple probability one in 20 would come off just by chance. You really ought to apply a much more rigorous statistical test.

Silk: I certainly agree. However in the perfusion technique used, a statistical significance of $P < 0.05$ means that we have consistently seen the same observation for an amino acid residue on six separate occasions in six different subjects, which with this technique must be considered significant.

Parsons: You would have to do an analysis of variance, with comparisons between amino acids and comparisons between patients; this would enable you to examine the variation between subjects and also between the effects of the different amino acids.

Silk: I shall certainly consider doing that.

Smyth: Cobalt stimulates dipeptidase activity greatly in homogenates of intestine, but not in everted sacs. This suggests that cobalt has easier access to dipeptides in homogenates than it has in everted sacs, and hence suggests that the peptidase activity is not at the cell surface (Cheeseman & Smyth 1971, 1975).

Semenza: To the best of my knowledge cobalt only activates that glycylglycinase which is water-soluble and probably located within the cell.

Smyth: That is why we use it.

Silk: I agree that this could be an exciting approach, especially as it could be done *in vivo*. However, studies in patients with cystinuria strongly suggest that surface hydrolysis has a role to play in peptide transport. In a recent study 1 mM-arginylleucine was perfused through the jejunum of six patients with homozygous cystinuria (Silk *et al.* 1975). About a third of dipeptide-bound arginine that disappeared from the gut lumen was recovered in intestinal aspirates in the free form. As there is a complete transport defect for free arginine in these patients, cystinuria provides one of the best models with which to estimate the degree of surface hydrolysis during transport of a dipeptide in man.

Smyth: I am not sure if the problem is not a semantic one, i.e. the meaning of cell surface. We did experiments with everted sacs in which glycylglycine was present in the mucosal fluid with or without methionine and found that methionine did not affect the final distribution of glycine in the different fluid compartments (Cheeseman & Smyth 1971). This however should have happened if hydrolysis had been at the surface (in our sense of the word) because the methionine would have inhibited the uptake of the liberated glycine. Another phenomenon which complicates the picture is the effect of amino acids on the activity of peptides (Cheeseman & Smyth 1975; Cheeseman *et al.* 1970). Some amino acids inhibit the action of one peptidase and stimulate the action of another, while other amino acids inhibit the action of both. This must greatly complicate the picture of what is happening during peptide digestion.

Silk: Of course with our technique we are studying the very artificial situation of uptake of individual peptides in a pure system. We have now realized that the interactions at the surface of the cell during absorption of a protein meal, for example, are extremely complex. Such knowledge is going to initiate a vast amount of work on the interactions of peptides and amino acids, and indeed of interactions of lipids and carbohydrates on absorption of free and peptide-bound amino acids.

Adibi: Your experiment raises several questions. First, in your studies of amino acid absorption, you related the concentration of polyethyleneglycol (PEG) in the intestinal aspirate to PEG concentration in the meal. Although this technique has been used previously, we and others have found that the concentration of PEG in the intestinal aspirate after a test meal is highly variable and therefore produces unpredictable absorption data. For this reason, we have abandoned this technique in our studies of absorption after an oral meal. I believe absorption studies should be performed when a steady state in PEG concentration has been maintained. This can be done by slow infusion of PEG before the meal is given. After the meal, the changes in amino acid or peptide concentration can then be related to a constant PEG ratio.

Secondly, you found that amino acid absorption was the same after the peptide and the amino acid meals. The greater rates of amino acid absorption from peptide solutions compared with amino acid solutions were shown to occur during the perfusion of a relatively small intestinal segment. It might be rather difficult to demonstrate this phenomenon over a large intestinal segment unless the concentrations of peptides and amino acids are sufficiently high. The kinetic advantage should become more apparent when the absorptive surface is reduced.

Thirdly, the peptide composition of your peptide meal is not determined. For example, it is not known how much hexapeptide, tetrapeptide or tripeptide was contained in the peptide meal. Knowledge of the composition is important for determining that amino acid absorption from a peptide meal is greater than from a free amino acid meal. The rate of amino acid uptake from an oligopeptide is not always greater than that from a corresponding amino acid mixture. In fact, in some situations, the rate of amino acid uptake from an oligopeptide could be either similar to or smaller than from a corresponding amino acid mixture (S. A. Adibi & E. L. Morse, unpublished work).

Silk: You criticize our oral feeding experiment. All I can say is that it was specifically designed to investigate normal physiological events. People eat protein meals and one is beginning to realize that in the intact human, gastric emptying is probably a major controlling factor. For example, the gastric emptying rate determines the concentrations of the protein digestion products that are presented to the mucosa. The experiment showing that 60% uptake took place by 120 cm merely confirms the original observation that absorption is extremely rapid (Borgström *et al.* 1957). However, you may remember that we did show great plasma α-amino nitrogen increments 30 min and 1 h after ingestion of the peptide meal compared with after ingestion of the equivalent free amino acid meal. At later times, though, the curves crossed over, so by 3 h there was no significant difference between total areas under the two curves.

Baker: Does your statement that you demonstrated a higher plasma α-amino nitrogen mean anything in physiological terms?

Silk: At present we are uncertain about this. When small gut function is compromised, for example after resection, these results may indicate that, early on, peptides are taken up more rapidly than free amino acids. Professor J. N. Hunt at Guy's Hospital is investigating whether the significant increases in plasma α-amino nitrogen increments after the peptide meal are due to increased gastric emptying of the peptide meal.

Regarding my thesis that more than one mode of uptake of peptides exists—Dr Adibi claims that dipeptides are transported intact. As I mentioned, I do not believe that a unified concept is any longer plausible. Thus, it has always been possible to detect free amino acids in the gut lumen, during dipeptide and tripeptide perfusion. For this to be possible, these free amino acids must have moved back across the unstirred water layer. It follows that the free amino acids detected in the gut lumen represent at the best only 50% produced by surface hydrolysis, as one must take into account efficient recapture and the inhibitory effect of the unstirred water layer on back diffusion into the bulk phase of the lumen. In summary, therefore, I agree that significant components of dipeptides may be transported intact, but I believe it is wrong to deny the possibility that surface brush border enzymes also play a role.

Semenza: A fourth possible explanation for independent peptide transport is essentially a very efficient capture mechanism, as put forward by Cheeseman & Parsons (1974). However, I shall suggest a modification of this mechanism later (Wacker & Semenza, this volume, pp. 109-116). Another transport agency, independent of the Na^+-dependent amino acid carrier, could be located in the immediate neighbourhood of the peptidase hydrolytic site or is perhaps even a part of it.

Dr Silk, you have been making these comparisons in normal humans fed with an artificial partial digest of free amino acids. What happens in pathological conditions where alterations in either transport or amino acid hydrolysis occur?

Silk: At present I cannot answer this question. This aspect is being studied now but we will have to wait for analysis of the sample. Both our group and Dr Adibi's group have shown that the kinetic advantage conferred by dipeptides on the uptake of amino acid residues is maintained in untreated coeliac disease, a condition in which there is a very severe mucosal lesion (Adibi *et al.* 1974; Silk *et al.* 1974).

Fottrell: It is very important that we clearly define what we mean by brush border peptidases. Criteria have already been established for other brush border enzymes such as disaccharidases which continue to increase in specific activity with continued purification of the brush border membranes. Other brush-border-associated enzymes, such as a peptide hydrolase recently studied in our laboratory (Piggott *et al.* 1976), may represent cytoplasmic contamination. This is very important, in my opinion, because in pathological conditions where the intestinal mucosa is damaged it is the brush border peptide hydrolases which will be more severely affected.

Dr Silk, when you compared the absorption of a tryptic hydrolysate of casein with a mixture of free amino acids did you consider that casein contains phosphate and carbohydrate? Side-chain substituents on the amino acids in the casein hydrolysate might therefore contribute to the differences in absorption rates which were found.

Payne: What is actually meant by tryptic hydrolysis? Pure trypsin acting on casein clearly couldn't generate 66% dipeptides and 33% amino acids, and even if it did, the free amino acids could be only lysine and arginine. The problem you raised could be resolved by finding out what the free amino acids are in that mixture, and that could readily be done. Four of the six amino acids which showed the biggest differences in relative absorption are ones that would be most modified by acid hydrolysis: namely Asn, Gln, Ser, Thr. So I would support Professor Fottrell's reservations.

Silk: The free amino acid test solution contained synthetic L-amino acids to the same pattern and molar concentration as the tryptic hydrolysate of casein. Thus the solutions were equivalent in terms of amino acid residue content. They were not equivalent in terms of carbohydrate and metal binding. Professor Matthews might be able to say more about the actual methods used for trypsin hydrolysis.

Matthews: It was done with crude trypsin, which is a mixture of pancreatic enzymes.

From this discussion, one might think that there was a controversy about brush border and intracellular hydrolysis. I don't think this is true. Both the processes occur (see Matthews 1975) and the extent of brush border hydrolysis as opposed to intracellular hydrolysis varies from one peptide to another. It may be more important for tripeptides than dipeptides. It also varies along the length of the gut, brush border hydrolysis probably being more prominent in the ileum. And it may well vary with the load. With a tremendous load more peptides may get through the hydrolase barrier and enter the cell.

Different usage by different people of terms such as 'intracellular hydrolysis' or 'membrane digestion' is beginning to cause much confusion. It would be an excellent thing if a few of us working in the field could agree on the usage of these terms.

References

ADIBI, S. A., FOGEL, M. R. & AGRAWAL, R. M. (1974) Comparisons of free amino acid and dipeptide absorption in the jejunum of sprue patients. *Gastroenterology 67*, 586-591

BORGSTRÖM, B., DAHLQVIST, A., LUNDH, G. & SJÖVALL, J. (1957) Studies of intestinal digestion and absorption in the human. *J. Clin. Invest. 36*, 1521-1536

CHEESEMAN, C. I. & PARSONS, D. S. (1974) Intestinal absorption of peptides. Peptide uptake by small intestine of *Rana pipiens*. *Biochim. Biophys. Acta 373*, 523-526

CHEESEMAN, C. I. & PARSONS, D. S. (1976) The role of some small peptides in the transfer of amino nitrogen across the wall of vascularly perfused intestine. *J. Physiol. (Lond.) 262*, 459-476

CHEESEMAN, C. I. & SMYTH, D. H. (1971) Site of intestinal hydrolysis of glycylglycine. *J. Physiol. (Lond.) 218*, 39-40

CHEESEMAN, C. I. & SMYTH, D. H. (1975) Interaction of amino acids, peptides and peptidases in the small intestine. *Proc. R. Soc. Lond. B Biol. Sci. 190*, 149-163

CHEESEMAN, C. I., NEWEY, H. & SMYTH, D. H. (1970) Effect of amino acids on the hydrolysis of dipeptides by rat small intestine. *J. Physiol. (Lond.) 212*, 25-26

MATTHEWS, D. M. (1975) Intestinal absorption of peptides. *Physiol. Rev. 55*, 537-608

PIGGOTT, C. O., O'CUINN, G. & FOTTRELL, P. F. (1976) Similarities between a dipeptide hydrolase from brush border and cytosol fractions of guinea-pig intestinal mucosa. *Biochem. J. 159*, 715-717

SILK, D. B. A., KUMAR, P. J., PERRETT, D., CLARK, M. L. & DAWSON, A. M. (1974) Amino acid and peptide absorption in patients with coeliac disease and dermatitis herpetiformis. *Gut 15*, 1-8

SILK, D. B. A., PERRETT, D. & CLARK, M. L. (1975) Jejunal and ileal absorption of dibasic amino acid on arginine-containing dipeptide in cystinuria. *Gastroenterology 68*, 1426-1432

Intestinal dipeptidases and dipeptide transport in the monkey and in man

A. N. RADHAKRISHNAN*

Wellcome Research Unit, Christian Medical College Hospital, Vellore

Abstract Among the powerful dipeptidases of the cytosol fraction the well-characterized 'master' dipeptidase of broad substrate specificity is an example of a 'true' dipeptidase. There are only a limited number of peptidases which together can hydrolyse the total range of the theoretically possible dipeptides. In the lumen, the dipeptides of dietary origin apparently enter the cell through a single transport system of broad specificity in the monkey and possibly also in man. Perfusion studies indicated impaired absorption of glycine and Gly-Gly in Indian compared to English subjects, but the kinetic advantage of dipeptide uptake over that of the free amino acid was maintained. In patients with tropical sprue, apart from decreased peptide absorption, there was an increased 'back-flow' of the constituent amino acids. The back-flow assumes nutritional significance since dipeptide uptake is competitively inhibited by amino acids of a certain specificity. Inhibition of brush border glycylglycine hydrolase (glycylglycine dipeptidase, EC 3.4.13.1) by L-leucine is non-competitive while that of cytosol enzyme is competitive. Certain other amino acids have no effect on glycylleucine hydrolase (glycylleucine dipeptidase, EC 3.4.13.2) from either fraction but inhibit Gly-Leu uptake. Thus, the inhibition of dipeptide uptake appears to be a consequence of interaction of amino acids directly with the dipeptide transport system and not with dipeptidases.

In the small intestine the hydrolysis of a variety of small peptides arising from the proteolytic digestion of dietary proteins is catalysed by enzymes present in the mucosa, both in the cytosol and in the brush border region. With Gly-Leu as a test substrate, the bulk of the activity (90–95%) was found to be in the cytosol fraction in several animal species. These two activities can be differentiated even in crude extracts since *p*-hydroxymercuribenzoate inhibits the cytosol but not the brush border enzyme (Heizer *et al* 1972; Das & Radhakrishnan 1972). When our work started several years ago, it was not known whether there were distinct dipeptidases, tripeptidases and oligo-

* *Present address:* School of Life Sciences, University of Hyderabad, Hyderabad-500001, India.

TABLE 1
Substrate specificity of the 'master' dipeptidase (from Das & Radhakrishnan 1973)

C-Terminal *N*-Terminal	Gly	Ala	Ile, Leu Val	Ser, Thr	Phe, Tyr, Trp, His	Met	Asp, Glu	Lys, Arg
Gly	**Gly-Gly**	Gly-Ala	Gly-Ile Gly-Leu Gly-Val	Gly-Thr	Gly-Trp Gly-Phe **Gly-His**	Gly-Met	Gly-Glu	Gly-Lys
Ala	Ala-Gly	Ala-Ala	Ala-Leu Ala-Val	Ala-Ser	Ala-Phe	Ala-Met	Ala-Asp	Ala-Arg Ala-Lys
Ile, Leu, Val	Leu-Gly	Leu-Ala	Leu-Leu Val-Leu	Leu-Ser	Leu-Phe	Leu-Met		
Ser, Thr	Ser-Gly	Ser-Ala	Ser-Leu His-Leu	Ser-Ser	Ser-Phe	Thr-Met		
Phe, Tyr Trp, His	Phe-Gly	Phe-Ala	Phe-Leu Tyr-Leu Trp-Leu		Phe-Phe	Phe-Met	Tyr-Glu	Tyr-Lys
Met	Met-Gly	Met-Ala	Met-Leu Met-Val	Met-Ser Met-Thr	Met-Phe Met-Tyr	Met-Met	Met-Glu	
Asp, Glu	Glu-Glu **Asp-Gly**	Glu-Ala	Glu-Val		Glu-Tyr		Glu-Glu **Glu-Asp**	Glu-Lys **Asp-Lys** Lys-Lys
Lys, Arg	Lys-Gly **Arg-Gly**	Lys-Ala **Arg-Ala**	Lys-Leu		Lys-Phe		Lys-Asp **Arg-Asp**	**Arg-Arg** **Arg-Lys**

All the peptides contained amino acids of the *laevo* configuration. The peptides that were inactive as substrates are shown in bold type.

peptidases. Also, there was no information on whether there was a fleet of enzymes to take care of the theoretically possible number of peptides from the digestion of proteins. There would be well over 400 dipeptides, over 8000 tripeptides, and so on. We therefore made a systematic effort to isolate and characterize at least one enzyme and this study led to the discovery of the so-called 'master' dipeptidase which shows an extremely wide range of action (Table 1). It is one of the most active enzymes in the body, with a molecular activity of 10^6 mol (mol dimer enzyme)$^{-1}$ min^{-1}. A free terminal $-NH_2$ group and a free terminal $-COOH$ group were essential for its activity and it would thus qualify to become the first 'true' dipeptidase to be isolated from the intestine (Das & Radhakrishnan 1973). It has no action at all on the following dipeptides: Gly-Gly, Gly-His, Gly-Pro, Pro-Gly and some of the dipeptides containing either basic or acidic amino acids. Our preliminary data would indicate that the above four dipeptides are hydrolysed by separate enzymes in the cytosol

(Krishna Baksi & A. N. Radhakrishnan, unpublished work). It is possible that some of the dipeptides with acidic amino acids are hydrolysed by a separate enzyme, as suggested by the work of Andria *et al.* (1976). All these results would strongly support the idea that there are only a limited number of 'true' dipeptidases in the cytosol with their characteristic substrate specificity but capable of efficiently handling the full range of luminal peptides.

Information on brush border enzymes is meagre and so far no 'true' dipeptidase has been characterized. One of the brush border enzymes isolated in a homogeneous state (Kim & Brophy 1976; Kim *et al.* 1976) acts on dipeptides as well as longer peptides, and appears to be more like an aminopeptidase. Other than in the hydrolysis of the terminal products of protein digestion, the role of brush border peptidases and the powerful cytosol peptidases is not clear. Perhaps they allow rapid entry of peptides into the mucosal cell by the establishment of a steep chemical gradient, or they may also serve as defence mechanisms, destroying certain peptides which may have deleterious effects on the body.

In the last few years there has been a significant change in our ideas on the terminal stages of protein digestion. The classical concept that amino acids are the only end-products of protein digestion which are actively absorbed has become untenable. Peptide absorption was shown to be distinctive in many animal species, including man, and is separate from amino acid absorption (for review see Matthews 1975; Das & Radhakrishnan 1976). An elegant proof for the entry of simple peptides into the enterocyte was provided by studies in Hartnup disease with a defect in neutral amino acid transport (Asatoor *et al.* 1970; Leonard *et al.* 1976) and in cystinuria with a defect in basic amino acid transport (Hellier *et al.* 1970, 1972). The 'affected' amino acids were absorbed efficiently when presented as dipeptides. Studies of the inhibition of uptake *in vitro* of [^{14}C]Gly-Leu by a variety of unlabelled dipeptides indicated the existence of a dipeptide uptake system of very broad specificity (Table 2). The inhibition of [^{14}C]Gly-Leu uptake was shown to be competitive with limited but judiciously selected dipeptides of widely differing chemical structure. These results gave us enough confidence to suggest that in the monkey intestine, and possibly also in man, there could be a general transport system for dipeptides (Das & Radhakrishnan 1975). Although we found inhibition of [^{14}C]Gly-Leu uptake by the tripeptides, Gly-Gly-Leu and Gly-Leu-Gly, the results were not unequivocal since the inhibition may well be due to the formation of dipeptides, Gly-Leu and Leu-Gly, by intestinal amino tripeptidase action and these dipeptides might be expected to cause the inhibition. More recently, Addison *et al.* (1975) have provided evidence indicating that the dipeptide transport system may also be shared by tripeptides.

TABLE 2

Specificity of the monkey intestinal glycyl-L-leucine uptake system (from Das & Radhakrishnan 1975)

	% inhibition of uptake of			% inhibition of uptake of	
Peptide	*[^{14}C]Glycyl-L-leucine*	*Glycyl-L-[^{14}C] leucine*	*Peptide*	*[^{14}C]Glycyl-L leucine*	*Glycyl-L-[^{14}C] leucine*
Gly-Gly	24	20	Gly-Asp	38	50
Gly-Ala	42	50	Gly-Glu	49	42
Gly-Val	70	65	Gly-Asn	50	63
Gly-Ile	80	78	Ala-Asp	52	62
Gly-Leu	82	86	Met-Glu	62	64
Gly-Ser	53	70	Asp-Gly	43	40
Gly-Phe	51	73	Glu-Ala	60	54
Gly-Trp	44	67	Glu-Val	79	58
Gly-His	45	63	Glu-Glu	57	54
Gly-Met	70	86	Gln-Gln	70	76
Ala-Gly	82	70	Tyr-Lys	72	77
Ala-Ala	73	73	Lys-Phe	70	67
Ala-Leu	89	92	Lys-Lys	64	62
Ala-Ser	88	74	Gly-D-Leu	12	10
Ala-Phe	80	86	Z-Gly-Leu	15	30
Ala-Met	85	86	Gly-β-Ala	15	20
Leu-Gly	74	87	Gly-γ-Abu	25	32
Ser-Leu	84	85	Gly-Gly-Leu	60	58
Met-Leu	70	91	Gly-Leu-Gly	62	58
Gly-Pro	76	69	Gly-Gly-Gly	26	20
Pro-Gly	63	51			

In these experiments, the concentration of ^{14}C-labelled glycyl-L-leucine was 0.5 mM and the concentration of the unlabelled peptides was 10 mM in all cases.
All the peptides, except glycyl-D-leucine, contained amino acids of the *laevo* configuration.
Z, *N*-benzyloxycarbonyl.
Gly-γ-Abu: glycyl-γ-aminobutyric acid.

Perfusion studies in man (Hellier *et al.* 1976*a*) showed impaired absorption of glycine and Gly-Gly in Indian subjects compared to English subjects, but the kinetic advantage of dipeptide uptake over that of the free amino acid was maintained (Fig. 1). In patients with tropical sprue (Hellier *et al.* 1976*b*) an apparent decrease in peptide absorption (but with the retention of kinetic advantage) and an increase in 'back-flow' of the constituent amino acids were noticed (Fig. 2). The latter result assumes nutritional significance since we subsequently found a direct interaction of certain amino acids with peptide uptake. We consider that a greater efflux of inhibitory amino acids into the lumen may affect the already reduced peptide uptake in patients with sprue, thus forming a vicious circle.

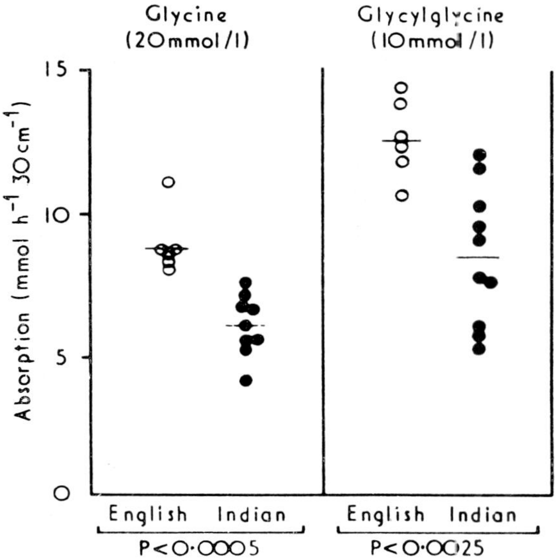

FIG. 1. Absorption of glycine (20 mM) and glycylglycine (10 mM) in normal English (○) and Indian (●) people. Horizontal lines represent means.

INTERACTION OF AMINO ACIDS WITH DIPEPTIDASES AND DIPEPTIDE UPTAKE

Our preliminary studies indicated that certain amino acids (e.g. leucine) inhibited dipeptide uptake. Earlier workers (Cheeseman & Smyth 1971; Addison *et al.* 1972) had shown that dipeptidases may also be inhibited by amino acids. Proline inhibited Gly-Pro hydrolase from horse erythrocytes (Adams & Smith 1952) and from pig intestine (Sjöström 1974), and this was ascribed to product inhibition. Intestinal leucine aminopeptidase is inhibited by leucine (Delange & Smith 1971). Rubino *et al.* (1971) reported inhibition of Gly-Pro influx by leucine, methionine and phenylalanine. Inhibition of Gly-Gly uptake by glycine, leucine, proline and methionine was observed by Caspary (1973). In human perfusion studies, inhibition of Gly-Gly uptake by leucine was also noted (Adibi 1971; Adibi & Soleimanpour 1974). It could be argued that inhibition of peptide uptake may be due to enzyme inhibition which would lead to accumulation of the dipeptide inside the cell with a consequent reduction in the rate of peptide entry (Matthews 1975). We have therefore made a systematic survey of the effect of various amino acids on mucosal peptidases (both cytosol and brush border), and on the kinetic nature of inhibition of the enzymes and of peptide uptake.

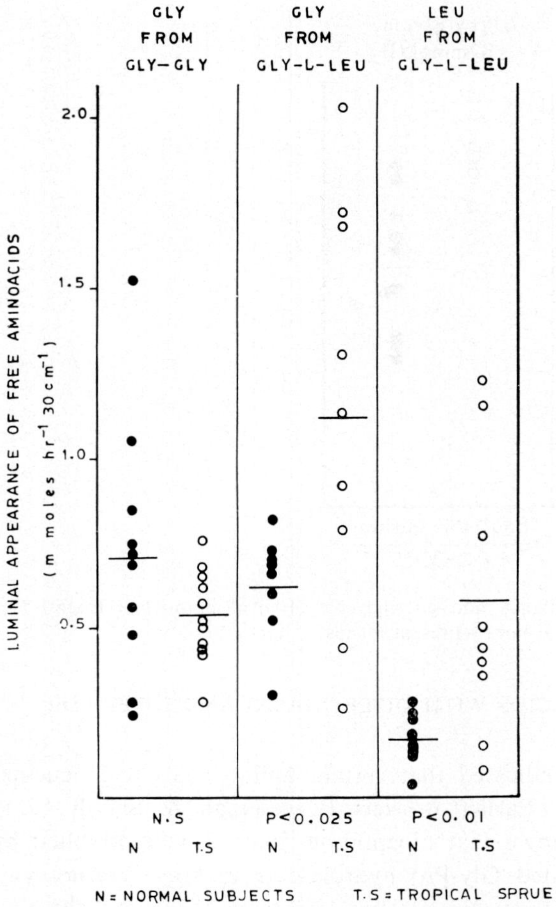

FIG. 2. Luminal appearance of glycine from glycylglycine (10 mM) and glycine and leucine from glycyl-L-leucine (10 mM) in controls and in patients with tropical sprue. Horizontal lines represent means.

Studies in vitro

The uptake of both Gly-Gly (Fig. 3) and Gly-Leu (Fig. 4) was inhibited by leucine and the nature of inhibition was found to be competitive by a reciprocal plot of the data. Since it is possible that this type of inhibition may also be obtained by amino acid inhibition of the peptidases, an effort was made to study the effect of various amino acids on mucosal peptidases. The cytosol fraction and the particulate fraction were effectively separated from each other

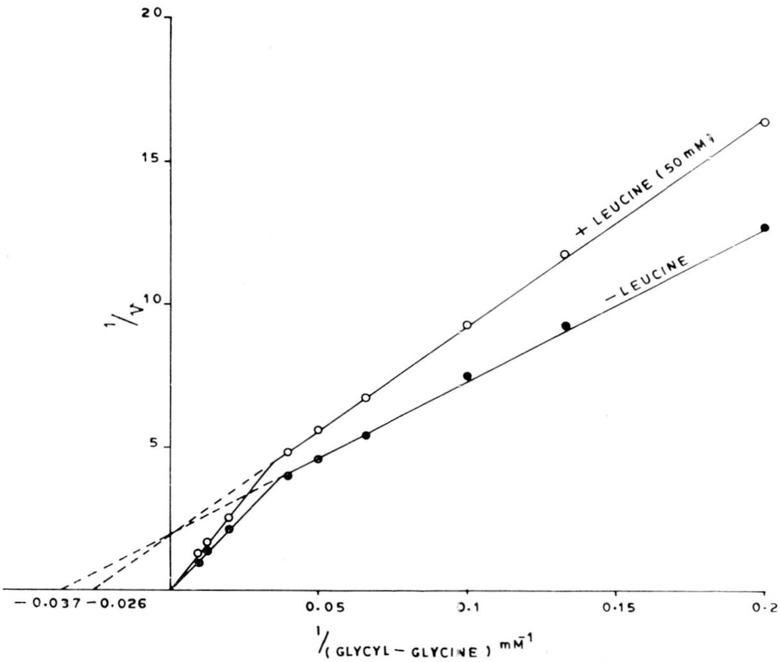

FIG. 3. Lineweaver-Burk plot showing competitive inhibition of uptake of [^{14}C]glycylglycine by L-leucine, using monkey intestinal strips. Inhibitor concentration, 50 mM.
v: 10^{-3} counts min^{-1} (mg tissue)$^{-1}$

and cross-contamination was monitored by the use of *p*-hydroxymercuribenzoate as a specific inhibitor of the cytosol enzyme (Heizer *et al.* 1972; Das & Radhakrishnan 1972). The data in Table 3 clearly show enzyme inhibition by a variety of amino acids. Both cytosol and brush border peptidases exhibit this phenomenon although the degree of inhibition may vary.

The kinetic nature of the inhibition was next studied. The inhibition by leucine of Gly-Gly hydrolysis is non-competitive with the brush border enzyme (Fig. 5) and is competitive with the cytosol enzyme (Fig. 6). We would like to think that these results provide reasonably strong proof that the amino acids may directly interfere with the transport systems responsible for peptide entry and that the brush border peptidases have no role in this phenomenon. The cytosol enzyme and the peptide uptake are both competitively inhibited by leucine, but it is unlikely that these are related since an additional parameter, the brush border barrier, is also involved. It was subsequently found that a variety of amino acids inhibit Gly-Leu uptake *in vitro* in monkey intestine (Table 4).

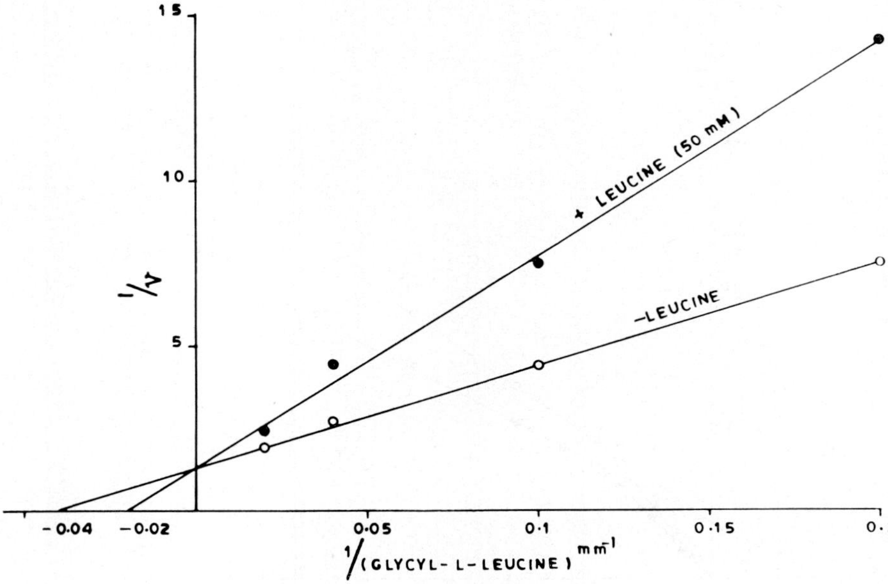

FIG. 4. Lineweaver-Burk plot showing competitive inhibition of uptake of glycyl-L-[^{14}] leucine by L-leucine, using monkey intestinal strips. Inhibitor concentration, 50 mM. v: 10^{-3} counts min^{-1} (mg tissue)$^{-1}$

TABLE 3

Effect of amino acids on Gly-Leu hydrolase of monkey intestine

L-*Amino acid (40 mM)*	*(Inhibition %)* Cytosol	*(Inhibition %)* Brush border	L-*Amino acid (40 mM)*	*(Inhibition %)* Cytosol	*(Inhibition %)* Brush border
Cys	100	100	Asp	45	39
His	95	74	Lys	34	0
Met	73	31	Val	31	16
Leu	64	36	Dopa	26	38
Nor-Leu	63	25	Thr	21	43
Trp	63	25	Hyp	12	0
Phe	62	47	Arg	10	23
Ile	59	22	Ser	8	19

Gly, Ala, Pro and Glu had no effect; Gly-Leu, 8 mM; Dopa, 20 mM

Gly-Gly hydrolysis by both the brush border and cytosol fractions of monkey intestine was markedly inhibited at high levels of the substrate, when the commercial Gly-Gly (Sigma) was used. But with the same sample of the

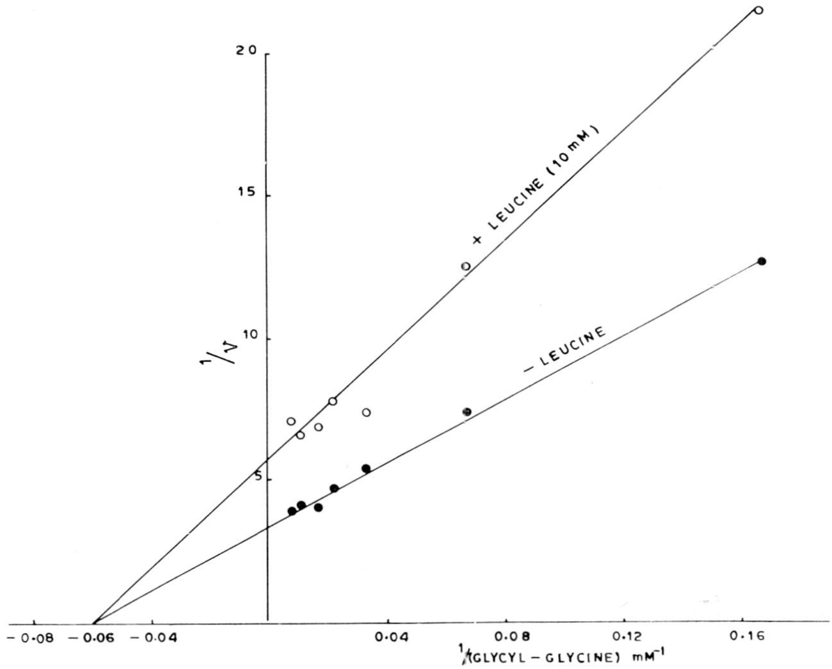

FIG. 5. Lineweaver-Burk plot showing non-competitive inhibition of glycylglycine hydrolysis by L-leucine (brush border enzyme from monkey intestine).
v: (μmol Gly-Gly hydrolysed) (30 min)$^{-1}$ (30 μg protein)$^{-1}$

TABLE 4

Effect of amino acids on Gly-[^{14}C]Leu uptake by monkey intestinal strips

L-*Amino acid (50 mM)*	*Inhibition (%)*	L-*Amino acid (50 mM)*	*Inhibition (%)*
Phe	62	Trp	31
Ala	59	Arg	30
Met	58	Pro	30
Leu	56	Ser	28
Cys	53	Hyp	27
Lys	49	Gly	27
Thr	47	Glu	7
Val	43	Asp	5
His	42		

Gly-Leu, 2 mM

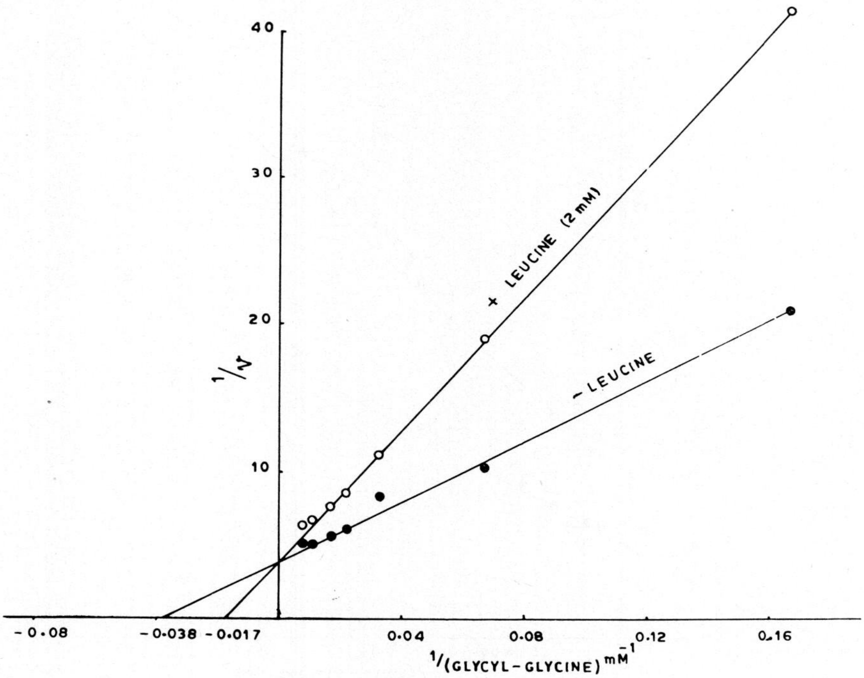

FIG. 6. Lineweaver-Burk plot showing competitive inhibition of glycylglycine hydrolysis by L-leucine (cytosol enzyme from monkey intestine).
v: (μmol Gly-Gly hydrolysed) $(30 \text{ min})^{-1}$ $(3\ \mu\text{g protein})^{-1}$

dipeptide no inhibition of Gly-Gly uptake by monkey intestinal strips was noted, even at concentrations inhibitory to the enzyme. However, crystallization from aqueous ethanol in the presence of 1 mM-EDTA eliminates this inhibitory effect, suggesting the removal of a heavy metal from Gly-Gly. We did not find any correlation between the effects of various amino acids on Gly-L-Leu hydrolysis and on Gly-L-Leu uptake. For example, glycine, proline, alanine and hydroxyproline have no effect on the hydrolysis of Gly-L-Leu (both cytosol and brush border) but they show considerable inhibition of Gly-L-Leu uptake. The above findings, summarized in Table 5, indicate that the brush border dipeptidase does not play a direct role in the dipeptide transport and that the site of action of inhibitory amino acids on dipeptide uptake is probably at the transport step.

TABLE 5

Comparison of inhibition of Gly-Leu hydrolases and Gly-[^{14}C]Leu uptake

L-*Amino acid*	*Inhibition (%)*		
	Enzyme		*Uptake*
	Brush border	*Cytosol*	
Ala	0	0	59
Lys	0	34	49
Pro	0	0	30
Hyp	0	12	27
Gly	0	0	27
Glu	0	0	7
Asp	39	45	5
Cys	100	100	53

Studies in vivo

Amino acid inhibition of dipeptide uptake was also observed in perfusion studies in man. The uptake of Gly-Gly (50 mM) was unaffected by the presence of glycine (30 mM) but was inhibited by about 20% in the presence of a mixture of L-leucine (30 mM) and glycine (30 mM), thus suggesting *in vivo* inhibition of Gly-Gly uptake by leucine (V. Ganapathy, M. D. Hellier & A. N. Radhakrishnan, unpublished work).

KINETICS OF GLY-GLY AND GLY-LEU UPTAKE

Our earlier studies (Das & Radhakrishnan 1975) had shown that the K_t values for Gly-Leu were of similar magnitude in both monkey (4 mM) and man (2.5 mM), when the range of concentration of the dipeptide varied from 0.5 to 5.0 mM. Subsequent experiments in which the concentration of the dipeptide ranged from 0.5 to 50.0 mM revealed an additional mode of uptake of Gly-Leu with a K_t of about 30 mM (Fig. 7). Similar careful studies with Gly-Gly (0.5–100 mM) confirmed these findings of two saturable transport systems with K_t values of 5 mM and 30 mM. At high concentrations of the dipeptide there is significant passive diffusion with Gly-Gly (Fig. 8), but not with Gly-Leu. In man, perfusion studies with Gly-Gly did not show a biphasic uptake pattern, but since the concentration intervals (10, 25, 50 and 100 mM) were far apart further experiments *in vivo* are necessary to substantiate the *in vitro* findings. The possible existence of two saturable systems for dipeptide uptake is also suggested by the work of Rubino *et al.* (1971) with Gly-Pro (with K_t values of

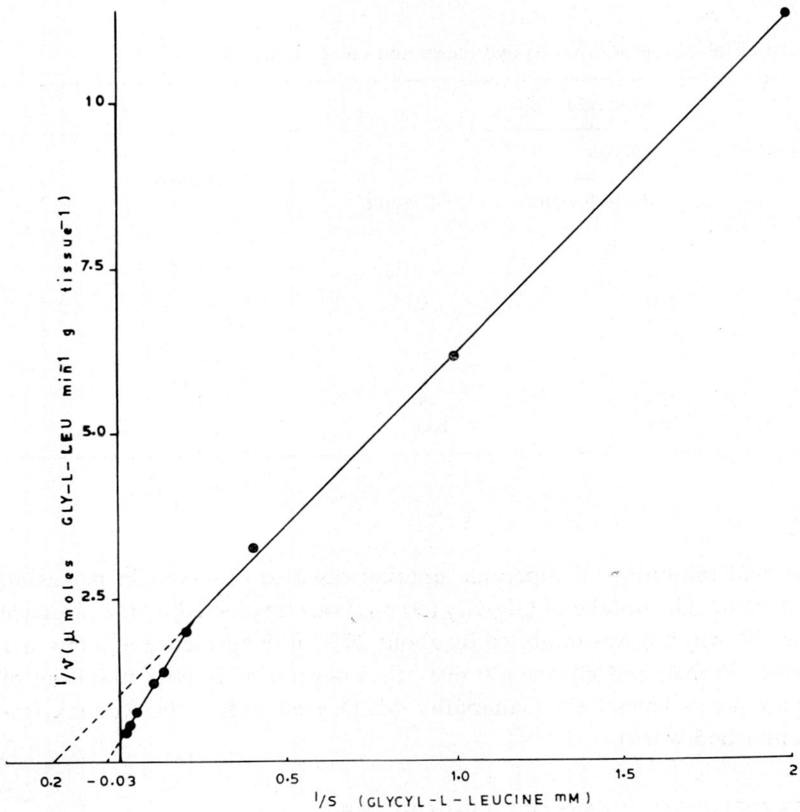

FIG. 7. Lineweaver-Burk plot for the uptake of glycyl-L-[^{14}C]leucine over a concentration range of 0.5 mM–50 mM, with monkey intestinal strips. K_t values: 5 mM and 30 mM.

0.9 mM and 57 mM). Rubino *et al.* (1971) suggested that the low affinity transport system was involved in the inhibition of Gly-Pro influx by leucine, methionine and phenylalanine. However, our studies indicate that the inhibition of Gly-Leu uptake by leucine involves both the low and high affinity transport systems, since the inhibition can be demonstrated with low and high dipeptide concentrations.

AN OVERVIEW OF PROTEIN DIGESTION

The available information on protein digestion is represented diagrammatically in Fig. 9. We assume that the following mechanisms may operate in the terminal stages of protein digestion.

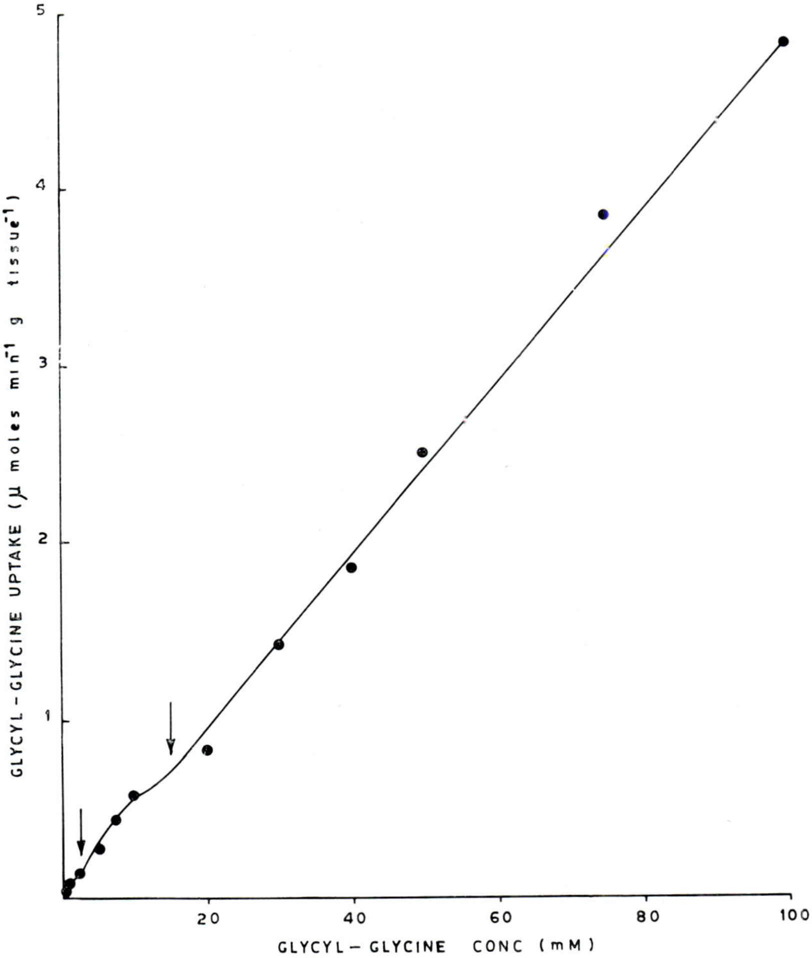

FIG. 8. Uptake of [^{14}C]glycylglycine (monkey intestinal strips) as a function of concentration of the peptide. The arrows show the breaks in the curve corresponding to K_t values of 5 mM and 30 mM. Above 20 mM, the uptake is predominantly due to diffusion.

(1) Proteins are hydrolysed in the lumen by the concerted action of pepsin, pancreatic enzymes and the mucosal surface proteases to a mixture of free amino acids and small peptides (dipeptides to hexapeptides), among which the peptides would predominate.

(2) The amino acids enter the cell through the different group-specific transport systems which would also normally act as efficient mechanisms for the recapture of effluxing amino acids.

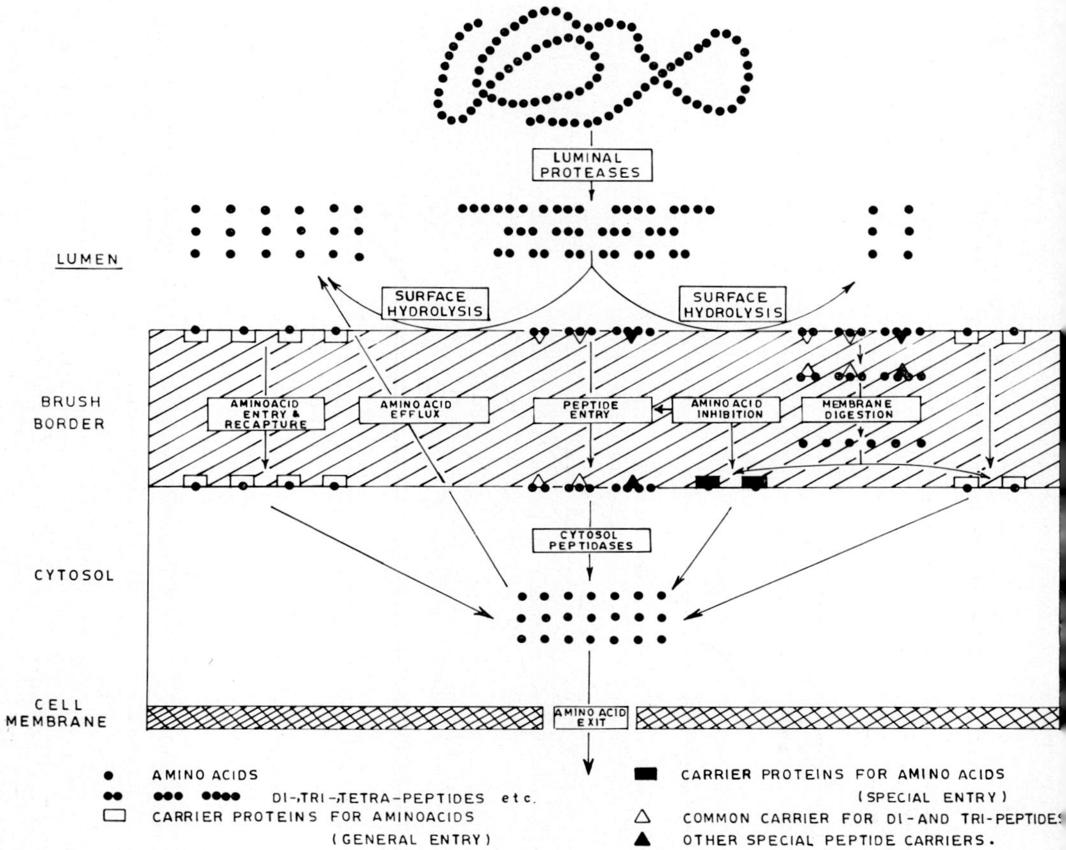

FIG. 9. A general scheme of protein digestion and absorption.

(3) It is assumed that peptides are 'absorbed' by three different mechanisms:

(*a*) intact peptides enter the mucosal cell by peptide transport systems, followed by intracellular hydrolysis into amino acids. This system may be directly affected by certain amino acids but without exhibiting group specificity for amino acid transport;

(*b*) peptides are converted by membrane digestion (or surface hydrolysis) into amino acids which are then transported into the cell by the usual amino acid transport systems. This component of peptide absorption would then be susceptible to amino acid inhibition of the amino acid transport systems. This should also exhibit amino acid group-specificity of transport;

(*c*) peptides are converted by membrane digestion (or surface hydrolysis) into amino acids which are then transported into the cell by special amino acid transport systems, not hitherto recognized. The hydrolytic products would not be in equilibrium with the luminal amino acids. This system would be kinetically rather similar to that given under 3*a*, regarding inhibition by other peptides and certain amino acids.

All three mechanisms for peptide transport postulated above (3*a*-*c*) may operate in physiological conditions. With existing knowledge it is difficult to speculate which mechanism is quantitatively most important, and obviously further work is necessary in this area. The entry system for intact peptides may allow transport of certain peptides of dietary origin which are resistant to luminal hydrolysis but which might have pronounced physiological effects.

ACKNOWLEDGEMENTS

I am deeply indebted to my colleagues Dr Manjusri Das and Mr V. Ganapathy who have collaborated in this work, and to Dr M. D. Hellier for his contributions and stimulating discussions during his sabbatical year in Vellore. I would like to place on record the help and encouragement of Professor S. J. Baker who was Head of our Unit during the project. My thanks are due to Dr Annie Gammon and Dr V. I. Mathan for their collaboration, to Mr V. N. Guiri for the charts, Mr J. Prabhakaran for the photographs, and Mr S. Krishnamurthy for his help in the preparation of the manuscript. Financial assistance from the Wellcome Trust, London and C.S.I.R. (India) is gratefully acknowledged.

References

ADAMS, E. & SMITH, E. L. (1952) Peptidases of erythrocytes. II. Isolation and properties of prolidase. *J. Biol. Chem. 198*, 671-682

ADDISON, J. M., BURSTON, D. & MATTHEWS, D. M. (1972) Evidence for active transport of the dipeptide glycyl sarcosine by hamster jejunum *in vitro*. *Clin. Sci. (Oxf.) 43*, 907-911

ADDISON, J. M., BURSTON, D., DALRYMPLE, J. A., MATTHEWS, D. M., PAYNE, J. W., SLEISENGER, M. H. & WILKINSON, S. (1975) A common mechanism for transport of di- and tri-peptides by hamster jejunum *in vitro*. *Clin. Sci. Mol. Med. 49*, 313-322

ADIBI, S. A. (1971) Intestinal transport of dipeptides in man. Relative importance of hydrolysis and intact absorption. *J. Clin. Invest. 50*, 2266-2275

ADIBI, S. A. & SOLEIMANPOUR, H. R. (1974) Functional characterization of dipeptide transport system in human jejunum. *J. Clin. Invest. 53*, 1368-1374

ANDRIA, G., MARZI, A. & AURICCHIO, S. (1976) α-Glutamyl-β-naphthylamide hydrolase of rabbit small intestine. Localization in the brush border and separation from other brush border peptidases. *Biochim. Biophys. Acta 419*, 42-50

ASATOOR, A. M., CHENG, B., EDWARDS, K. D. G., LANT, A. F., MATTHEWS, D. M., MILNE, M. D., NAVAB, F. & RICHARDS, A. J. (1970) Intestinal absorption of two dipeptides in Hartnup disease. *Gut 11*, 380-389

CASPARY, W. F. (1973) Interaction of dipeptide absorption and transport of free amino acids, in *Biochemical and Clinical Aspects of Peptide and Amino Acid Absorption* (Rommel, K. & Goebell, H., eds.), pp. 29-40, Schattauer, Stuttgart

CHEESEMAN, C. I. & SMYTH, D. H. (1971) Site of intestinal hydrolysis of glycylglycine. *J. Physiol. (Lond.) 218*, 39P-40P

DAS, M. & RADHAKRISHNAN, A. N. (1972) Substrate specificity of a highly active dipeptidase purified from monkey small intestine. *Biochem. J. 128*, 463-465

DAS, M. & RADHAKRISHNAN, A. N. (1973) Glycyl-L-leucine hydrolase, a versatile 'master' dipeptidase from monkey small intestine. *Biochem. J. 135*, 609-615

DAS, M. & RADHAKRISHNAN, A. N. (1975) Studies on a wide spectrum intestinal dipeptide uptake system in the monkey and in the human. *Biochem. J. 146*, 133-137

DAS, M. & RADHAKRISHNAN, A. N. (1976) Role of peptidases and peptide transport in the intestinal absorption of proteins. *World Rev. Nutr. Diet. 24*, 58-87

DELANGE, R. J. & SMITH, E. L. (1971) Leucine amino peptidase and other N-terminal exopeptidases, in *Enzymes*, vol. 3: *Hydrolysis: Peptide-bonds*, 3rd edn. (Boyer, P. D., ed.), pp. 82-102, Academic Press, New York

HEIZER, W. D., KERLEY, R. L. & ISSELBACHER, K. J. (1972) Intestinal peptide hydrolases. Difference between brush border and cytoplasmic enzymes. *Biochim. Biophys. Acta 264*, 450-461

HELLIER, M. D., PERRETT, D. & HOLDSWORTH, C. D. (1970) Dipeptide absorption in cystinuria. *Br. Med. J. 4*, 782-783

HELLIER, M. D., HOLDSWORTH, C. D., PERRETT, D. & THIRUMALAI, C. (1972) Intestinal dipeptide transport in normal and cystinuric subjects. *Clin. Sci. (Oxf.) 43*, 659-668

HELLIER, M. D., RADHAKRISHNAN, A. N., GANAPATHY, V., GAMMON, A. & BAKER, S. J. (1976*a*) Intestinal absorption in normal Indian and English people. *Br. Med. J. 24*, 186-188

HELLIER, M. D., RADHAKRISHNAN, A. N., GANAPATHY, V., MATHAN, V. I. & BAKER, S. J. (1976*b*) Intestinal perfusion studies in tropical sprue. I. Amino acid and dipeptide absorption. *Gut 17*, 511-516

KIM, Y. S. & BROPHY, E. J. (1976) Rat intestinal brush border membrane peptidases. I. Solubilization, purification, and physico-chemical properties of two different forms of the enzyme. *J. Biol. Chem. 251*, 3199-3205

KIM, Y. S., BROPHY, E. J. & NICHOLSON, J. A. (1976) Rat intestinal brush border membrane peptidases. II. Enzymatic properties, immunochemistry, and interactions with lectins of two different forms of the enzyme. *J. Biol. Chem. 251*, 3206-3212

LEONARD, J. V., MARRS, T. C., ADDISON, J. M., BURSTON, D., CLEGG, K. M., LLOYD, J. K., MATTHEWS, D. M. & SEAKINS, J. W. (1976) Intestinal absorption of amino acids and peptides in Hartnup disorder. *Pediatr. Res. 10*, 246-249

MATTHEWS, D. M. (1975) Intestinal absorption of peptides. *Physiol. Rev. 55*, 537-608

RUBINO, A., FIELD, M. & SHWACHMAN, H. (1971) Intestinal transport of amino acid residues of dipeptides. Influx of glycine residue of glycyl-L-proline across mucosal border. *J. Biol. Chem. 246*, 3542-3548

SJÖSTRÖM, H. (1974) Enzymatic properties of pig intestinal proline dipeptidase. *Acta Chem. Scand. B28*, 802-808

Discussion

Silk: When strips of intestine are incubated *in vitro* there is a very rapid efflux of cytoplasmic peptidases into the mucosal medium (Silk & Kim 1976). In your work with strips of monkey intestine have you considered the possibility, therefore, that inhibition of peptide uptake by amino acids occurs on the mucosal side of the strips?

Radhakrishnan: The incubation is very brief (2 min), and inhibition very strong, comparable to inhibition by peptides. In our experiments the amino acids are used at much higher concentrations, 40 or 50 mM, than the peptides, which are used at 2 mM.

Silk: Despite what you say, we showed that efflux of cytoplasmic enzymes is instantaneous (Silk & Kim 1976). What component of that 2 mM dipeptide substrate is in fact hydrolysed in the mucosal medium by effluxing cytoplasmic enzymes?

Radhakrishnan: In our earlier studies we measured the amino acids in the medium. The figure is very low with the 2-min incubation, although after 10 or 20 min there is extensive hydrolysis. There is an increase of both glycine and leucine in the lumen of patients with sprue and we are inclined to ascribe it to back-flow. Has anyone shown an actual kinetic advantage for a peptide which has more than one type of amino acid?

Matthews: I was surprised when you said that peptides with two different amino acid residues don't show a kinetic advantage over the equivalent mixture of free amino acids. Dr Adibi showed that there was a kinetic advantage with Gly-Leu (Adibi 1971), and David Silk found it with Gly-Ala, Ala-Gly and Ala-Gly-Gly (Silk *et al.* 1973, 1974). My own group initially described the phenomenon with mixed peptides of Gly and Met (Matthews *et al.* 1969) and later, using rat small intestine *in vitro*, we reported it with nearly 20 mixed dipeptides (Burston *et al.* 1972), so I think it does occur. In all instances I have quoted, there was much more rapid uptake of one or both amino acids from the peptides than from the equivalent amino acid mixtures.

Radhakrishnan: With Gly-Gly, for instance, if it goes a little faster than half the rate of the amino acid, more glycine will be absorbed because of the two residues. That is, if the rate constant of glycine is 1.0, and the rate of Gly-Gly transport is 0.6, there is an advantage of about 20%. With mixed peptides there may not be an actual kinetic advantage. With homologous peptides there is a clear kinetic advantage, but with other peptides I am not sure whether there is.

Matthews: Kinetic advantage is a confusing term (Matthews 1975). Do you mean by kinetic advantage that the constituent amino acids of a peptide are absorbed more rapidly from a peptide than from the free form?

Radhakrishnan: No; let me make it clear. If we assume two separate transport systems, we can say there is a kinetic advantage of a dipeptide, say Gly-Leu, only if the actual rate of Gly-Leu uptake is greater than that of either glycine or leucine, all experiments being done at saturating concentrations for the respective transport system.

Adibi: There is evidence that one doesn't need to infuse a dipeptide with the same amino acid residues for a kinetic advantage to occur (Adibi 1971; Hellier *et al.* 1972*a,b*; Silk *et al.* 1973).

Nicholson: One of the problems in interpreting peptide perfusion studies or *in vitro* peptide absorption studies is the appearance of free amino acids in the perfusate or in the incubation medium. Do the amino acids arise from surface hydrolysis of peptides or from efflux of amino acids after intracellular peptide hydrolysis? Professor Radhakrishnan, you ascribe the appearance of luminal amino acids in your experiments to efflux rather than to surface hydrolysis. What did you do to distinguish between these two possibilities?

Radhakrishnan: We have not done anything specific in the human perfusion studies. However, there is no difference in the rate of amino acid uptake; for example, the rate of absorption of glycine and leucine was the same in the patients with sprue and the control subjects. So we infer that the increased levels of amino acids may be due to back-flow.

Rhodes: That is a key point. One really needs some way of differentiating hydrolysis from efflux. In theory one way of doing this is to inhibit the enzyme. I wonder whether Professor Kim or Dr Maroux have any suggestions for perfusionists about inhibiting the membrane hydrolase?

Kim: We have thought about it, like everyone else. Unfortunately most of the agents which seem to inhibit the brush border enzyme also seem to affect the cytoplasmic enzyme as well. For instance, zinc very effectively reduces brush border activity but it also inhibits the cytoplasmic enzyme. We thought about using antibodies but I don't know how feasible that is.

De Laey: It might be possible to reduce the membrane enzyme activity by reducing oral nutrition and giving the substrate parenterally. It has been shown that true brush border enzymes do not increase on parenteral nutrition but only on oral nutrition (Levine *et al.* 1974).

Kim: Starvation and prolonged parenteral nutrition both reduce the brush border enzyme activity considerably, but they don't completely abolish it. Therefore it will be difficult to interpret the results.

De Laey: Invertase activity was quite recently shown to be almost abolished by parenteral nutrition (Castro *et al.* 1975).

Rubino: I found the data on the biphasic kinetics of Gly-Leu uptake very interesting in relation to the problem of whether there is more than one transport process for peptides.

When we studied glycylproline uptake in rabbit ileum the bimodal kinetics we observed suggested that there are two transport processes, one with a low K_m and the other with a high K_m (Rubino *et al.* 1971). More recently we did analogous studies in rabbit jejunum but we only found evidence for a single transport mechanism. Even in the intestine of newborn rabbits there is evidence for a single transport mechanism for Gly-Pro.

Is it possible that one of the two apparent saturable processes for Gly-Leu uptake is due to hydrolysis of the dipeptide at the membrane followed by transport of free amino acids, so that one of the two kinetic components reflects transport of free amino acids?

Radhakrishnan: We can show amino acid inhibition of both the high K_t and the low K_t processes. I don't think this is due to hydrolysis since the inhibition was very much higher than the observed hydrolysis. We have not seen this in the human, possibly because the concentration intervals were far apart.

Rubino: Do you think that inhibition of Gly-Leu uptake by amino acids is due to competition at the transport site?

Radhakrishnan: We think the inhibition is at the transport site and not at the enzyme level. One can get the same kind of data as a consequence of cytosol enzyme inhibition, but we don't believe that is what happened.

Meister: Has anyone else who has studied uptake of a dipeptide obtained results similar to those found by Professor Radhakrishnan—that is, non-equivalent appearance of the two amino acids within the cell? Does one need to consider dilution with endogenous amino acids?

Adibi: The amounts of endogenous amino acids are extremely small as compared to concentrations of amino acids in perfusion solutions (Adibi & Gray 1967). This is agreed by everyone who has looked into it. With Gly-Leu we used concentrations ranging from 10 to 100 mM and compared each concentration with an equimolar mixture. Both glycine and leucine were taken up at higher rates from the dipeptide than from free amino acid solutions (Adibi 1971). I think Silk *et al.* (1973) and Hellier *et al.* (1972*a*,*b*) have found this with other dipeptides. There is not much controversy over this. The observation is not due to dilution with endogenous amino acids.

Radhakrishnan: If the amino acid and the dipeptide transport systems are separate, it is unfair to compare the two systems in suboptimal conditions. For instance, it is not fair to compare the transport of Gly-Leu and an equimolar amino acid mixture. It is not like Gly-Gly. The saturating points are different for the peptide system and for the amino acid system. If you are measuring the uptake of an amino acid, it should be done at the saturating concentration of the amino acid transport system. Similarly, when you measure the peptide transport you should also use the saturating concentration of the peptide transport system.

Adibi: You mean we should not use equimolar Gly-Leu but use leucine and glycine separately at comparable molar concentrations? Indeed, we have measured kinetics of leucine uptake from leucine solutions. The V_{max} of leucine uptake is 414 μmol/(min 15 cm), and the V_{max} of leucine uptake from glycylleucine solutions is 600 μmol/(min 15 cm) (Adibi & Soleimanpour 1974). Therefore, in our experimental conditions the maximal rate of leucine uptake from glycylleucine is greater by 50%.

Gardner: Why do you use monkeys, Professor Radhakrishnan?

Radhakrishnan: We get them very cheaply!

Smyth: Did you try using Gly-Leu instead of Leu-Gly? In our experience a number of amino acids stimulate the hydrolysis of Leu-Gly but inhibit hydrolysis of Gly-Leu (Cheeseman & Smyth 1975).

Radhakrishnan: We have not done that.

Fottrell: Were the patients with tropical sprue receiving treatment at the time of the study?

Baker: They were untreated patients, but the mucosa is not usually flat in tropical sprue, unlike what is seen in coeliac disease.

Fottrell: Are there differences in the rates of absorption in treated and untreated patients?

Baker: Yes; as the patient recovers the intestinal lesion returns towards normal and the absorption improves.

Fottrell: When comparing stabilities between brush border and cytosol peptide hydrolases in crude extracts, do you find that the brush border enzymes retain their stabilities after solubilization from the membranes?

Radhakrishnan: That work was reported by Heizer *et al.* (1972). The difference is very marked. The cytosol enzyme is very sensitive to *p*-hydroxymercuribenzoate at very low concentrations. At the concentration we used we could not pick up any cytosol enzyme activity. No one has yet isolated a true dipeptidase from the brush border so I am only referring to Gly-Leu or Gly-Gly hydrolase activity.

Payne: I understand that the cytosol enzyme which you purified is multimeric. In view of its extremely broad specificity can you unambiguously say that it is a single enzyme? What makes me ask this are recent reports on multimeric peptidases which indicate that associations of the subunits in different proportions can alter the substrate specificities (Stoll *et al.* 1973).

Radhakrishnan: The enzyme I am talking about is a single enzyme with a dimeric subunit. The molecular weight is 107 000. The specificity, especially the lack of activity, that we showed is with very large amounts of enzyme. The published reports are very confusing, and there may be some species differences too.

Fottrell: Although there are some species differences between peptide hydrolases there are also many similarities between the enzymes from different species. Your 'master dipeptidase' from monkey has much in common with one of the peptidases we have studied (Donlon *et al.* 1975). In the guinea-pig, our studies have shown that there are multiple forms of these enzymes, with overlapping specificities in some cases.

Maroux: Have you compared the activities of the brush-border and cytosol enzymes towards your substrates?

Radhakrishnan: We have done this in about half a dozen animals. We have not separated the brush borders but we separated the particulate and the supernatant fractions. We washed the particulate fraction extensively and we assume that it represents the brush border portion. The activity in this fraction varies from about 5 to 10% of the total activity when Gly-Leu is the test substrate, but we may have a situation where a particular peptidase is more in the brush border and less in the cytosol.

Maroux: Do you think that the enzyme responsible for this activity is a specific enzyme of the brush border?

Radhakrishnan: I have no information on the brush border enzyme. The one we isolated from the cytosol fraction is very specific. The only enzyme from the brush border purified to homogeneity is the one Professor Kim isolated, but it seems to act on larger peptides as well.

Kim: How many tripeptides have you tried your enzyme on?

Radhakrishnan: Just two; we tried substitutions of Gly-Leu at the amino end and at the carboxyl end. The two tripeptides studied were Gly-Gly-Leu and Gly-Leu-Gly. The enzyme doesn't touch these at all, even at very high concentrations and with increased incubation times. So when we say it doesn't act, we really mean it.

Josefsson: There has been a lot of confusion over the location of the dipeptidases. The glycylleucine dipeptidase which has been purified from two different species is entirely located in the cytosol. With specific antibodies against this peptidase we have been able to show conclusively that there is no activity whatsoever of this enzyme in the brush border. Glycylleucine dipeptidase is a true dipeptidase and so far there is no indication of any true dipeptidases in the brush border. We know there are enzymes with dipeptidase activity but we have to be very careful when we talk about dipeptidases.

Radhakrishnan: I was talking about glycylleucine dipeptidase activity and not a specific enzyme when I was talking about the brush border.

Parsons: If an amino-oligopeptidase hydrolyses a tripeptide it can't be a dipeptidase.

Radhakrishnan: Not by my definition.

Adibi: Did you observe inhibition of dipeptide uptake by amino acids in monkey intestine only or have you observed the same in other systems?

Radhakrishnan: We observed this in the human perfusion studies, in one experiment.

Adibi: We looked for this but didn't find it (Adibi & Soleimanpour 1974). Has anyone else seen it?

Smyth: We found that amino acids could stimulate or inhibit digestion of peptides, but did not affect uptake (Cheeseman & Smyth 1975).

Matthews: David Silk and I have some perfusion data in man which are relevant (P. D. Fairclough *et al.*, unpublished work). We perfused the jejunum with a mixture of peptides and amino acids, a partial enzymic hydrolysate, in the presence of an extremely high concentration (100 mM) of phenylalanine. We got such extensive inhibition of uptake of nearly all amino acids that we could only conclude that the amino acid was to some extent inhibiting peptide uptake.

Silk: These experiments were extended to compare the inhibitory effect of high concentrations of phenylalanine on uptake of amino acid residues from both the tryptic hydrolysate of casein and its equivalent amino acid mixture. Much to our surprise, uptake of peptide-bound residues was inhibited more than uptake of free amino acids.

Discussion

ADIBI, S. A. (1971) Intestinal transport of dipeptides in man: relative importance of hydrolysis and intact absorption. *J. Clin. Invest. 50*, 2266-2275

ADIBI, S. A. & GRAY, S. J. (1967) Intestinal absorption of essential amino acids in man. *Gastroenterology 52*, 837-845

ADIBI, S. A. & SOLEIMANPOUR, M. R. (1974) Functional characterization of dipeptide transport system in human jejunum. *J. Clin. Invest. 53*, 1368-1374

BURSTON, D., ADDISON, J. M. & MATTHEWS, D. M. (1972) Uptake of dipeptides containing basic and acidic amino acids by rat small intestine *in vitro*. *Clin. Sci. (Oxf.) 43*, 823-837

CASTRO, G. A., COPELAND, E. M., DUDRICK, S. J. & JOHNSON, L. R. (1975) Intestinal disaccharidase and peroxidase activities in parenterally nourished rats. *J. Nutr. 105*, 776-781

CHEESEMAN, C. I. & SMYTH, D. H. (1975) Interaction of amino acids, peptides and peptidases in the small intestine. *Proc. R. Soc. Lond. B Biol. Sci. 190*, 149-163

DONLON, J., O'CUINN, G., PIGGOTT, C. O. & FOTTRELL, P. F. (1975) Comparative studies of multiple peptide hydrolases in guinea-pig small intestinal mucosa and the implications of these studies for the 'missing peptidase' hypothesis for coeliac disease, in *Isozymes*, vol. II (Markert, C. L., ed.), pp. 275-287, Academic Press, New York

HEIZER, W. D., KERLEY, R. L. & ISSELBACHER, K. J. (1972) Intestinal peptide hydrolases. Differences between brush border and cytoplasmic enzymes. *Biochem. Biophys. Acta 264*, 450-461

HELLIER, M. D., HOLDSWORTH, C. D., PERRETT, D. & THIRUMALAI, C. (1972*a*) Intestinal dipeptide transport in normal and cystinuric subjects. *Clin. Sci. (Oxf.) 43*, 659-668

HELLIER, M. D., HOLDSWORTH, C. D., MCCOLL, I. & PERRETT, D. (1972*b*) Dipeptide absorption in man. *Gut 13*, 965-969

LEVINE, G. M., DEREN, J. J., STEIGER, E. & ZINNO, R. (1974) Role of oral intake in maintenance of gut mass and disaccharide activity. *Gastroenterology 67*, 975-982

MATTHEWS, D. M. (1975) Absorption of peptides by mammalian intestine, in *Peptide Transport in Protein Nutrition* (Matthews, D. M. & Payne, J. W., eds.), pp. 61-146, North-Holland, Amsterdam

MATTHEWS, D. M., LIS, M. T., CHENG, B. & CRAMPTON, R. F. (1969) Observations on the intestinal absorption of some oligopeptides of methionine and glycine in the rat. *Clin. Sci. (Oxf.) 37*, 751-764

RUBINO, A., FIELD, M. & SHWACHMAN, H. (1971) Intestinal transport of amino acid residues of dipeptides. I. Influx of the glycine residue of glycyl-L-proline across mucosal border. *J. Biol. Chem. 246*, 3542-3548

SILK, D. B. A. & KIM, Y. S. (1976) Release of peptide hydrolases during incubation of intact intestinal segments *in vitro. J. Physiol. (Lond.) 258*, 489-497

SILK, D. B. A., PERRETT, D. & CLARK, M. L. (1973) Intestinal transport of two dipeptides containing the same two neutral amino acids in man. *Clin. Sci. Mol. Med. 45*, 291-299

SILK, D. B. A., PERRETT, D., WEBB, J. P. W. & CLARK, M. L. (1974) Absorption of two tripeptides by the human small intestine: a study using a perfusion technique. *Clin. Sci. Mol. Med. 46*, 393-402

STOLL, E., ERICSSON, L. H. & ZUBER, H. (1973) The function of the two subunits of thermophilic aminopeptidase I. *Proc. Natl. Acad. Sci. U.S.A. 70*, 3781-3784

Dipeptide transport in the intestinal mucosa of developing rabbits

A. RUBINO and S. GUANDALINI

Institute of Child Health, Department of Paediatrics, 2nd School of Medicine, Naples

Abstract Influxes of glycyl-L-proline (a dipeptide which is not hydrolysed in the membrane and is transported intact across the brush border) and of glycyl-L phenylalanine (a dipeptide which has affinity for the peptide transport system and is hydrolysed at the brush border membrane) have been studied in the small intestine of fetal, newborn and suckling rabbits. For glycyl-L-phenylalanine, transport as the intact dipeptide and 'membrane hydrolysis + amino acid transport' have been measured separately by using glycyl-L-proline and L-leucine as selective inhibitors of each pathway. For comparison, uptake of free glycine and of free phenylalanine has also been studied.

The intestine of newborn rabbits is shown to have a translocation process for intact dipeptides which is saturable with a low K_t and stimulated by sodium ions, and which is not shared by free amino acids. This process resembles that described in adult animals, except that the maximal velocity is much higher in newborns. The developmental pattern of this uptake process for dipeptides differs markedly from that of free glycine, thus providing a new type of evidence for the distinction between amino acid and dipeptide transport processes. The developmental pattern of the free phenylalanine uptake process also differs from the development of the 'superficial hydrolysis + amino acid transport' component of glycyl-L-phenylalanine uptake.

These data suggest that the advantage of mucosal uptake of peptides, compared to the uptake of free amino acids, is much greater in the early stages of postnatal life than in the adult.

It has become well established that a carrier mechanism selective for dipeptides and tripeptides occurs in the intestinal brush border (Matthews & Adibi 1976). There is evidence that this carrier mechanism is distinct and independent of the transport systems for free amino acids. This evidence is of three main types. First, amino acids do not compete with peptides for mucosal uptake, whereas peptides do compete with one another (Adibi 1971; Rubino *et al.* 1971; Addison *et al.* 1972, 1974, 1975; Caspary 1973; Cook 1973, 1974; Das & Radhakrishnan 1974, 1975; Adibi *et al.* 1975; Sleisenger *et al.* 1976). Secondly, in some congenital defects of amino acid transport the 'affected' amino acids are absorbed normally or almost normally when presented to the mucosa as dipeptides (Asatoor *et al.* 1970; Hellier *et al.* 1972; Tarlow *et al.* 1972; Milne & Asatoor 1975; Silk *et al.* 1975). Thirdly, during transport peptides and free amino acids show different types of interaction with sodium ions (Rubino *et al.* 1971; Cheeseman & Parsons 1974).

Additional evidence allowing individual transport systems to be distinguished may come from ontogenetic studies, provided that there are differences in the developmental patterns. In this paper we discuss transport of dipeptides across the intestinal brush border of developing rabbits. The particular dipeptides referred to are glycyl-L-proline and glycyl-L-phenylalanine. These substrates were chosen for the developmental study for the reasons given in the next section.

MECHANISMS OF UPTAKE OF GLYCYL-L-PROLINE AND GLYCYL-L-PHENYLALANINE

Glycyl-L-proline influx has been studied in the ileum (Rubino *et al.* 1971) and more recently in the jejunum (A. Rubino, S. Guandalini & S. Auricchio, unpublished work) of adult rabbits. Although in the ileal mucosa there was evidence for two entry mechanisms—one with a low K_t and a low V_{max}, the other with a high K_t and a high V_{max} (Rubino *et al.* 1971)—only one entry process (low K_t and low V_{max}) was observed in the jejunum. The influx process with a low K_t was studied further as to its interaction with sodium ions and molecular specificity (Rubino *et al.* 1971). The maximal velocity of this influx is stimulated by sodium and it therefore differs from the influxes of free amino acids, where sodium affects the affinity constants and not the maximal velocity (Schultz & Curran 1970). There is no competition at the influx site between glycyl-L-proline and its constituent amino acids or other free amino acids, though there is competition with various dipeptides and tripeptides. Indeed, provided that they had a free amino group, all the dipeptides and tripeptides we tested competed with glycyl-L-proline.

Two further observations made in our laboratory are relevant to the mechanism of glycyl-L-proline uptake. First, glycyl-L-proline dipeptidase of the rabbit small intestine has been shown to be entirely intracellular and to be absent from the brush border (A. Rubino, M. Field & S. Auricchio, unpublished work). Secondly, influxes of the glycine and proline residues of glycyl-L-proline were measured simultaneously in rabbit small intestine by using glycyl-L-proline labelled with a different isotope in each residue: over a glycyl-L-proline concentration range from 0.5 to 30 mM, the ratio between the two influxes was not significantly different from 1. Glycyl-L-proline therefore appears to be transported intact across the brush border by the system selective for dipeptides and tripeptides, and to be hydrolysed subsequently within the cell. Since this dipeptide appears to follow only this absorptive pathway, it can serve as a useful substrate in studies of the development of the carrier system for peptides.

It has become clear, however, that there is at least one other mechanism for peptide absorption, namely hydrolysis at the brush border followed by transport of the released amino acids (Matthews & Adibi 1976). In general, it would appear that there are only a few peptides that are not hydrolysed at the brush border and that are therefore transported intact; most peptides can follow both pathways for absorption, since they both have affinity for the peptide transport system and are hydrolysed at the brush border. Glycyl-L-phenylalanine may be an example of such a dipeptide, since it is a competitor of glycyl-L-proline at the influx site, but is also hydrolysed by the brush border (Andria *et al.* 1976).

DEVELOPMENT OF DIPEPTIDE UPTAKE

In our studies of the intestinal uptake of glycyl-L-proline and of glycyl-L-phenylalanine in fetal, newborn and suckling rabbits, segments of small intestine were preincubated in an oxygenated Ringer solution (Rubino *et al.* 1971) and then incubated in the presence of the radioactively labelled substrate. Glycyl-L-proline was labelled in the glycine residue, while glycyl-L-phenylalanine was labelled in the phenylalanine residue. In other words, we measured the uptake of the glycine residue of glycyl-L-proline and that of the phenylalanine residue of glycyl-L-phenylalanine. Incubation was less than 60 s and during this time uptake was linear, indicating that unidirectional influx was measured. Uptake was calculated from the amount of radioactivity in the tissue at the end of incubation and it was corrected for the extracellular space, as given by the addition of inulin to the incubation medium.

Since the entire segment of small intestine was suspended in the medium, both mucosal and serosal sides were exposed, but preliminary results indicate that uptake by the serosal side was always less than 10% of uptake by the mucosa.

Fig. 1 shows the uptake of glycyl-L-proline by rabbit jejunum during development and, for comparison, the uptake of free glycine. There was no significant change in the uptake of free glycine during the whole period under study, while glycyl-L-proline uptake showed a striking perinatal peak, with a gradual increase from the 25th day of gestation. Uptake of glycylproline

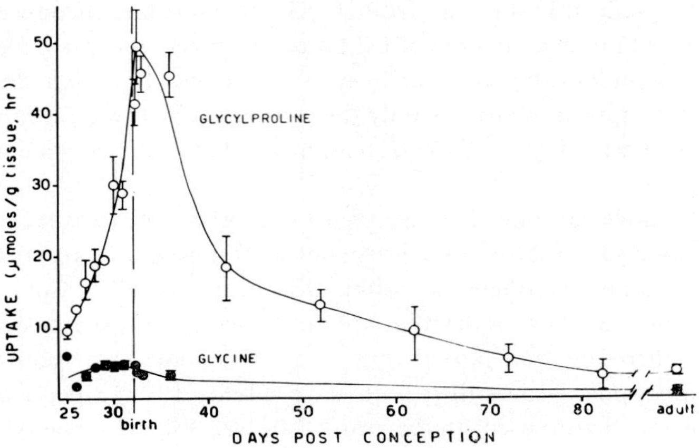

FIG. 1. Uptake of glycyl-L-proline (○) and free glycine (●) in fetal, newborn and suckling rabbits. Uptake values are means ± 1 S.E.M. Concentrations of both substrates in the incubation medium were 0.5 mM.

remained constant during the first six postnatal days and then declined steadily, reaching adult values at about 50 days of postnatal age. Analogous results were obtained in the ileum of developing rabbits. The prenatal period during which glycyl-L-proline uptake increased corresponds to the stage at which villi appear and enterocytes increase in number and differentiate into mature cells (Deren *et al.* 1965). It is noteworthy, however, that the uptake of free glycine did not show any prenatal increase and neonatal peak, thus differing from glycyl-L-proline uptake.

Fig. 1 shows that the influx of dipeptide-bound glycine has a kinetic advantage over the influx of free glycine, and this advantage is much more striking in newborn than in adult animals.

Some properties of the glycyl-L-proline influx process were studied when this influx was at its peak, i.e. in newborn rabbits aged one to six days. When measured in the presence of an excess of either amino acids or dipeptides, glycyl-L-proline influx is not inhibited, or is only slightly inhibited, by free amino acids, including glycine, L-proline, L-lysine and L-leucine, although it is strongly inhibited by all dipeptides tested (Fig. 2).

When glycyl-L-proline influx was measured at various substrate concentrations in the small intestine of newborn animals, no evidence for bimodal kinetics was obtained; in other words there was no evidence for more than one carrier mechanism. Table 1 compares the kinetic constants in newborn rabbit jejunum to the values found in adult animals. The maximal velocity is greater

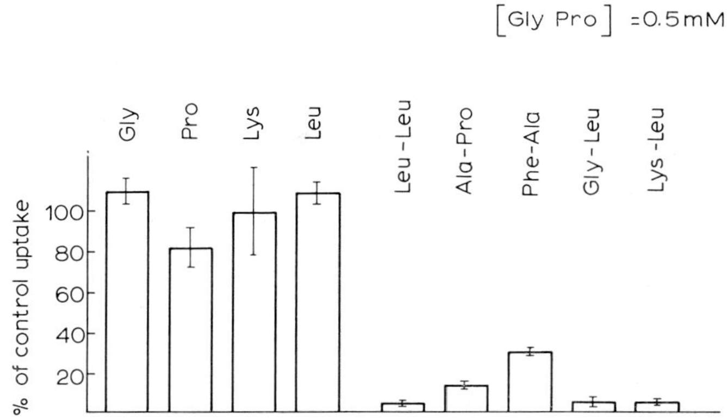

FIG. 2. Effect of free amino acids and dipeptides on glycyl-L-proline influx in the jejunum of newborn rabbits. The amino acids and dipeptides tested as inhibitors were added to the incubation medium at concentrations of 20 mM and 5 mM respectively. The incubation media for control tissues were adjusted with mannitol to the same osmolarity as the media containing the inhibitors. The glycyl-L-proline concentration was 0.5 mM in all cases. Influxes measured in the presence of inhibitors are given as percentages of control influxes (means ± 1 S.E.M.).

in newborn animals by a factor of 13, while there is no difference between the affinity constants in neonates and adults.

Under standard conditions, the influx was measured with 140 mM-sodium in both the preincubation and the incubation solutions. When choline was substituted for sodium in both solutions, the influx was reduced. The main effect of sodium was on the maximal velocity, which was reduced by almost two-thirds under sodium-free conditions, while there was no significant change in K_t (Guandalini *et al.* 1974). This type of effect of sodium ions is similar to that observed previously for glycyl-L-proline influx in adult rabbits (Rubino *et al.* 1971) and it differs from the behaviour of the influxes of free amino acids, where only the affinity constants are affected by sodium (Schultz & Curran 1970).

TABLE 1

Kinetic constants of glycyl-L-proline uptake in the rabbit jejunum (mean ± 1 S.E.M.)

	Newborn	*Adult*
V_{max} (μmol/cm^2 h)	11.2 ± 2.5	0.81 ± 0.25
K_t (mM)	0.95 ± 0.26	0.91 ± 0.35

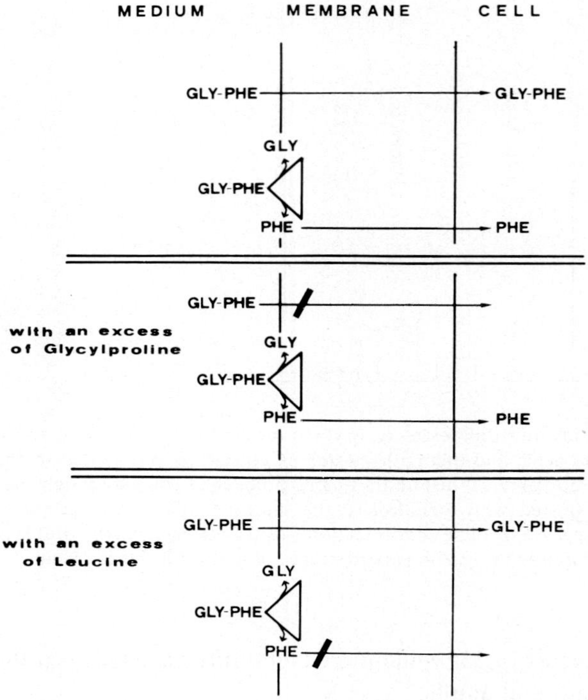

FIG. 3. Mechanisms for entry of the phenylalanine residue of glycyl-L-phenylalanine across the intestinal brush border of rabbits. Horizontal arrows and triangles represent transport systems and glycyl-L-phenylalanine dipeptidase, respectively. The top of the figure outlines two uptake mechanisms for the Phe residue: one is transport as intact dipeptide, the other is hydrolysis at the brush border membrane followed by uptake of the released Phe through the carrier system for neutral free amino acids. When measured in presence of an excess (20 mM) of either glycyl-L-proline or L-leucine in the incubation medium, the carrier mechanisms for dipeptide and for free Phe, respectively, are blocked, thus permitting uptake to be measured through either pathway separately.

In summary, the glycyl-L-proline influx process in newborn rabbits does not appear to differ fundamentally from the influx in adult animals, except that the maximal velocity is much bigger in newborn animals than in adults.

The uptake of glycyl-L-phenylalanine can be viewed as the resultant of two components, namely influx as intact dipeptide and hydrolysis at the brush border followed by transport of the released phenylalanine through the carrier system for neutral amino acids (Fig. 3). In our attempt to measure these two components separately, we determined glycyl-L-phenylalanine influx in the presence of a 40-fold excess of either glycyl-L-proline or leucine. The principle was that, in the presence of glycyl-L-proline, the influx of intact glycyl-L-phenyl-

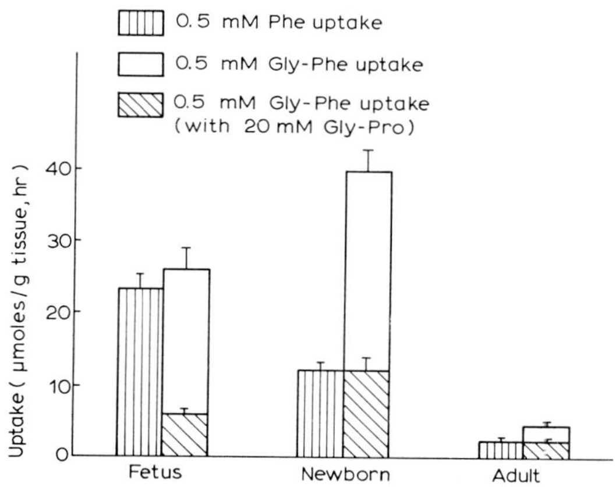

FIG. 4. Uptake of free phenylalanine and of the Phe residue of glycyl-L-phenylalanine by rabbit jejunum. Fetuses from 25 to 30 days of gestation, newborns 1 to 5 days old. Concentrations of substrates and the inhibitor (glycyl-L-proline) are given in the figure. Errors are 1 S.E.M.

alanine might have been abolished, so that the residual uptake could have been due entirely to hydrolysis followed by phenylalanine transport (Fig. 3). In contrast, in the presence of an excess of L-leucine, transport of free phenylalanine might have been eliminated and therefore the residual uptake could have been due entirely to transport as intact peptide. The following observations made in adult rabbits indicate that glycyl-L-proline and L-leucine work as selective inhibitors of either of the absorptive mechanisms for glycyl-L-phenylalanine, thus permitting each pathway to be measured separately (Guandalini *et al.* 1976). (1) The inhibitions by glycyl-L-proline and by L-leucine were additive and both were of the competitive type. (2) In the presence of L-leucine, the maximal velocity of glycyl-L-phenylalanine influx was similar to that described previously for glycyl-L-proline influx (Rubino *et al.* 1971). (3) In the presence of L-leucine the only significant effect of sodium on glycyl-L-phenylalanine influx was a stimulation of the maximal velocity, i.e. the type of effect described previously for the peptide carrier system (Rubino *et al.* 1971), while in the presence of glycyl-L-proline the only significant effect of sodium was on the affinity constant, i.e. the type of interaction with sodium described previously for the phenylalanine influx process (Schultz & Curran 1970).

In Fig. 4, [^{14}C-Phe]glycyl-L-phenylalanine uptake is compared with the uptake of free phenylalanine in the jejunum of fetal, newborn and adult rabbits. Fetuses ranged from 25 to 30 days of gestation and the results have been grouped as there were no significant changes in uptake over this period.

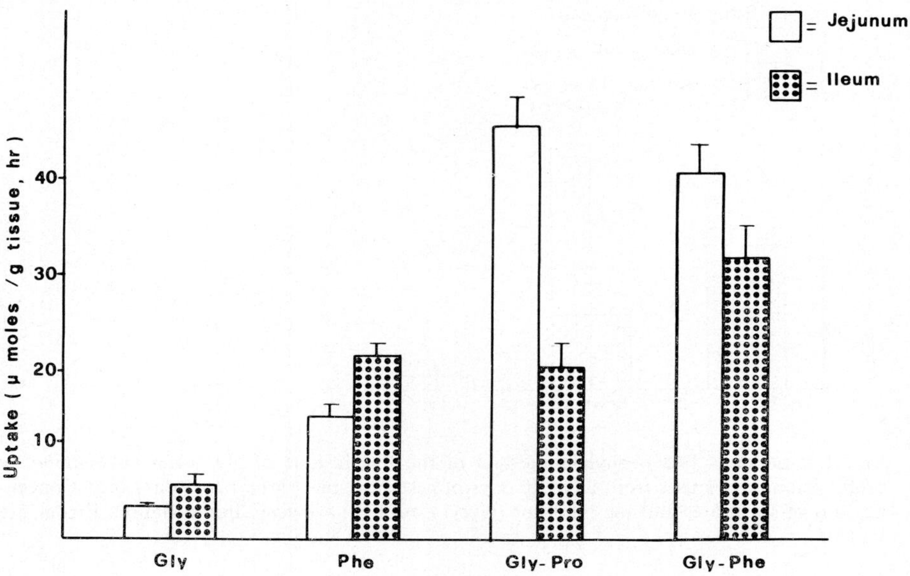

FIG. 5. Uptake of amino acids and dipeptides in jejunum and ileum of newborn rabbits. All four substrates were present in the incubation media in a concentration of 0.5 mM. Errors are 1 S.E.M.

Likewise, results from newborns aged from one to five days have been grouped as there were no significant changes in uptake during this period. The uptake of free phenylalanine is less in newborns than in fetuses, and it decreases further in adults. On the other hand, the uptake of glycyl-L-phenylalanine increases after birth as compared to prenatal levels. In other words, uptake of free phenylalanine differs from that of dipeptide-bound phenylalanine during these developmental stages.

Glycyl-L-phenylalanine uptake measured in the presence of an excess of glycyl-L-proline (i.e. the component due to superficial hydrolysis and transport of the released phenylalanine) follows the same pattern as the total glycyl-L-phenylalanine uptake: that is, it is lower in fetuses than in newborn rabbits, thus differing from the uptake of free phenylalanine. This difference in developmental patterns between uptake of free phenylalanine and uptake of the dipeptide through hydrolysis + phenylalanine transport suggests that in fetuses the hydrolytic enzyme might not yet have developed and that it might develop completely only after birth.

The glycyl-L-phenylalanine uptake measured in the presence of glycyl-L-proline is 50% of the total uptake in adult animals, while it is only 30% in newborns and 24% in fetuses. These findings suggest that transport as intact peptide is relatively more important during fetal and neonatal periods than in adults.

In Fig. 5 the uptake of glycine, phenylalanine and dipeptides in the jejunum of newborn rabbits is compared with that in the ileum. The uptake of glycine and of phenylalanine is larger in the ileum, while uptake of dipeptides is larger in the jejunum, suggesting that the distribution of amino acid transport mechanisms along the neonatal small bowel may differ from that of the dipeptide transport process. This observation is in keeping with previous *in vitro* observations (Crampton *et al.* 1973) in adult rats; however *in vivo*, jejunum and ileum of rats absorb glycyl-L-proline at the same rate (Lane *et al.* 1975).

CONCLUSIONS

The transport mechanism selective for intact peptides and the absorptive mechanism due to hydrolysis at the brush border and transport of the released amino acids both occur in the small intestine of fetal and newborn rabbits.

In newborn rabbits, the dipeptide carrier mechanism shows kinetic properties, molecular specificity and type of interaction with sodium similar to what has been described previously in adult animals, except that the maximal velocity is higher in newborns (by a factor of 13 in the jejunum and by a factor of 5 in the ileum).

The developmental pattern of the influx mechanism for dipeptides differs from that of the influx mechanism for neutral amino acids, thus providing additional evidence for a distinction between these two processes. Also, dipeptide influx and the influx of neutral free amino acids in newborn animals differ from each other in their distribution through the length of the small intestine.

Finally, due to the larger maximal velocity of the dipeptide carrier mechanism in newborns, its relative importance compared to the carrier mechanisms for free amino acids seems to be much bigger in neonatal than in adult small intestine. The significance of this observation in relation to the composition of elemental diets for premature as well as full-term infants should be further investigated.

ACKNOWLEDGEMENTS

We gratefully acknowledge the financial support provided by the Consiglio Nazionale delle Ricerche, Italy, and the technical assistance of Eduardo De Gregorio.

References

ADDISON, J. M., BURSTON, D. & MATTHEWS, D. M. (1972) Evidence for active transport of the dipeptide glycylsarcosine by hamster jejunum *in vitro*. *Clin. Sci. (Oxf.) 43*, 907-911

ADDISON, J. M., MATTHEWS, D. M. & BURSTON, D. (1974) Competition between carnosine and other peptides for transport by hamster jejunum *in vitro*. *Clin. Sci. Mol. Med. 46*, 707-714

ADDISON, J. M., BURSTON, D., DALRYMPLE, J. A., MATTHEWS, D. M., PAYNE, J. W., SLEISENGER, M. H. & WILKINSON, S. (1975) A common mechanism for transport of di- and tripeptides by hamster jejunum *in vitro*. *Clin. Sci. Mol. Med. 49*, 313-322

ADIBI, S. A. (1971) Intestinal transport of dipeptides in man: relative importance of hydrolysis and intact absorption. *J. Clin. Invest. 50*, 2266-2275

ADIBI, S. A., MORSE, E. L., MASILAMANI, S. S. & AMIN, P. M. (1975) Evidence for two different modes of tripeptide disappearance in human intestine: uptake by peptide carrier systems and hydrolysis by peptide hydrolases. *J. Clin. Invest. 56*, 1355-1363

ANDRIA, G., MARZI, A. & AURICCHIO, S. (1976) α-Glutamyl-β-naphthylamide hydrolase of rabbit small intestine. Localization in the brush border and separation from other brush border peptidases. *Biochim. Biophys. Acta 419*, 42-50

ASATOOR, A. M., CHENG, B., EDWARDS, K. D. G., LANT, A. P., MATTHEWS, D. M., MILNE, M. D., NAVAB, F. & RICHARDS, A. J. (1970) Intestinal absorption of two dipeptides in Hartnup disease. *Gut 11*, 380-389

CASPARY, W. F. (1973) Interaction of dipeptide absorption and transport of free amino acids, in *Biochemical and Clinical Aspects of Peptide and Amino Acid Absorption* (Rommel, K. & Goebell, A., eds.), pp. 29-40, Schattauer, Stuttgart and New York

CHEESEMAN, C. I. & PARSONS, D. S. (1974) Intestinal absorption of peptides. Peptide uptake by small intestine of *Rana pipiens*. *Biochim. Biophys. Acta 373*, 523-526

COOK, G. C. (1973) Independent jejunal mechanisms for glycine and glycylglycine transfer in man *in vivo*. *Br. J. Nutr. 30*, 13-19

COOK, G. C. (1974) Effect of glycylglycine and glycine on jejunal absorption rate of L-histidine in man *in vivo*. *J. Physiol. (Lond.) 237*, 187-194

CRAMPTON, R. F., LIS, M. T. & MATTHEWS, D. M. (1973) Sites of maximal absorption and hydrolysis of two dipeptides by rat small intestine *in vivo*. *Clin. Sci. (Oxf.) 44*, 583-594

DAS, M. & RADHAKRISHNAN, A. N. (1974) Studies on the uptake of glycyl-L-leucine by strips of monkey small intestine. *Indian J. Biochem. Biophys. 11*, 12-16

DAS, M. & RADHAKRISHNAN, A. N. (1975) Studies on a wide-spectrum intestinal dipeptide uptake system in the monkey and in the human. *Biochem. J. 146*, 133-137

DEREN, J. J., STRAUSS, E. M. & WILSON, T. H. (1965) The development of structure and transport systems of the fetal rabbit intestine. *Dev. Biol. 12*, 467-489

GUANDALINI, S., RUBINO, A. & AURICCHIO, S. (1974) Intestinal transport of dipeptides by newborn rabbits. *Acta Pediatr. Scand. 63*, 165-166

GUANDALINI, S., MAGAZZU, G., CONTI, A. & RUBINO, A. (1976) A dual mechanism for peptide uptake by the small intestine of rabbits. *Pediatr. Res. 10*, 895

HELLIER, M. D., HOLDSWORTH, C. D., PERRETT, D. & THIRUMALAI, C. (1972) Intestinal dipeptide transport in normal and cystinuric subjects. *Clin. Sci. (Oxf.) 43*, 659-668

LANE, A. E., SILK, D. B. A. & CLARK, M. L. (1975) Absorption of two proline containing peptides by rat small intestine *in vivo*. *J. Physiol. (Lond.) 248*, 143-149

MATTHEWS, D. M. & ADIBI, S. A. (1976) Peptide absorption. *Gastroenterology 71*, 151-161

MILNE, M. D. & ASATOOR, A. M. (1975) Peptide absorption in disorders of amino acid transport, in *Peptide Transport in Protein Nutrition* (Matthews, D. M. & Payne, J. W., eds.), pp. 167-182, North-Holland, Amsterdam

RUBINO, A., FIELD, M. & SHWACHMAN, H. (1971) Intestinal transport of amino acid residues of dipeptides. 1. Influx of the glycine residue of glycyl-L-proline across mucosal border. *J. Biol. Chem. 246*, 3542-3548

SCHULTZ, S. G. & CURRAN, P. F. (1970) Coupled transport of sodium and organic solutes. *Physiol. Rev. 50*, 637-704

SILK, D. B. A., PERRETT, D. & CLARK, M. L. (1975) Jejunal and ileal absorption of dibasic amino acids and arginine-containing dipeptide in cystinuria. *Gastroenterology 68*, 1426-1432

SLEISENGER, M. H., BURSTON, D., DALRYMPLE, J. A., WILKINSON, S. & MATTHEWS, D. M. (1976) Evidence for a single common carrier for uptake of a dipeptide and a tripeptide by hamster jejunum *in vitro*. *Gastroenterology 71*, 76-81

TARLOW, M. J., SEAKINS, J. W. T., LLOYD, J. K., MATTHEWS, D. M., CHENG, B. & THOMAS, A. J. (1972) Absorption of amino acids and peptides in a child with a variant of Hartnup disease and coexistent coeliac disease. *Arch. Dis. Child. 47*, 798-803

Discussion

Parsons: One point of importance here is the stage of development of the exocrine digestive enzymes for protein. In rabbit stomach the development of pepsin secretion is delayed, isn't it?

Hemmings: There is very little secretion in the stomach of the fetal rabbit. The meconium is practically undisturbed throughout fetal life. Immediately after birth gastric secretion is still very small.

Parsons: What happens to pancreatic secretion?

Hemmings: I think again that there is very little pancreatic secretion until some time after birth.

Parsons: Professor Rubino, will you not therefore have to consider other factors in addition to those in your list?

Rubino: Zoppi *et al.* (1972) have shown that in humans pancreatic secretion is low during the first few postnatal days, but it develops rapidly and is stimulated by feeding. The development of amylase can be induced by starch and that of trypsin is stimulated by a protein-rich diet.

Parsons: Where are these peptides coming from if the pancreatic secretion isn't working?

Rubino: Even if maturation is not yet complete there is enough pancreatic proteolytic activity for alimentary proteins to be digested to a considerable extent, so that oligopeptides can be produced even in newborns.

Josefsson: The glycylproline system is an excellent comparative system because the only enzyme known to be capable of splitting amino acyl-proline peptide bonds is the proline dipeptidase in the intestinal cytosol. Therefore this dipeptide seems to be exclusively one which can be hydrolysed inside the cell, and hydrolysis occurs only in the cytosol. In contrast to the other dipeptides that have been talked about so far, Gly-Pro can therefore only be hydrolysed intracellularly. In general it seems that quantitatively most dipeptide hydrolysis is intracellular, that is, in the cytosol.

In studies where you try to inhibit the system with various kinds of amino acids in high concentrations you have to keep in mind that the cytosol-located dipeptidases we are talking about are very heavily product-inhibited. For example, proline dipeptidase is inhibited by proline, and glycylleucine dipeptidase is inhibited by a lot of different amino acids, amongst which leucine is one of the strongest inhibitors. When you interpret the inhibition data it is therefore possible that the inhibition by proline in your studies depends on inhibition of the enzyme activity, while the inhibition obtained with the dipeptides is an effect on the transport system. That means that there is a possibility of selective inhibition when either dipeptides or proline are added, because the glycylproline activity can't be inhibited more than by proline. So all the inhibition with dipeptides seems to be an inhibition of the transport system used for the uptake of Gly-Pro.

Your studies on the rabbit fetus, the newborn and the adult are in complete parallel with our studies on humans (Lindberg 1966), pigs (Lindberg & Karlsson 1970), and rats (Lindberg & Owman 1966). We found that both the proline dipeptidase activity and the main dipeptidase activity followed exactly the same curves. We got extremely high values just around the birth of the animals, then these values came down more or less slowly, though rather more rapidly than your uptake figures indicate, until the adult state. I think it is a general phenomenon that the cytosol dipeptidases have extremely high activity at the time of birth. The nutritional implications will have to be studied later but it is very important that amongst all the enzymes involved in proteolytic hydrolysis only the dipeptidases show these extremely high concentrations at the time of birth.

Rubino: Glycylproline is a convenient substrate to use in studying the dipeptide transport system because, as far as I know, it is one of the very few dipeptides not hydrolysed at the brush border. However, after incubation in the presence of glycylproline, only free amino acids can be found within the cell. The activity of glycylproline dipeptidase in the cytosol is so high that there is no way of showing intact glycylproline in the cell. This is in contrast with other dipeptides, such as glycylsarcosine, which have been found intact within the cell.

I have no explanation for the inhibition of glycylproline influx by proline other than the assumption that in newborn rabbits the transport process for intact glycylproline has a slight affinity for proline. As glycylproline dipeptidase is within the cell, and therefore operates after the influx of the dipeptide, I don't see how inhibition of the influx can be explained by inhibition of the enzyme activity inside the cell.

Josefsson: The normal functioning enterocyte, without proline in that high concentration, has enough capacity to hydrolyse all the glycylproline that enters the cells immediately. But when you add proline in such a high concentration and saturate the enterocyte with proline, proline dipeptidase is strongly inhibited and you can expect an accumulation of glycylproline into the cell. The capacity for hydrolysis is becoming so low that you really get an effect. The reduction in uptake was about 20%, so it was rather small, but I think this is the only way of explaining it. There is no glycylproline activity whatsoever anywhere else.

Rubino: What we measured in our experiments was influx, and the tissue was incubated with both substrate and inhibitors for less than 1 min. Do you think that in these conditions inhibition of influx can be explained by an effect on the intracellular enzyme?

Josefsson: Yes, because you have such a high concentration of proline.

Smyth: Isn't it likely that if intracellular hydrolysis is altered, this will affect the intracellular concentration of peptide and hence the gradient of peptide concentration across the intestine?

Josefsson: No, I don't think so. This is an excellent system for selecting between the dipeptide transport system and the hydrolysis of the dipeptide which occurs separately. The uptake of glycylproline is a pure transport phenomenon. This is why you get a strong inhibition when you add high concentrations of different dipeptides to your system. The main point is that this dipeptide differs from the other dipeptides in the system.

Semenza: Professor Rubino, you measured glycylproline uptake activity around birth. Do you know of any data on the development of brush-border-bound leucylaminopeptidase?

Rubino: I don't know of any study on the development of brush border peptidases.

De Laey: You probably know the results of Koldovsky and his group in 1966. They studied the development of enzyme activities in the small intestine of the rat, mouse and rabbit. Their main conclusion was that the developmental curves of tripeptidase and leucylaminopeptidase were similar low at birth, passing through a maximum in the 21-day-old rat (weaning), followed by a decrease (Noack *et al.* 1966). They observed an analogous pattern for most brush border enzymes not only in rats but also in mice and rabbits (Koldovsky 1966). The brush border enzyme findings seem to be slightly different from what you find with cytosol enzymes.

Rhodes: How much free glycine or free proline effluxes back into the lumen in the glycylproline transport experiments?

Rubino: At the end of incubation with glycylproline no free glycine or free proline are detectable in the medium.

Rhodes: One minute is rather short for that efflux. Have you done longer incubations?

Rubino: No. All I can say is that after 1-min incubations there is no free glycine or free proline in the medium. With glycylphenylalanine the story is different: at the end of incubation with this dipeptide there is free glycine in the medium.

Fottrell: The development of intestinal peptidase activity from either the brush border or cytosol fraction has important applications in the nutrition of the infant. Wall & Peters (1971) and our own group (P. F. Fottrell & R. Keane, unpublished work) found that cortisone has a marked influence on these enzymes. We found that a rat cytosol peptide hydrolase with relatively broad substrate specificity increased in activity when cortisone is injected about a day after birth. Steroids such as prednisolone are sometimes used in the treatment of coeliac disease, and one of the effects of such steroids may be the enhancement of intestinal peptide hydrolase activity. It would be interesting, therefore, to compare the effects of steroids such as cortisone on both the peptide transport systems and the peptide hydrolases.

Kenny: It might be interesting to look at the activity of dipeptidyl peptidase IV (EC 3.4.14. –) during development in the rabbit gut. This enzyme, which is present in the brush border of the gut, as it is in the kidney, hydrolyses substrates with a Gly-Pro- N-terminal sequence. *In vivo*, this enzyme is likely to be the means by which glycyl-L-proline is generated. More interesting is the possibility that the transport protein for glycyl-L-proline is identical with this dipeptidyl peptidase. If it showed the same development pattern as that observed by Professor Rubino, there would be some support for this idea.

Matthews: In connection with the question of why the newborn gut transports the peptide so very well, Professor Mellander mentioned to me that peptides had been found in milk.

Mellander: In human milk three peptide fractions have been isolated in our laboratory.

Lloyd: Does the composition of the diet affect development, Professor Rubino? You are showing a developmental pattern which *in utero* is based on intravenous food with the protein consisting of amino acids; immediately after birth it is based on milk containing peptides as well as protein; and then the adult rabbit presumably eats grass and other laboratory food. In talking about the implications of your findings for the nutrition of human infants it would be important to know whether the developmental pattern is actually influenced by the type of feeding, otherwise unjustified conclusions might be drawn from the results.

Rubino: I agree. We have not yet studied the effect of type of feeding on the developmental pattern of the dipeptide transport process. The controlling mechanisms for the increase in uptake during development and for the decrease after birth are both unexplored. One possibility is to study the effect of different diets. One other point to be considered is the concentrations of amino acids in plasma, which decrease immediately after birth and might influence the rate of dipeptide uptake by the intestine. The effect of hormones will also have to be explored. Anyway I fully agree that it is too early to draw conclusions about nutrition in newborns.

Parsons: The fetus swallows amniotic fluid; it would therefore be most interesting to see whether there are any peptides in the amniotic fluid.

Nützenadel: We should be aware that mucosal transfer as measured by Professor Rubino does not necessarily reflect intestinal absorption. The developmental pattern of the amino acid and glucose transport has been observed in many tissues. Different metabolic requirements of the intestinal epithelial cell during ontogeny could be the cause of the changes that are observed. In the rat, enhanced intracellular accumulation of glucose into immature intestinal mucosa was found *in vitro* (Fitzgerald *et al.* 1971), but *in vivo* the disappearance rates of glucose from the intestinal lumen were lower in newborn than weanling animals (Younoszai & Cynek 1975). Therefore the interpretation of these *in vitro* results in terms of nutrition of the newborn may be misleading.

Rubino: I agree. *In vitro* studies provide one kind of information but *in vivo* studies give other kinds of information and they are necessary for drawing conclusions about absorption and nutrition.

Radhakrishnan: You mentioned that you find no dipeptide inside the cell, Professor Rubino, but I am not clear about your double-labelling experiments. Wouldn't you expect glycine and proline to have different intracellular-pool sizes, which would alter the ratio?

Rubino: We were measuring the initial rate of unidirectional influx of the dipeptide across the brush border. When we did incubations with glycylproline labelled in both the glycine and the proline residues, the ratio of the glycine residue influx to the proline residue influx was 1. In other words there was no difference at all in the rates of uptake of the two amino acid residues, suggesting that what was transported across the brush border was the intact dipeptide.

Mellander: Proteins such as egg proteins and casein, with resistant peptide linkages, are used in the nutrition of very young individuals. Later I would like to go into the question of substrates (see p. 294).

Hemmings: The rabbit gut is rather inert proteolytically but there is a fierce exchange of protein across the highly proteolytic membrane of the yolk sac. We have found much catheptic activity in the rabbit yolk sac (Jones &

Hemmings 1971). I am sure a lot of interesting peptidase activity would be found in this organ.

Kenny: I have looked for peptidase activity in the yolk sac. The peptidases known to be present in the microvillus membrane of the gut and kidney are detectable in the yolk sac membrane but the activities are very low, about 1% or less of those in the kidney microvillus.

Burston: Why do you preincubate the tissue, Professor Rubino?

Rubino: Schultz *et al.* (1967) first described the technique of measuring influx across the intestinal brush border, as applied to influx of free amino acids. When we decided to measure influx of glycylproline, we were trying to compare our results with those on the influx of free amino acids reported by Schultz's group. Therefore we used the same experimental conditions, including a 30-min preincubation. Let me add that if preincubation lasts 10 min the results for glycylproline influx are exactly the same as if it lasts 30 min.

Burston: Dr Silk, would the cytosol enzyme keep coming out after 30 min?

Silk: In our experiments there was a linear efflux of cytoplasmic enzymes for up to an hour (Silk & Kim 1976).

Professor Rubino, is there a 70% uptake of Gly-Phe in the presence of Gly-Pro in adult animals?

Rubino: Yes.

Silk: Do you interpret these results as indicating that 70% of Gly-Phe in those experiments was hydrolysed at the surface of the cell?

Rubino: Yes. But this was only for ileum, not for jejunum. In relation to Dr Burston's question, I should stress that the tissue was washed out at the end of preincubation before being exposed to the labelled substrate. Also, at the end of incubation we could not find any free glycine or free proline in the medium, suggesting that no hydrolytic enzyme was in the medium itself.

Rhodes: Have you measured the enzyme in the preincubation medium?

Rubino: No, we haven't.

Semenza: Free amino acids have been used for inhibition. We do not know yet what the kinetic mechanism for these dipeptidases is, but it is likely to turn out to be either ordered uni bi or iso ordered uni bi in Cleland's (1963) nomenclature, by analogy with other peptidases. If this is so, we must remember that the first product will be a much poorer inhibitor than the second one liberated. In other words, with glycylleucine, for example, I would expect that leucine would be a much poorer inhibitor of glycylleucine peptidase activity than glycine. In fact, leucine is likely not to inhibit at all.

Norén: It really is the reverse. Leucine is the stronger inhibitor of glycylleucine dipeptidase.

Semenza: In terms of K_i?

Norén: Yes.

References

CLELAND, W. W. (1963) The kinetics of enzyme-catalyzed reactions with two or more substrates or products. I. Nomenclature, and rare equations. *Biochim. Biophys. Acta 67*, 104-137

FITZGERALD, S. F., REISER, S. & CHRISTIANSEN, P. A. (1971) Developmental pattern of sugar and amino acid transport in postnatal rat small intestine. *Pediatr. Res. 5*, 698-703

JONES, R. E. & HEMMINGS, W. A. (1971) Proteolytic enzymes of the yolk-sac splanchnopleur. *Biochim. Biophys. Acta 242*, 278-287

KOLDOVSKY, O., HERINGOVA, A., HOSKOVA, J., JIRSOVA, V., NOACK, R., FRIEDRICH, M. & SCHENK, G. (1966) The postnatal development of enzyme activities of the small intestine. *Biol. Neonat. 9*, 33-43

LINDBERG, T. (1966) Intestinal dipeptidases: characterization, development & distribution of intestinal dipeptidases of the human foetus. *Clin. Sci. (Oxf.) 30*, 505-515

LINDBERG, T. & KARLSSON, B. (1970) Changes in the intestinal dipeptidase activities during fetal and neonatal development of the pig as related to the ultrastructure of mucosal cells. *Gastroenterology 59*, 247-256

LINDBERG, T. & OWMAN, C. (1966) Intestinal dipeptidases. Development of dipeptidase activity in the small intestine of the rat as related to the development of the intestinal mucosa. *Acta Physiol. Scand. 68*, 141-151

NOACK, R., KOLDOVSKY, O., FRIEDRICH, M., HERINGOVA, A., JIRSOVA, V. & SCHENK, G. (1966) Proteolytic and peptidase activities of the jejunum and ileum of the rat during postnatal development. *Biochem. J. 100*, 775-778

SCHULTZ, S. G., CURRAN, P. F., CHEZ, R. A. & FUISZ, R. E. (1967) Alanine and sodium fluxes across mucosal border of rabbit ileum. *J. Gen. Physiol. 50*, 1241-1260

SILK, D. B. A. & KIM, Y. S. (1976) Release of peptide hydrolases during incubation of intact intestinal segments *in vitro*. *J. Physiol. (Lond.) 258*, 489-497

WALL, A. J. & PETERS, T. J. (1971) Changes in structure and peptidase activity of rat small intestine induced by prednisolone. *Gut 12*, 445-448

YOUNOSZAI, M. K. & CYNEK, A. (1975) *In vivo* D-glucose absorption in developing rat small intestine. *Pediatr. Res. 9*, 130-133

ZOPPI, G., ANDREOTTI, G., PAJNO-FERRARA, F., NJAI, D. M. & GABURRO, D. (1972) Exocrine pancreas function in premature and full term neonates. *Pediatr. Res. 6*, 880-886

Mechanisms of peptide transport

D. BURSTON, T. C. MARRS, M. H. SLEISENGER*, T. SOPANEN† & D. M. MATTHEWS

Department of Experimental Chemical Pathology, Vincent Square Laboratories of Westminster Hospital, London

Abstract In order to see whether the mammalian intestine possesses distinct dipeptide and oligopeptide systems, as *E. coli* does, the characteristics of uptake of the series Gly-Sar, Gly-Sar-Sar and Gly-Sar-Sar-Sar were investigated, using rings of everted hamster intestine *in vitro*. The dipeptide and the tripeptide were taken up by an active Na^+-dependent mechanism, but there was no appreciable uptake of the tetrapeptide. Uptake of Gly-Sar was inhibited in a competitive way by Gly-Sar-Sar, and the degree of inhibition was reasonably similar to that predicted from the K_t values of the two peptides. Gly-Sar-Sar, at high concentrations, could cause virtually complete inhibition of uptake of Gly-Sar, and vice versa. Gly-Sar-Sar-Sar had no effect on uptake of Gly-Sar. It was concluded that the tetrapeptide had little or no affinity for the uptake mechanism and that the intestine did not have distinct dipeptide and oligopeptide uptake systems. The simplest explanation of the results is that the dipeptide and the tripeptide share a single uptake system, though the possibility that they share multiple uptake systems which do not differ greatly in their kinetic characteristics cannot be excluded.

The hypothesis that neutral, basic and acidic peptides are taken up by three separate systems has been shown to be an over-simplification. Both basic and acidic peptides can inhibit uptake of neutral peptides. Glu-Glu inhibits uptake of Gly-Sar in a competitive way. There is a large component in uptake of Glu-Glu which cannot be inhibited by Gly-Sar, and this may indicate that Glu-Glu is taken up by two systems, one shared with Gly-Sar and one not available to the latter peptide.

Ehrlich ascites carcinoma cells and liver and kidney slices from the hamster can all take up dipeptides, apparently by a mediated process. Peptide uptake is however considerably less rapid than that of free His. Recent experiments indicate that the scutellum of germinating barley is capable of taking up dipeptides on a large scale, apparently as the result of active transport.

---

* Present address: University of California Service, Veterans Administration Hospital, San Francisco, California, USA

† Present address: Department of Biochemistry, University of Helsinki, Finland

Much of the recent work in our laboratory has been directed to the problem of whether there is only one or more than one peptide uptake system in mammalian small intestine. We have also carried out some preliminary experiments on peptide uptake by Ehrlich ascites carcinoma cells, and by liver and kidney slices, to investigate the ability of mammalian cells other than the absorptive cells of the intestine to take up small peptides, and we have just begun an investigation of peptide uptake by the scutellum of germinating barley. In this paper we shall discuss results in these areas. Standard three-letter abbreviations will be used for amino acid residues.

INTESTINAL PEPTIDE UPTAKE–ARE THERE MULTIPLE SYSTEMS?

Rubino *et al.* (1971) showed that intestinal uptake of Gly-Pro was inhibited by several dipeptides of neutral amino acids and by Gly-Gly-Gly, and during the last five years it has become clear that Gly-Pro, Car (β-Ala-His), Gly-Sar, Gly-Sar-Sar and a large number of other dipeptides and tripeptides behave as if they share, to some extent at any rate, the same uptake system or systems (see Matthews 1975*a,b*). In our work on competition for peptide uptake we have made extensive use of Car, Gly-Sar and Gly-Sar-Sar as reference peptides, since they are exceptionally resistant to hydrolysis and are accumulated largely intact in the intestinal wall *in vitro*; such hydrolysis as occurs appears to be almost entirely intracellular. In consequence, complications due to competition for transport involving free amino acids liberated in the brush border are avoided, and results are much easier to interpret. We have approached two problems. The first is that of whether the intestine, like *E. coli*, has distinct dipeptide and oligopeptide uptake systems. The results of our investigation suggest that it does not. The second problem is that of whether there are several distinct peptide uptake systems favouring particular classes of dipeptide, as some authors have suggested. Investigations of this problem are still far from complete—but it seems to us that the existence of more than one peptide uptake system is distinctly possible.

ARE THERE DISTINCT DIPEPTIDE AND OLIGOPEPTIDE UPTAKE SYSTEMS IN THE INTESTINE?

In *E. coli* (Payne 1975, 1976), the dipeptide uptake system includes among its structural requirements a terminal NH_2 group, a terminal COOH group, and a peptide bond. The oligopeptide system is distinguished by its lack of requirement for the terminal COOH group, which gives it the ability to accept peptides of several (up to seven) amino acid residues; probably it could accept larger peptides than this if they were not prevented from reaching it by the limitation in molecular size imposed by the pores in the cell wall surrounding the plasma membrane. Dipeptides can utilize the oligopeptide system, but oligopeptides cannot utilize the dipeptide uptake system.

In investigating the question of whether the intestine possesses dipeptide and oligopeptide uptake systems of this kind, we have relied on the behaviour of the series Gly-Sar, Gly-Sar-Sar, and Gly-Sar-Sar-Sar. All investigations were carried out *in vitro*, using rings of everted hamster jejunum incubated under O_2 at 37 °C in Krebs–Ringer phosphate saline, pH 7.2, either for 20 min, at which time accumulation is near-maximal, or for 2 min, which approximates conditions of influx. Preliminary experiments using Gly-Sar showed that uptake of this peptide was most extensive in a region about 8–10 cm from the pylorus, and rings were prepared from this region. In experiments on accumulation, unlabelled peptides were used and, after extraction from the rings, peptides and amino acids were estimated by ion-exchange chromatography; in experiments in which influx was measured, peptides labelled with [^{14}C]Gly were used and, after extraction, radioactivity was measured by scintillation counting.

The earliest findings (Addison *et al.* 1972, 1975*a*) bearing on the question of separate dipeptide and oligopeptide uptake systems were that both the dipeptide Gly-Sar and the tripeptide Gly-Sar-Sar were concentrated by the rings by a Na^+-dependent, active process. On the other hand the tetrapeptide Gly-Sar-Sar-Sar was very poorly taken up and was not concentrated (Fig. 1). Such uptake of the tetrapeptide as there was, was unaffected by Na^+-replacement and by 2,4-dinitrophenol (DNP), so that it did not appear to be the result of an active process.

In the next series of experiments (Addison *et al.* 1975*b*), accumulation of Gly-Sar-Sar was inhibited by a wide range of other dipeptides and tripeptides including Gly-Sar, but not by Gly-Sar-Sar-Sar. These findings suggested that if Gly-Sar, Gly-Sar-Sar and Gly-Sar-Sar-Sar were representative, the ability of hamster jejunum to take up peptides was confined to dipeptides and tripeptides, and did not extend to tetrapeptides. Thus it seemed unlikely that the gut possessed an oligopeptide transport system similar to that of *E. coli*.

The possibility remained, however, that either Gly-Sar or Gly-Sar-Sar might be taken up to some extent by a system unavailable to the other peptide. This was explored by an investigation of the kinetics of influx of these peptides, and of competition for influx between them (Sleisenger *et al.* 1976). Influx, V, of each peptide was measured over a range of concentrations, S, from 0.25 to 20 μmol/ml. Lineweaver-Burk plots ($1/V$ against $1/S$) were linear over the entire range (Fig. 2), as were Hanes plots (S/V against S) and Hofstee plots (V against V/S), suggesting that influx of both peptides conformed to simple Michaelis–

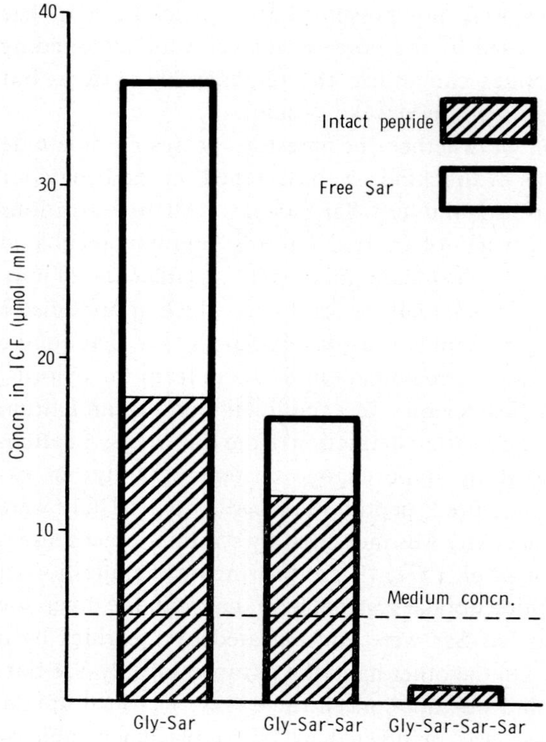

FIG. 1. Uptake (20 min) by rings of everted hamster jejunum of Gly-Sar, Gly-Sar-Sar and Gly-Sar-Sar-Sar, from 5 µmol/ml solutions. Final concentrations in the medium were not measurably different from the initial concentration. With Gly-Sar and Gly-Sar-Sar, some free Sar and free Gly appeared in the tissue as the result of hydrolysis which was apparently largely, if not entirely, intracellular. Thus total peptide uptake is represented by the sum of intact peptide and free Sar (or free Gly) appearing in the tissue, rather than by appearance of intact peptide alone. ICF: intracellular fluid. (Data from Addison *et al.* 1975*a*.)

Menten kinetics. Thus, this part of the investigation provided no evidence that influx of either peptide involved more than one transport system—though it certainly did not rule out such a possibility since, when more than one transport system exists, kinetic plots may fail to provide evidence of heterogeneity unless the kinetic parameters of the individual systems differ widely (Christensen 1975). All methods of plotting provided very similar estimates of K_t and V_{max}; those from the Lineweaver–Burk plots are given in Table 1. It appears that Gly-Sar-Sar has a lower apparent affinity for transport than Gly-Sar, and also a lower V_{max}.

In the next part of the investigation, influx of Gly-Sar was measured over a

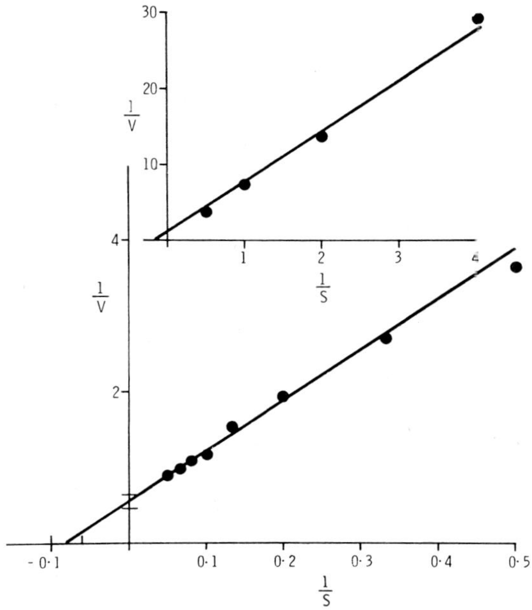

FIG. 2. Lineweaver–Burk plot of influx of [^{14}C]Gly-Sar-Sar into rings of everted hamster jejunum. Each point is derived from measurements of influx into six rings, two from each of three animals. The line plotted is a regression line obtained by the method of least squares and weighted for the variance of $1/V$. V is in μmol (g initial wet w)$^{-1}$ min^{-1} and S is in μmol/ml. 95% confidence limits for intercepts on both axes are indicated. (Reproduced from Sleisenger *et al.* 1976.)

range of concentrations in the presence of a constant concentration (10 μmol/ml) of Gly-Sar-Sar. The inhibition of influx of Gly-Sar was apparently competitive, and the degree of inhibition observed was slightly greater than, though not significantly different from, the inhibition predicted from the K_t values of the two peptides. This would not be expected if one of the peptides were taken up by two systems, one of which is not available to the other peptide—a situation in which the observed inhibition would be expected to be less than that predicted. The existence of the unstirred layer causes an error in estimations of K_t values, apparent K_t being larger than the true K_t of the transport mechanism. This effect has been discussed in detail elsewhere (Sleisenger *et al.* 1976); it seemed unlikely that the errors introduced by the unstirred layer were large enough, under our conditions, to invalidate the conclusions. However, arguments based on measurements of transport kinetics are seldom overwhelmingly convincing, so in a final series of experiments the problem was approached in a simpler way.

TABLE 1

K_t and V_{max} for Gly-Sar and Gly-Sar-Sar

	K_t (*μmol/ml*)	V_{max} (*μml (g initial wet wt)$^{-1}$ min^{-1}*)
Gly-Sar	7.6(6.5-9.4)	2.7(2.4-3.0)
Gly-Sar-Sar	12.6(10.0-16.8)	1.9(1.6-2.3)

In the Lineweaver–Burk plots, regression lines were plotted by the method of least squares after weighting for the variance of $1/V$. Values for K_t and V_{max} are given with 95% confidence limits in parentheses. (Data from Sleisenger *et al.* 1976.)

If two substrates are transported exclusively by a single shared system, or by multiple shared systems none of which is available to one substrate while being unavailable to the other, it should be possible to approach 100% inhibition of transport of each substrate by the other, if the concentration of the inhibitor substrate is sufficiently high. Experiments were therefore carried out in which influx of Gly-Sar (1 μmol/ml) was measured in the presence of Gly-Sar-Sar (5, 10, 40 and 60 μmol/ml), and influx of Gly-Sar-Sar (1 μmol/ml) was measured in the presence of Gly-Sar (5, 10, 40 and 60 μmol/ml). Plots of $1/1-R$ (where $R = V$ inhibited/V non-inhibited) against the reciprocal of the inhibitor concentration (Preston *et al.* 1974) intersected the ordinate in the region of 1.0 (Fig. 3), suggesting that at infinitely high concentrations of Gly-Sar-Sar, inhibition of influx of Gly-Sar would be virtually complete ($> 90\%$) and vice versa. Control experiments were carried out in which influx of each of the peptides (1 μmol/ml) was measured in the presence of Gly (60 μmol/ml) to exclude the possibility of some non-specific element in inhibition at very high concentrations of inhibitor. No effect was observed. Finally, influx of Gly-Sar (1 μmol/ml) and influx of Gly-Sar-Sar (1 μmol/ml) were studied in the presence of Gly-Sar-Sar-Sar (40 μmol/ml). There was no effect, showing that even at this high molar ratio the tetrapeptide did not inhibit influx of the dipeptide or the tripeptide.

The simplest hypothesis which will explain the results of this investigation is that the dipeptide Gly-Sar and the tripeptide Gly-Sar-Sar share a single uptake system which has little or no affinity for the tetrapeptide Gly-Sar-Sar-Sar. The results do not exclude the possibility that more than one system is involved in uptake of the dipeptide and tripeptide, but they do indicate that if multiple systems are involved none of these systems is available to one peptide while being unavailable to the other.

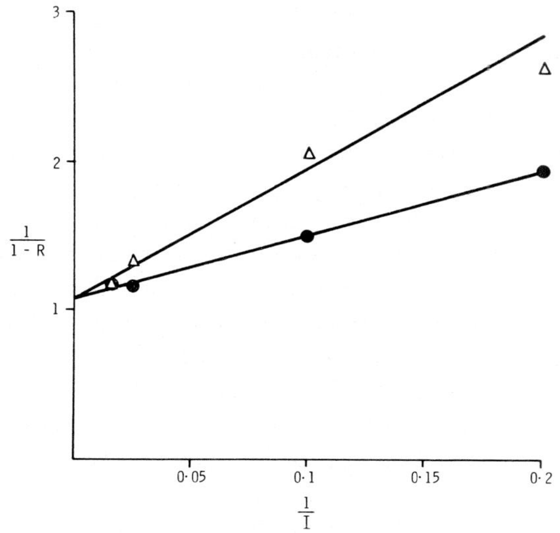

FIG. 3. Preston-Schaeffer-Curran plots (Preston *et al.* 1974) of influx of [^{14}C]Gly-Sar (1 μmol/ml) in the presence of Gly-Sar-Sar (5–60 μmol/ml) △——△ and influx of [^{14}C]Gly-Sar-Sar (1 μmol/ml) in the presence of Gly-Sar (5–60 μmol/ml) ●——●. (Reproduced from Sleisenger *et al.* 1976.)

ARE PEPTIDES TAKEN UP BY DIFFERENT SYSTEMS ACCORDING TO THE NET CHARGE ON THE AMINO ACID SIDE-CHAINS?

At an early stage in the investigation of the transport of amino acids by the small intestine, evidence was obtained suggesting that amino acids might be grouped into three main transport groups, containing neutral, basic and acidic acids (Wiseman 1955). In spite of the description of unexpected cross-inhibitions, the recognition that neutral amino acids are apparently transported by multiple systems, and that any given amino acid may be transported by more than one system, this early broad classification has not lost all validity. Accordingly, it seemed reasonable to be alert to the possibility that analogous transport groups might also exist for peptides.

A preliminary investigation of the ability of various dipeptides and tripeptides to inhibit uptake of Car by rings of everted hamster jejunum *in vitro* (Addison *et al.* 1974) showed that uptake of this peptide (5 μmol/ml) was inhibited by equimolar Gly-Pro, Pro-Hyp, Met-Met, Gly-Gly, Gly-Gly-Gly and Gly-Sar—but not by equimolar Glu-Glu or Lys-Lys. Further, there was no significant inhibitory effect of equimolar Glu-Glu on uptake of Lys from Lys-Lys (5 μmol/ml), or equimolar Lys-Lys on uptake of Glu from Glu-Glu (5 μmol/ml). (With readily hydrolysed peptides such as Glu-Glu, Lys-Lys and many others, peptide uptake is represented only by an increase in the free amino acids of the intestinal tissue. Uptake of amino acids released from the peptide in the brush border and in the incubation medium also contributes to this increase to a variable extent. 'Uptake of Lys from Lys-Lys', 'uptake of Glu from Glu-Glu' and similar phrases are intended to cover uptake of both peptide-bound and free amino acids.) These results led us to suggest the possibility that Car, Glu-Glu and Lys-Lys were each transported by separate systems. It was of course recognized that no conclusions could be drawn from the behaviour of peptides studied at one arbitrary concentration only, and we made it clear that our suggestion was based on only one possible interpretation of the results. In fact, at the time this suggestion was made we were rather prejudiced in favour of the view that there might be only one uptake system for all dipeptides and tripeptides, since we already suspected that dipeptides and tripeptides shared a system, and we knew that in *E. coli* the peptide uptake systems were said to be indifferent to the charge on the amino acid side chains. A little later, Das & Radhakrishnan (1975), having shown that influx of Gly-Leu into monkey small intestine was inhibited by a very wide range of dipeptides, and that for Ala-Ala, Ala-Leu, Lys-Lys and Glu-Glu the inhibition was competitive, put forward the hypothesis that all dipeptides of dietary origin might be taken up by the same system.

A second preliminary investigation (Addison *et al.* 1975*b*), this time of the ability of various dipeptides and tripeptides to inhibit uptake of Gly-Sar-Sar, showed that the hypothesis that neutral, basic and acidic peptides were taken up by separate systems was, at best, an over-simplification. Uptake of Gly-Sar-Sar (which apparently shares an uptake system with Car and many other neutral dipeptides and tripeptides) was fairly strongly inhibited by equimolar (5 μmol/ml) Lys-Lys, though not, curiously, by equimolar Lys-Lys-Lys; and it was also inhibited by equimolar Glu-Glu-Glu, though there was no significant inhibition by equimolar Glu-Glu. Internally neutralized peptides (Lys-Glu, Glu-Lys and Lys-Glu-Gly) also inhibited uptake of Gly-Sar-Sar (Fig. 4). The results have not by any means been summarized in full, but there would be little point in doing so, since it soon became clear that no further progress could be made without systematic and detailed investigations, including measurements of transport kinetics so that the concentrations used in competition experiments could take account of K_t values. More than a year ago we embarked on the first such investigation. It is turning out to be slow work and, though interesting observations have been made and some interim conclusions arrived at, much more remains to be done before the results can be adequately interpreted.

It was decided that in the first investigation we should keep the situation as

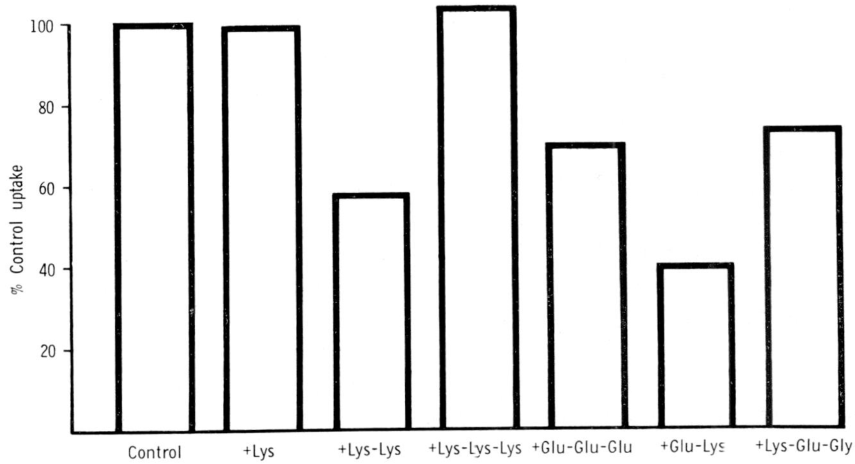

FIG. 4. Uptake (20 min) by rings of everted hamster jejunum of Gly-Sar-Sar (5 μmol/ml) alone and in the presence of free Lys and of several peptides of Lys and Glu. (Data from Addison *et al.* 1975*b*).

simple as possible by basing it on one neutral dipeptide which is resistant to hydrolysis, and one acidic dipeptide, and we set out to find out what we could about transport interactions between Gly-Sar and Glu-Glu. Rings of everted hamster jejunum were used, the experimental procedure being similar to that employed in the investigation of the behaviour of the series Gly-Sar, Gly-Sar-Sar and Gly-Sar-Sar-Sar (pp. 81-85). In the initial experiments, we studied 20-min uptake of unlabelled peptides. At a later stage, 20-min uptake and 2-min influx were measured, using [^{14}C]Gly-Sar and [^{14}C]Glu-Glu. When necessary, the pH of peptide-containing solutions was readjusted to the initial pH of the buffer used.

The first experiment gave encouraging results. Uptake (20 min) of Gly-Sar was measured in the presence of Glu-Glu in concentrations of from 5 to 30 μmol/ml, and a Preston–Schaeffer–Curran plot (Preston *et al.* 1974) suggested that at infinitely high concentrations of Glu-Glu uptake of Gly-Sar would be completely inhibited. Subsequently, under conditions of influx and at pH 5 (see below), it was shown that the inhibition was competitive. In the next experiment, we tried the effect of Gly-Sar on 20-min uptake of Glu from Glu-Glu. (Glu was measured as usual as the sum of Glu + Ala [Burston *et al.* 1972] since the tissue transaminates a proportion of Glu, which is represented by Ala.) Inhibitions were fairly substantial—uptake of Glu from Glu-Glu (5 μmol/ml) was reduced to 69% of control values by Gly-Sar (10 μmol/ml) and to 52% by Gly-Sar (60 μmol/ml). However, the Preston–Schaeffer–Curran plot gave irregular distribution of points, so that no extrapolation to infinitely high concentrations of Gly-Sar could be made. In spite of this, the results suggested that Gly-Sar was unlikely to be capable of completely inhibiting uptake of Glu from Glu-Glu, however high the inhibitor concentration. At this stage, an obvious possibility had to be explored: that the failure of Gly-Sar to inhibit uptake of Glu-Glu completely was because a substantial component in uptake of Glu from Glu-Glu was uptake of free Glu released in the incubation medium and brush border—this would not be expected to be inhibited by Gly-Sar owing to the independence of uptake of peptides and amino acids. We thought that if the free Glu component in uptake could be minimized, uptake of Glu from Glu-Glu might become much more susceptible to the inhibitory effect of Gly-Sar. The work of Fogel & Adibi (1974) suggested that brush border hydrolysis might be decreased by a reduction in pH. It was soon found that the concentration of free Glu appearing in the incubation medium at the end of 20 min could be reduced to about one-third if pH was reduced from the standard value of about 7 to 5. This reduction was accompanied by an *increase* in total uptake of Glu from the peptide, which was approximately doubled (Fig. 5). Subsequent experiments were carried out at pH 5 instead of 7.2, using a Tris-phosphate buffer (Reiser & Christiansen 1973). It is interesting to compare the effects of pH on hydrolysis and uptake in the case of Glu-Glu with the effects on hydrolysis and uptake in the case of Gly-Leu, one of the peptides studied by Fogel & Adibi (Fig. 6).

At pH 5, the final concentration of free Glu in the medium, after 20 min incubation in Glu-Glu (10 μmol/ml), was about 0.2 μmol/ml. Uptake of Glu from this concentration of free Glu was found to be unmeasurably low at 20 min, whereas uptake of Glu from Glu-Glu (10 μmol/ml) was more than 10 μmol/g wet weight. This suggested that unless the effective concentration of free Glu in the brush border during incubation with the peptide was much higher than the final concentration of free Glu appearing in the medium, uptake of free Glu was only a trivial component of total uptake at pH 5. A ten times higher (2.0 μmol/ml) concentration of free Glu in the medium was needed to give an uptake of free Glu of only 3 μmol/g wet weight.

In spite of the reduction in uptake of free Glu presumably associated with the reduction in hydrolysis at pH 5, it was found that the ability of Gly-Sar to inhibit uptake of Glu from Glu-Glu was not appreciably enhanced, and in some experiments appeared to be diminished. A pilot experiment showed that uptake of Glu from Glu-Glu (5 μmol/ml) was reduced to 48% of the control value by Gly-Sar (60 μmol/ml) at pH 5, as against 52% at pH 7.2. Measurement of uptake of Glu from Glu-Glu (10 μmol/ml) in the presence of a range of concentrations of Gly-Sar suggested (though again the Preston–Schaeffer–

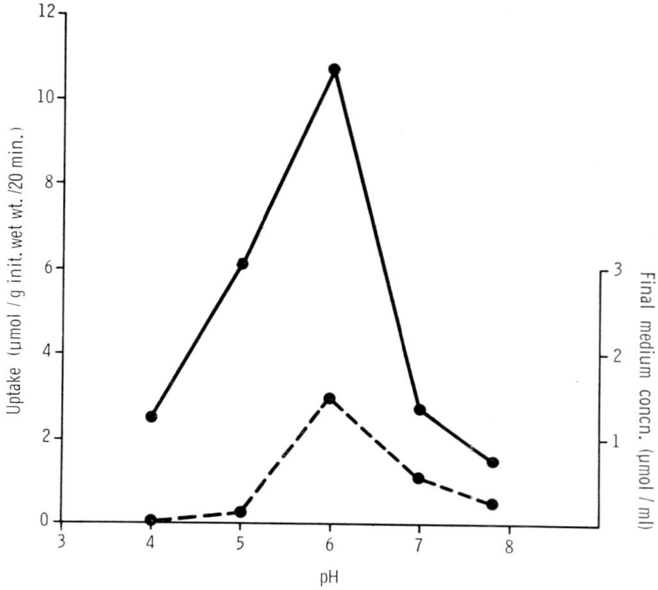

FIG. 5. Effect of pH on uptake of [^{14}C]Glu from [^{14}C]Glu-Glu and on hydrolysis of Glu-Glu by rings of everted hamster jejunum. Uptake ●—●; final concentration of Glu in medium ●---●. The incubation time was 20 min and the concentration of Glu-Glu 5 μmol/ml. Uptake, which is expressed as Glu-Glu, was estimated by measurements of radioactivity in the tissue. The final concentration of free Glu in the incubation medium was measured on the amino acid analyser. Each flask contained as usual a single ring weighing between 10 and 20 mg. The medium (3 ml) was buffered with Tris-phosphate buffer (Reiser & Christiansen 1973). Each point is the mean of six observations.

Curran plot was unsatisfactory) that an approach to complete inhibition was still unattainable. In the highest inhibition observed in this experiment, in the presence of Gly-Sar (60 μmol/ml), inhibited uptake was still 84% of the control value.

Since the possibility could not be ruled out that the concentration of free Glu released in the brush border was much higher than that appearing in the medium, we next made an attempt to minimize uptake of free Glu by the use of Met, still working at pH 5. We already knew that Met was a powerful inhibitor of uptake of free Glu, and it was shown that Met (20 μmol/ml) reduced uptake of free Glu (10 μmol/ml) to only 11% of the mean control value. Uptake of Glu from Glu-Glu (10 μmol/ml) was reduced by Met (20 μmol/ml) to 67% of the mean control value. If we assume that Met inhibits only uptake of free Glu, and not that of Glu-Glu, this suggests that something over 33% of the uptake of Glu from Glu-Glu (10 μmol/ml) was accounted for by uptake of

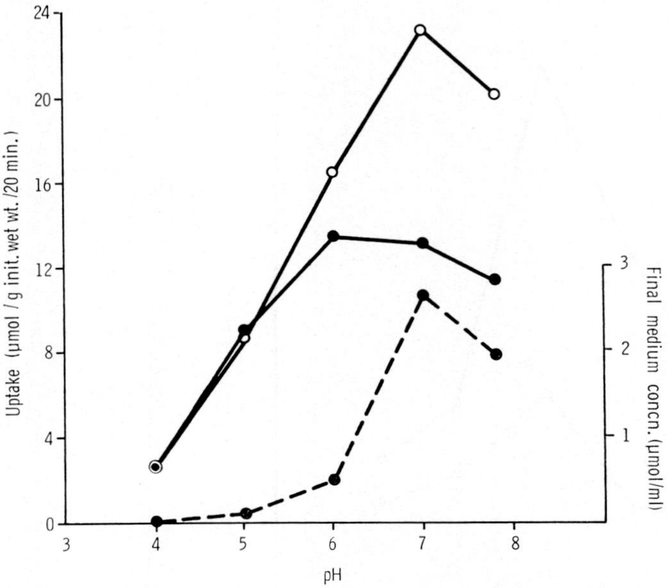

FIG. 6. Effect of pH on uptake of Gly and Leu from Gly-Leu and on hydrolysis of Gly-Leu by rings of everted hamster jejunum. Uptake of Gly ●——●; uptake of Leu ○——○; final concentration of Leu in medium ●------●. The peptide concentration was 5 μmol/ml and experimental conditions were as in the legend to Fig. 5. Both uptake and the final concentrations of amino acids in the medium were estimated on the amino acid analyser. Each point is the mean of six observations. The concentrations of Gly in the medium, which are similar to those of Leu, are omitted for the sake of clarity. Maximal total uptake and hydrolysis is at pH 7, not pH 6 as with Glu-Glu. Note that at pH 4 and 5, where hydrolysis is minimized, uptake of Leu and Gly from the peptide becomes equal. This is consistent with the hypothesis that the inequality in uptake occurring at higher pH values is because hydrolysis followed by uptake of free amino acids contributes to total uptake, in addition to uptake of intact peptide, and uptake of free Gly and Leu occurs at unequal rates.

free Glu. This estimate is surprisingly high, and it might be an overestimate. However, if the failure of Gly-Sar to cause more than weak inhibition of uptake of Glu from Glu-Glu at pH 5 is due to a substantial component of uptake of free Glu in uptake of the peptide, the presence of Met would certainly be expected to enhance the inhibitory effect of Gly-Sar. In fact, it did not. Uptake of Glu from Glu-Glu (10 μmol/ml) in the presence of Gly-Sar (40 μmol/ml) plus Met (20 μmol/ml) was 68% of the mean control value—which was as though, in this experiment, Gly-Sar had no inhibitory effect at all. These were puzzling results, but they did suggest that there was a substantial component in uptake of Glu from Glu-Glu which was not the result of uptake of free Glu and which was not susceptible to inhibition by Gly-Sar. At this

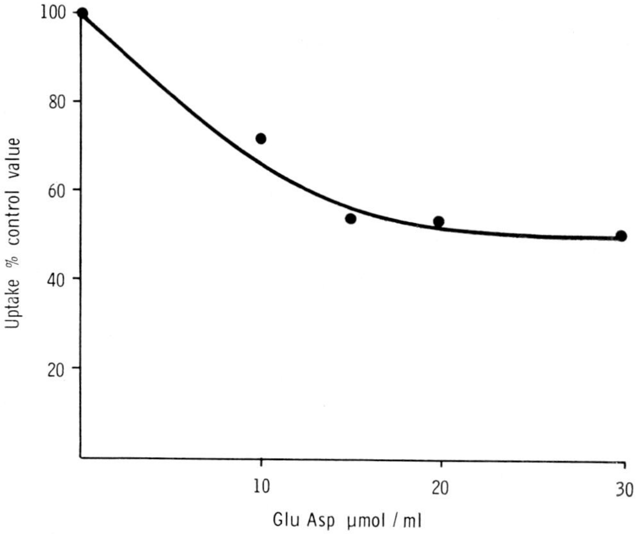

FIG. 7. Inhibition of 20-min uptake of [^{14}C]Glu from [^{14}C]Glu-Glu (10 µmol/ml) by increasing concentrations of Glu-Asp. The pH of the incubation medium was 5. Though the Preston-Schaeffer-Curran plot was not satisfactory, it seems unlikely that Glu-Asp is capable of causing complete inhibition of uptake of Glu from Glu-Glu.

stage we began to entertain the hypothesis that uptake of Glu-Glu might involve two peptide uptake systems, one shared with Gly-Sar, and one available to Glu-Glu and probably to other acidic peptides, but not available to Gly-Sar.

Up to this time we had had no [^{14}C]Glu-Glu, which put severe limitations on the work. When we obtained the labelled peptide, it became possible to test the effect of unlabelled Glu-Asp on uptake of Glu from labelled Glu-Glu. We thought it was highly likely that Glu-Asp, another acidic dipeptide, would share the same uptake systems postulated for Glu-Glu, and should be able to inhibit uptake of the latter peptide completely. To our surprise, no more than about 50% inhibition of 20-min uptake of [^{14}C]Glu from [^{14}C]Glu-Glu could be obtained with Glu-Asp (Fig. 7). It might be argued that the component in uptake which cannot be inhibited by Glu-Asp is uptake of free Glu, though one might expect uptake of free (labelled) Glu released from Glu-Glu in the brush border to be inhibited to some extent by (unlabelled) Glu and Asp released from the inhibitor peptide. A pilot experiment on 20-min uptake of Glu from Glu-Glu (10 µmol/ml) in the presence of Glu-Asp (40 µmol/ml) plus Met (20 µmol/ml) showed that uptake of Glu from Glu-Glu was more strongly inhibited than by Glu-Asp alone, but still remained at 28% of the mean control value. A similar experiment in the presence of Glu-Asp (20 μmol/ml) plus Gly-Sar (20 μmol/ml) plus Met (20 μmol/ml) showed that this mixture reduced uptake to 35 % of the mean control value.

The results available at the time of writing, though obviously still very incomplete, give the impression that there may be some component in uptake of Glu from Glu-Glu which is uptake of peptide, not uptake of free Glu, and which is not completely susceptible to inhibition by Gly-Sar or by Glu-Asp, alone or in combination. If this proves to be correct, it will be a quite unexpected finding. The results of a recent investigation of the kinetics of influx of Glu from [^{14}C]Glu-Glu over the concentration range 0.125–20 μmol and at pH 5 are at least consistent with the hypothesis that there is more than one component in uptake of this peptide, since Lineweaver-Burk and other linearizing plots are clearly biphasic. However, the possibility that one of the components is influx of free Glu has not yet been excluded.

The work reported here has been described because even at this stage it may be of some interest to other investigators—but we must repeat that this is very much an interim report and so far we have not reached any firm conclusions.

Note added after completion of original manuscript: Recent experiments indicate that at pH 5 influx (as opposed to 20-min uptake at pH 7.2) of [^{14}C]Gly-Sar is not completely inhibitable by Glu-Glu, Gly-Gly or 'cold' Glu-Sar. A substantial component of influx remains even at infinitely high concentrations of inhibitor, and there are good reasons for believing that this component is taken up by simple diffusion rather than by mediated peptide transport. It is unlikely to be influx of free [^{14}C]Gly, since brush border and intramedium hydrolysis of Gly-Sar seems to be negligible. When this diffusion component is taken into account, Preston-Schaeffer-Curran plots show that Glu-Glu can completely inhibit Gly-Sar, suggesting they share a common uptake system. Whether Gly-Sar can completely inhibit Glu-Glu has not yet been determined under these conditions.

UPTAKE OF DIPEPTIDES BY EHRLICH ASCITES CARCINOMA CELLS AND BY LIVER AND KIDNEY SLICES

An investigation still in its early stages has shown that Ehrlich ascites carcinoma cells and liver and kidney slices from the hamster all possess the ability to take up dipeptides; higher peptides have not yet been tested. Details of experimental techniques will be published elsewhere. Incubations were at 37 °C under O_2 in Krebs-Ringer phosphate buffer. Analyses were by ion-exchange chromatography.

Fig. 8 shows 20-min uptake of His, Gly-Gly and Car by Ehrlich ascites

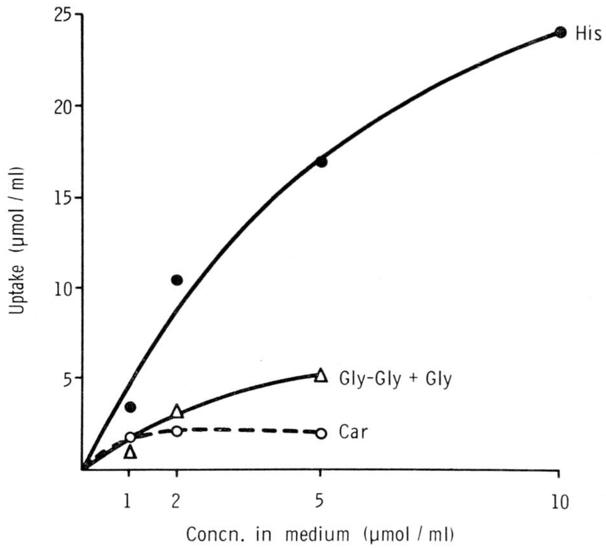

FIG. 8. Uptake (20 min) of His, Gly-Gly and Car by Ehrlich ascites carcinoma cells. Each point is the mean of six observations. The increase in free Gly on incubation with Gly-Gly has been expressed as Gly-Gly and added to intact Gly-Gly appearing in the cells, on the assumption that hydrolysis of this peptide is entirely intracellular.

carcinoma cells. Both peptides appeared in the cells intact, and with Gly-Gly there was also an increase in free Gly in the cells, amounting to about 60% of the total Gly uptake. Uptake of the peptides was poor compared to that of His. Peptide uptake appeared to be a carrier-mediated process, as suggested by the shape of the concentration curves for uptake and the observation that uptake of Car was strongly inhibited by Gly-Sar. 2,4-Dinitrophenol (0.0005 M) inhibited uptake of Car, but replacement of medium Na^+ by K^+ had no effect at any concentration of Car studied. The ability of Ehrlich ascites carcinoma cells to take up intact Gly-Gly was noted many years ago by Christensen & Rafn (1952).

Uptake (20 min) of His, Car, Gly-Gly and Gly-Sar by liver and kidney slices is shown in Figs. 9 and 10. Both tissues took up the peptides, which appeared in them intact. With Gly-Sar, a small amount of free Sar also appeared. At some concentrations, the concentrations of Car in the intracellular fluid slightly exceeded those in the incubation medium, but not enough to provide convincing evidence of the ability of the tissue to concentrate this peptide against a gradient. Replacement of Na^+ by K^+ in the medium had no effect on uptake of Car (5 μmol/ml) by either liver or kidney. The ability of kidney slices (and skeletal muscle) to take up Car by a carrier-mediated process

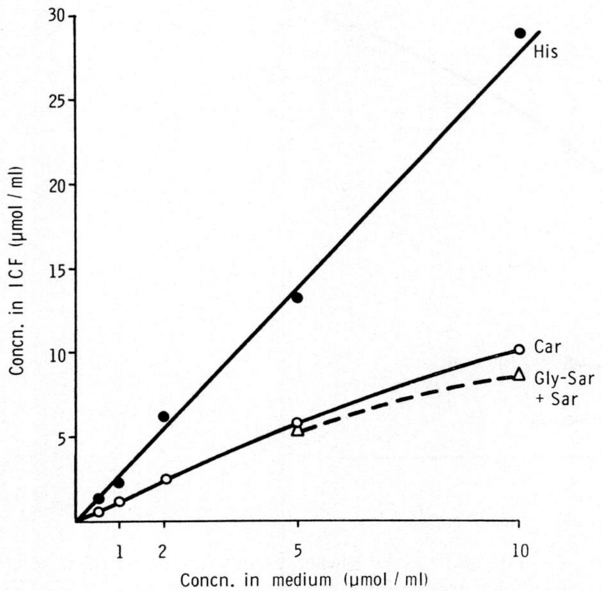

FIG. 9. Uptake (20 min) of His, Car and Gly-Sar by hamster liver slices. Each point is the mean of six observations. The free Sar appearing on incubation with Gly-Sar has been added to intact Gly-Sar appearing in the tissue, on the assumption that hydrolysis of this peptide is entirely intracellular. Below a concentration of 5 μmol/ml, uptakes of Gly-Sar are so similar to those of Car that the lines cannot be plotted separately.

was previously reported by Nützenadel & Scriver (1976): they found that Na^+ depletion did inhibit uptake. Adibi (this volume) has shown that intact Gly-Sar appears in rat kidney *in vivo* after intravenous administration.

The results of these experiments add to evidence suggesting that the ability to take up small peptides may be possessed by many different types of animal cell.

PEPTIDE UPTAKE BY THE SCUTELLUM OF GERMINATING BARLEY

Investigations of the distribution of peptidases in germinating barley (see Enari & Mikola, this volume) suggested that complete hydrolysis of protein to amino acids was unlikely to occur in the endosperm (food store), and that the scutellum (the absorptive organ responsible for transferring digestion products from the endosperm to the seedling) might therefore possess the ability to take up peptides as well as free amino acids, these peptides subsequently undergoing hydrolysis in the scutellum and seedling. The following is a short

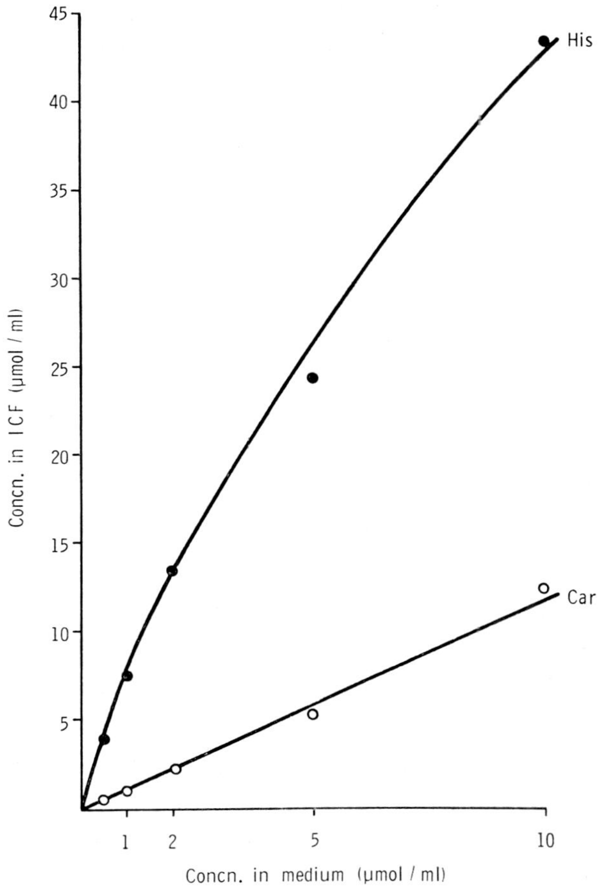

FIG. 10. Uptake (20 min) of His and Car by slices of hamster renal cortex. Each point is the mean of six observations.

account of investigations (by Dr Sopanen) of the possibility of peptide uptake by the scutellum.

Barley grains (*Hordeum vulgare* L.cv. Himalaya) were germinated for three days under aseptic conditions in the dark at room temperature, and the scutella were dissected out. The scutella were preincubated for 4 h at room temperature in water stirred by a stream of oxygen and then incubated with shaking for 1 h at 30 °C under O_2 in the test solution (three scutella, weighing 5–7 mg each, in 3 ml). In early experiments, the test solutions consisted of a peptide dissolved in water without buffer; later, pH was adjusted to 5.2 by acetic acid or Na_2HPO_4.

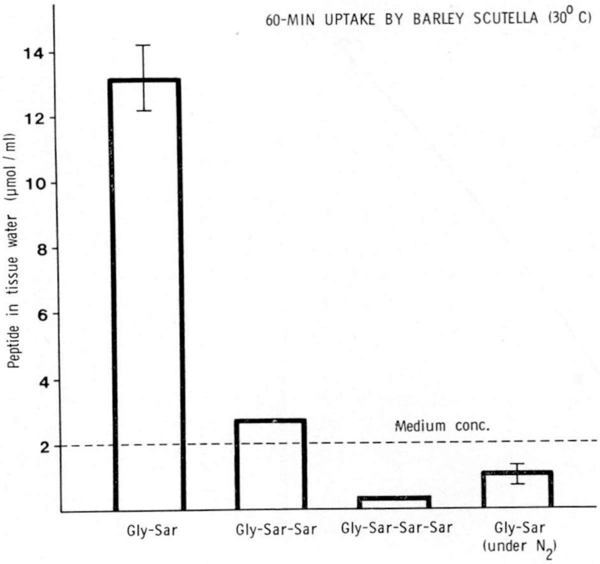

Fig. 11. Uptake of Gly-Sar, Gly-Sar-Sar and Gly-Sar-Sar-Sar by barley scutella. For Gly-Sar, S.E.M. is indicated. For Gly-Sar-Sar and Gly-Sar-Sar-Sar, the results shown are from single experiments, but subsequent work has shown them to be representative. The initial medium concentration of each peptide is shown. The final concentrations were either slightly lower than this, or not measurably different.

After incubation, the scutella were rinsed with water and after extraction with sulphosalicylic acid (60 g/l) for 5 min at 100 °C, peptides and amino acids were estimated by ion-exchange chromatography.

The initial results are shown in Fig. 11. Gly-Sar (2 μmol/ml) was extensively taken up with no measurable hydrolysis in either the incubation medium or the scutellum, an apparent concentration gradient (tissue:medium) of over 6:1 being achieved; uptake was greatly reduced by anoxia. The two-hour uptake was approximately double that at one hour. Uptake of Gly-Sar-Sar (2 μmol/ml) was much less extensive than that of the dipeptide, and uptake of Gly-Sar-Sar-Sar was negligible.

Uptake of Gly-Sar (2 μmol/ml) was reduced to 35% of control values by Gly-Gly (10 μmol/ml) and to 15% by Gly-Gly (40 μmol/ml), but Gly (40 μmol/ml) did not inhibit uptake of Gly-Sar (2 μmol/ml). These experiments also showed that both Gly and Gly-Gly were rapidly taken up, uptake of Gly from Gly-Gly (40 μmol/ml) being about four times as rapid as that of equimolar Gly. Little or no intact Gly-Gly was found in the scutella, uptake being represented almost entirely by an increase in Gly. With both Gly and

Gly-Gly, there was some conversion of Gly to another ninhydrin-reactive compound, probably glutamine.

The results show that barley scutella possess the ability to take up small peptides, apparently as the result of active transport. As in animal tissues and microorganisms, peptide transport is probably independent of that of amino acids, as suggested by the inhibition experiments with Gly and Gly-Gly. These results also show that peptide transport by the scutellum can be rapid in relation to that of an amino acid. The extensive uptake of Gly-Sar from aqueous solution suggests that peptide transport is unlikely to be Na^+-dependent. The results with Gly-Sar-Sar-Sar suggest that as in mammalian small intestine the ability to take up tetra- and higher peptides may be slight or absent. An interesting feature of the results is that uptake of Gly-Sar by the scutella, which have no villi, took place at 30 °C at more than half the rate of uptake (influx) of Gly-Sar by hamster jejunum at 37 °C.

The work described provides what we believe to be the first demonstration of active transport of small peptides in a higher plant.

ACKNOWLEDGEMENTS

A grant from the Medical Research Council is gratefully acknowledged. Part of the work reported was carried out while Professor Sleisenger was a Macy Faculty Scholar, 1974 to 1975. Dr Sopanen is in receipt of support from the Finnish Cultural Foundation and the Emil Aaltonen Foundation.

References

ADDISON, J. M., BURSTON, D. & MATTHEWS, D. M. (1972) Evidence for active transport of the dipeptide glycylsarcosine by hamster jejunum *in vitro*. *Clin. Sci. (Oxf.) 43*, 907-911

ADDISON, J. M., MATTHEWS, D. M. & BURSTON, D. (1974) Competition between carnosine and other peptides for transport by hamster jejunum *in vitro*. *Clin. Sci. Mol. Med. 46*, 707-714

ADDISON, J. M., BURSTON, D., PAYNE, J. W., WILKINSON, S. & MATTHEWS, D. M. (1975*a*) Evidence for active transport of tripeptides by hamster jejunum *in vitro*. *Clin. Sci. Mol. Med. 49*, 305-312

ADDISON, J. M., BURSTON, D., DALRYMPLE, J. A., MATTHEWS, D. M., PAYNE, J. W., SLEISENGER, M. H. & WILKINSON, S. (1975*b*) A common mechanism for transport of di- and tripeptides by hamster jejunum *in vitro*. *Clin. Sci. Mol. Med. 49*, 313-322

ADIBI, S. A. (1977) Clearance of dipeptides from plasma: role of kidney and intestine, in this volume, pp. 265-280

BURSTON, D., ADDISON, J. M. & MATTHEWS, D. M. (1972) Uptake of dipeptides containing basic and acidic amino acids by rat small intestine *in vitro*. *Clin. Sci. (Oxf.) 43*, 823-837

CHRISTENSEN, H. N. (1975) *Biological Transport*, 2nd edn., pp. 136-142, Benjamin, New York

CHRISTENSEN, H. N. & RAFN, M. L. (1952) Uptake of peptides by a free-cell neoplasm. *Cancer Res. 12*, 495-497

DAS, M. & RADHAKRISHNAN, A. N. (1975) Studies on a wide-spectrum intestinal dipeptide uptake system in the monkey and in the human. *Biochem. J. 146*, 133-137

ENARI, T.-M. & MIKOLA, J. (1977) Peptidases in germinating barley grain: properties, localization and possible functions, in this volume, pp. 335-346

FOGEL, M. R. & ADIBI, S. A. (1974) Assessment of the role of brush border peptide hydrolases in luminal disappearance of dipeptides in man. *J. Lab. Clin. Med. 84*, 327-333

MATTHEWS, D. M. (1975*a*) Absorption of peptides by mammalian intestine, in *Peptide Transport in Protein Nutrition* (Matthews, D. M. & Payne, J. W., eds.), pp. 61-146, North-Holland, Amsterdam

MATTHEWS, D. M. (1975*b*) Intestinal absorption of peptides. *Physiol. Rev. 55*, 537-608

NÜTZENADEL, W. & SCRIVER, C. R. (1976) Uptake and metabolism of β-alanine and L-carnosine by rat tissues *in vitro*: role in nutrition. *Am. J. Physiol. 230*, 643-650

PAYNE, J. W. (1975) Transport of peptides in microorganisms, in *Peptide Transport in Protein Nutrition* (Matthews, D. M. & Payne, J. W., eds.), pp. 283-364, North-Holland, Amsterdam

PAYNE, J. W. (1976) Peptides and microorganisms, in *Advances in Microbial Physiology*, vol. 13 (Rose, A. H. & Tempest, O. W., eds.), pp. 55-113, Academic Press, London

PRESTON, R. L., SCHAEFFER, J. F. & CURRAN, P. F. (1974) Structure-affinity relationships of substrates for the neutral amino acid transport system in rabbit ileum. *J. Gen. Physiol. 64*, 443-467

REISER, S. & CHRISTIANSEN, P. A. (1973) The properties of Na^+-dependent and Na^+-independent lysine uptake by isolated intestinal epithelial cells. *Biochim. Biophys. Acta 307*, 212-222

RUBINO, A., FIELD, M. & SHWACHMAN, H. (1971) Intestinal transport of amino acid residues of dipeptides. 1. Influx of the glycine residue of glycyl-L-proline across mucosal border. *J. Biol. Chem. 246*, 3542-3548

SLEISENGER, M. H., BURSTON, D., DALRYMPLE, J. A., WILKINSON, S. & MATTHEWS, D. M. (1976) Evidence for a single common carrier for uptake of a dipeptide and a tripeptide by hamster jejunum *in vitro*. *Gastroenterology 71*, 76-81

WISEMAN, G. (1955) Preferential transference of amino acids from amino acid mixtures by sacs of everted small intestine of the golden hamster (*Mesocricetus auratus*). *J. Physiol. (Lond.) 127*, 414-422

Discussion

Gardner: Professor Matthews has raised a very important issue, particularly in asking where the organism might get its peptides from. I suspect that the answer is that intact peptides *do* cross the intestinal mucosa. This has become obscured by the fact that we now know that there are very active peptide hydrolases within the cells and we have therefore come to assume that absorbed peptides will be hydrolysed intracellularly and that free amino acids will be the only products of protein digestion entering the blood-stream. Ten to twenty years ago we assumed that just because there were proteases in the intestinal lumen we should look only for amino acid absorption. At that time the possibility of peptides entering the mucosal cells was not seriously entertained.

The experiments which I have done on the absorption of amino acids and peptides from complex mixtures probably represent reasonably faithfully the physiological conditions of protein digestion within the intestinal lumen. I have been using enzymic partial digests of casein, in particular. About 30%

of the amino-nitrogen fraction which crosses the mucosa is peptide-bound (Gardner 1975).

The total concentration of peptide-bound amino acids on the serosal surface can exceed by a considerable factor (about four times) the concentration of the peptide-bound amino acids in the intestinal lumen. I am not saying that the peptides on the serosal side have exactly the same composition as the peptides in the lumen; possibly we ought some time to discuss the difficult business of analysing peptides and complex mixtures of peptides.

The fact that the peptide-nitrogen concentration crossing the gut is greater than the concentration in the lumen rules out the possibility that this is a leak phenomenon. Other pieces of evidence convince me that I am not talking about a leak. For instance in experiments with a marker dye such as phenol red, which has a molecular weight of about 300, this dye does not cross the gut in appreciable concentrations. The experimental system was an *in vitro* one which Professor Fisher and I developed for other work on rat intestine (Fisher & Gardner 1974). We used the trick which Smyth & Taylor (1957) introduced whereby the fluid which is transported across the mucosa drips off from the outside of the perfused organ and exudes via the torn-off blood vessels and lymphatics so that there is direct access to the fluid and solutes which have crossed the mucosa without their dilution or contamination by the blood-stream. For detecting the peptides I measured all amino acids by ion-exchange chromatography before and after acid hydrolysis.

It is quite well known that parenteral nutrition mixtures such as Aminosol which contain peptides are well tolerated and well taken up by tissues (e.g. Lidström & Wretlind 1952). I think this supports Professor Matthews' view that peptides in the blood-stream can be utilized by other organs. It would be interesting to know whether other people find large molecules crossing the gut. I think the answer is yes.

Matthews: Professor M. H. Sleisenger did some experiments in our laboratory in which concentrations of amino acid in portal plasma were followed during and after intraduodenal infusion of either a partial enzymic hydrolysate of casein or the equivalent amino acid mixture (Sleisenger *et al.* 1977). The results were not easy to interpret but were most easily interpreted on the hypothesis that when the enzymic hydrolysate is absorbed, a substantial proportion of some amino acids enters the portal blood in peptide form.

Parsons: Dr Gardner, how big were the peptides that crossed the epithelium in your preparations?

Gardner: I don't know. Probably predominantly two to four residues in length. This is a difficult analytical problem which cannot be solved by simple techniques like gel filtration.

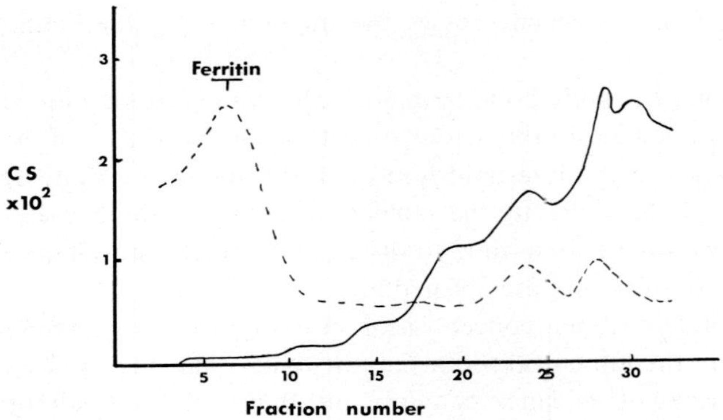

FIG. 1 (Hemmings). Sugar gradient ultracentrifugation profile (——) of clarified carcase macerate after ^{131}I-labelled ferritin was fed to suckling rats. ^{125}I-labelled markers of ferritin, bovine serum albumin and hog stomach pepsin are incorporated (------). Tubes were made up with a gradient from 5-40% sucrose, and centrifuged in the Sw 29 rotor of a Spinco model L centrifuge for 18 h at 39 000 rev./min.

Parsons: Do you only get those results when you have a complete mixture in the lumen, e.g. an enzymic hydrolysate, and not when you have just one or two oligopeptides?

Gardner: I cannot answer this definitely yet. I suspect that this may not be a general phenomenon happening to all peptides but just with specific peptides. Therefore casein hydrolysates, for instance, may be a fortuitous choice, with a particularly large quantity of intact peptides crossing the mucosa. In some other preliminary experiments with commercial bacteriological peptone, which is an enzymic digest of muscle extract, I have not observed so much peptide in the serosal secretion.

Hemmings: I have spent my working life dealing with the passage of macromolecules, IgG for the most part, through living cellular barriers, the fetal membranes and placenta of the rabbit, and the small intestine of the suckling rat and mouse. To me it is a commonplace that large molecules such as proteins should enter and pass through living cells. I would now like to suggest that this is not just a peculiarity restricted to a few specialized types of cell, but a general physiological principle.

The passage of immunoglobulin across the gut of the suckling rat occurs primarily across the duodenum, and continues for the first 21 days of life (Brambell 1970). It then stops suddenly. This is the 'cut-off'. We discovered that while this transport of undamaged IgG was taking place across the duodenum, a second process was occurring across the ileum where damaged or partially cleaved molecules were being absorbed. Thus when we put a dose of

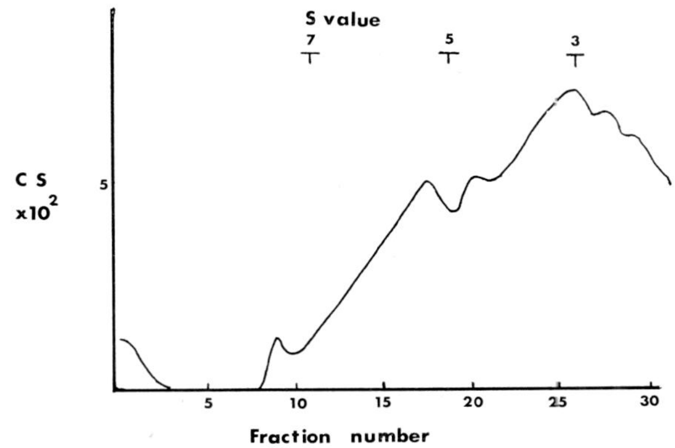

FIG. 2 (Hemmings). Sugar gradient ultracentrifugation profile of clarified carcase macerate of adult rats after they had been fed with bovine IgG labelled with ^{131}I. Spun in the Sw 50 rotor at 50 000 rev./min for 18 h.

labelled ferritin into the ileum of the suckling rat, 75% of the protein-bound activity in the circulation of that rat was no longer precipitable with specific immune serum against ferritin (Hemmings & Williams 1975). Fig. 1 shows the profile of the labelled protein in the skeletal muscle of such a rat, demonstrated by sugar-gradient ultracentrifugation (Hemmings 1975*a*). It can be seen that although there is no intact ferritin present, there is a lot of material of molecular weights in the range of albumin and pepsin.

We went on to ask whether this process was terminated at 21 days, as is the transport of intact IgG. In fact it continued (Hemmings 1975*b*) and it reaches a peak as late as 100 days after birth (C. Hemmings *et al.* 1977) When a dose of bovine IgG is fed at this age, 40% is present in the carcase as degradation products of high molecular weight. If one corrects this figure for the amount left in the gastrointestinal tract, the value is about 60% of the absorbed dose. The IgG is mostly in the 20 000–50 000 daltons range of molecular size. Fig. 2 presents a scan of the carcase after such an experiment.

Clearly we had to rethink the process of digestion, and we have come up with a hypothesis of what might be termed 'distributed digestion'. The large protein fragments are being taken through the gut wall into the circulation and distributed throughout the tissues, entering the individual cells and probably being finally broken down there. We therefore went on to consider other barriers in the body, such as the blood-brain barrier, and we looked at the brain content after feeding various proteins. The specific activity of the brain turned out to be much the same as that of skeletal muscle (Hemmings 1977). After porcine haemoglobin labelled with ^{131}I was fed to adult rats, the macerated

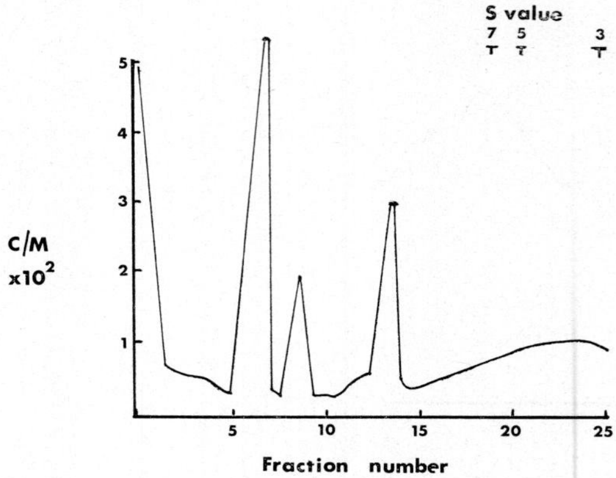

FIG. 3 (Hemmings). Sugar gradient ultracentrifugation profile of brain macerate of adult rats fed with ^{125}I-labelled porcine haemoglobin.

brain tissue, clarified by an initial spin to rid it of organelles, gave the profile illustrated in Fig. 3 (Hemmings 1976).

Few breakdown fragments are present, most of the activity being in the peaks to the left of the 5S marker, which is approximately the position of the original haemoglobin; so evidently the haemoglobin derivatives have formed complexes in the brain, all of which are heavier than the original molecule. This is a most interesting phenomenon which still puzzles us.

The ultracentrifuge diagrams are in effect profiles of the cell sap of these tissues. Basically this is the soluble cytosol, so we can assume that the derivatives are indeed entering the cells of those tissues. The proposition is that the cell membranes of virtually all the cells of the body are capable of permitting entry to these breakdown products, although their molecular weight is so high.

We have begun immunofluorescence studies in the attempt to find where these fragments are. Fig. 4 is a section of ileum after bovine IgG has been fed to an adult rat, developed with rabbit anti-bovine IgG-fluorescein isothiocyanate. You can see the fluorescence in the epithelial cells, and in the lymphatic drainage of the villus. We have tried this on the brain too, where there are much lower concentrations of antigen. Fig. 5 shows a typical section, somewhere in the forebrain (our knowledge of brain morphology is very elementary).

To summarize the position at present, we have carried out experiments of the type described with a variety of proteins—ferritin, as already mentioned (Hemmings 1975*b*), rat and bovine IgG (Hemmings 1975*c*), α, β and γ gliadin (C. Hemmings *et al.* 1976*a,b*) and porcine haemoglobin (Hemmings 1976).

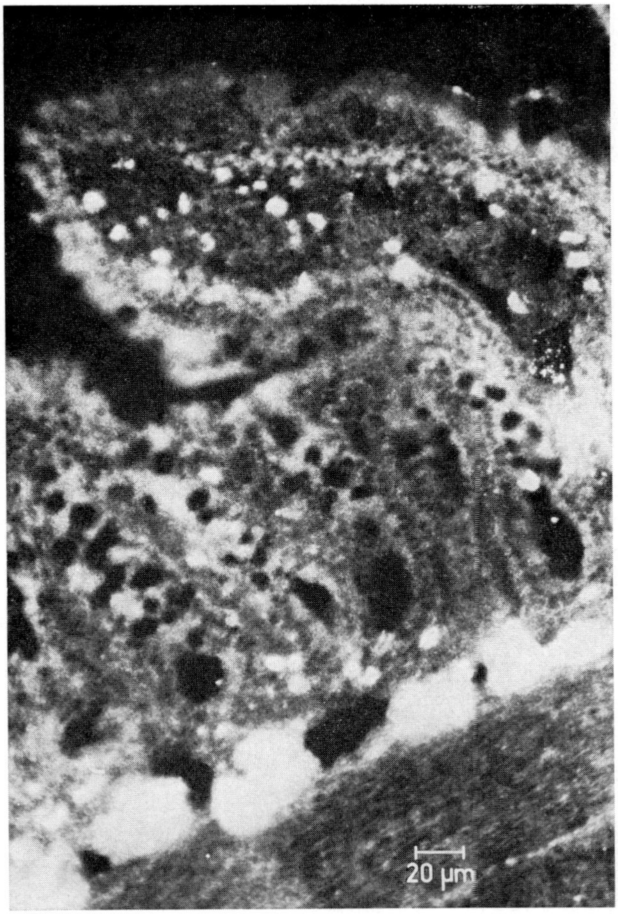

FIG. 4 (Hemmings). Fluorescence micrograph of adult rat intestine after the animal had been fed with bovine IgG, developed with rabbit IgG anti-bovine IgG-fluorescein isothiocyanate.

All these, in the adult, show a high proportion of the dose to be present in the carcase as large fragments.

There are of course many reports on the passage of immunologically significant amounts of intact protein across the gut, dating back to 1900 (Hemmings 1977). The present investigations extend this phenomenon quantitatively.

Baker: In your study of a protein macromolecule labelled with iodine, how do you know that what you are looking at finally is what went in and not just a shift of the label?

Hemmings: If the label is taken off the protein, we have monoiodotyrosine,

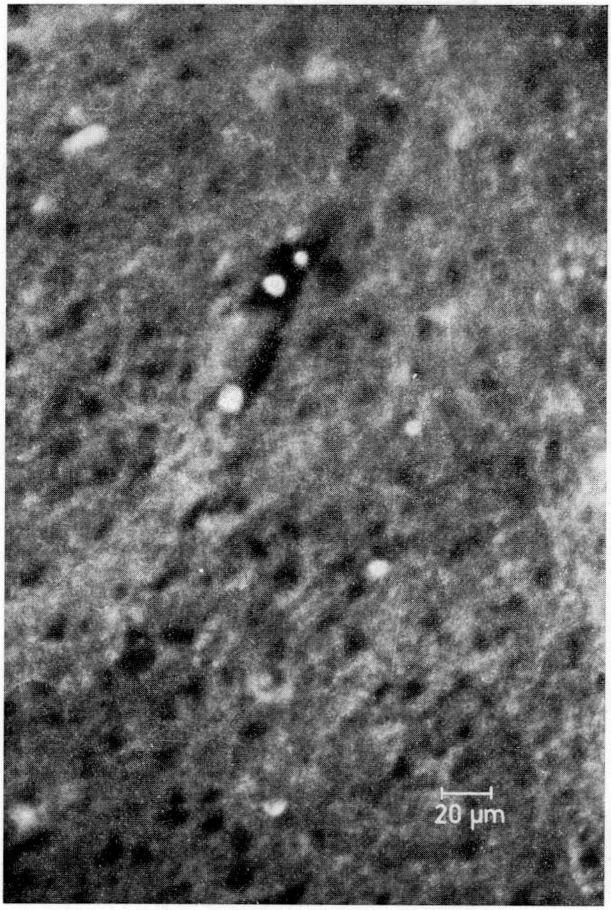

FIG. 5 (Hemmings). Fluorescence micrograph of forebrain of adult rat fed with bovine IgG, developed as for Fig. 4.

which is not resynthesized into protein. Iodide, a further step down the line, would not be resynthesized into protein either, except that in a long-term experiment, if the thyroid is not blocked, it gets incorporated into thyroglobulin (Jones 1977).

Josefsson: Have you done any electron microscope studies of the mechanism of absorption?

Hemmings: Yes. With molecules like ferritin the protein is probably engulfed into the cell pinocytically but it becomes distributed through the cytosol.

Josefsson: Is that by endocytosis?

Hemmings: There is a mechanism of endocytosis but it is not entirely vesicular. The protein becomes intimately distributed in the cytosol and one can see it leaving the cell by way of the intercellular spaces, apparently by some type of assisted diffusion process.

Josefsson: The membrane and the ferritin-containing molecule might interact and then fuse into the cell. Have you any indication of such fusion?

Hemmings: No. The pictures show indiscriminate scattering of ferritin through the cytosol.

Rhodes: What was the dose of ferritin you gave?

Hemmings: Both suckling and adult rats got about 20 mg ferritin into the intestinal lumen. It worked in the adult too (up to 100 days).

Rhodes: Then how much of the ferritin in the adult is hydrolysed by proteolytic enzymes from the pancreas and the small intestine?

Hemmings: We cheated a little in that experiment and put the ferritin into the ileal lumen, but the bovine IgG went through the whole gastrointestinal tract. About 60% of the absorbed dose was present as breakdown products in the adult.

Mellander: Your interesting contribution made me think about how the problem of intestinal absorption might be entirely avoided experimentally. Some years ago we injected cow's milk intravenously into calves and found that 1 litre of milk could be injected without any trouble (Mellander & Hedenstedt 1957). Later the same technique was used for giving human milk intravenously to infants, also without any side-effects. Apparently this is one way of investigating how macromolecules such as proteins can be utilized without first being hydrolysed in the intestinal tract.

Matthews: Dr Hemmings' interesting work suggests a view of protein absorption very similar to that held just over a century ago. Discussing this period, Folin (1905) wrote 'The usefulness of the proteolytic enzymes has always been admitted, but their usefulness, it has been thought, must be restricted to the mere solution of coagulated proteins'; and Van Slyke & Meyer (1913-14) wrote that it was held that '... the proteins are absorbed, with little or no chemical change, directly into the circulation, from which they are taken by the tissues and incorporated into their substance.' In other words, it was thought that proteins were brought into solution in the lumen of the gut and then absorbed with minimal if any breakdown. It was already known that protein digestion did produce some free amino acids, but these were thought to be too elementary for utilization, so that they were wasted. In recent years our views on protein absorption have reverted to those of 1912. Perhaps they are now going to revert to those of 1870!

Fottrell: I have some difficulty with the interpretation of Dr Hemmings' experiments. The possibility of isotopic exchange must be considered and it would also be useful to have some characterization of the labelled proteins in the tissues. Furthermore, although some uptake of intact proteins does occur in the intestine it is a relatively specialized phenomenon and most dietary protein is degraded before absorption.

Hemmings: The breakdown products of iodinated protein, iodotyrosine and iodide, are excreted rapidly without incorporation into protein synthesized *de novo* (Jones 1977). So in this sense iodoproteins are much more reliable indicators than, say, ^{14}C- or tritium-labelled proteins. In the studies where bovine IgG was fed to adult rats we located the intact protein immunologically rather than by virtue of any radioactive label. We also have confirmatory studies using quantitative serological precipitation with both bovine IgG and α-gliadin (C. Hemmings *et al.* 1976*c*).

As to the generality of the phenomenon of gut transport of intact protein, there are many reports of its occurrence in dog, guinea-pig and man, as well as in rat. It surely seems to be general.

Fottrell: Professor Matthews' paper on the existence of multiple transport systems for peptides is interesting in view of the studies on multiple peptide hydrolases. It would appear that nature allows for possible genetic errors by having multiple transport and hydrolysis systems which have overlapping functions.

Payne: When one considers a peptide undergoing transport, one should perhaps consider as distinct forms those which have either a protonated or non-protonated α-N-terminus and the influence of the p*K* of the amino group in determining the relative proportion of each species. When it appears impossible to completely inhibit the uptake of one peptide by another peptide, what one sees may arise because a certain proportion of one peptide may exist say in the non-protonated form and that proportion may be different for the other peptide, and if there are different transport systems they may differ in their specificity with respect to this particular characteristic (see Payne, this volume, pp. 305-325). Professor Matthews presented results of competition experiments that could be interpreted as evidence for different systems. As a separate point, we should like to report that quite independently we have demonstrated peptide transport in plants (Higgins & Payne 1977*a*, *b*). Using a dansyl detection method, Gly-Sar was found to be concentrated about 10-fold by germinating barley embryos suspended in Na^+/K^+ phosphate, pH 6.2. A similar accumulation occurred when the excised embryo was placed with the scutellum surface in contact with filter paper impregnated with aqueous Gly-Sar solution. Gly-Sar was found in the roots, shoots and scutellum residue. Peptide transport can be demonstrated after several hours' incubation, using seeds that have imbibed water for only a few hours. As accumulation of intact Gly-Sar was inhibited by anoxia, azide and dinitrophenol, it appears to be occurring by an active transport mechanism.

Matthews: The suggestion about protonation is most interesting and ought to be borne in mind.

Nützenadel: Obviously, one dipeptide which crosses the intestinal epithelium intact is carnosine (β-alanyl-L-histidine). It can be found in the urine of man after intake of meat which contains this dipeptide. Patients with carnosinaemia—a metabolic disorder with high urinary excretion of carnosine—seem to be unable to metabolize this compound. Thus in the normal and mutant state carnosine excretion is similar to that of urinary amino acids and we could conclude that unhydrolysed dipeptides appearing in the blood are nutrients.

Smyth: I was interested in Professor Matthews' use of non-hydrolysable peptides. Some years ago Rhodes *et al.* (1967) showed that D-leucylglycine was not hydrolysed by the brush border and we used D-leucylglycine as a non-hydrolysable peptide. We found that its uptake was inhibited by L-leucyl-L-alanine and by L-leucyl-L-leucine (Cheeseman & Smyth 1975). We also looked for dipeptides in the blood-stream after putting them into the intestinal lumen (Newey & Smyth 1959). We found glycylglycine in the blood, but no other dipeptide. We thought this might be because glycylglycine peptidase has a very weak activity, and hence this particular dipeptide escaped hydrolysis (Newey & Smyth 1960).

Gardner: Analytically the quantitation and characterization of peptides in complex mixtures is a very difficult thing to do.

Smyth: Our method of estimating blood peptides was a very simple one, depending on the increase in glycine on acid hydrolysis. We look forward to hearing of work with more refined techniques to determine whether any other peptides can get into the blood-stream.

References

BRAMBELL, F. W. R. (1970) *The Transmission of Passive Immunity from Mother to Young* (*Frontiers of Biology*, vol. 18), North-Holland, Amsterdam

CHEESEMAN, C. I. & SMYTH, D. H. (1975) Interaction of amino acids, peptides and peptidases in the small intestine. *Proc. R. Soc. Lond. B Biol. Sci. 190*, 149-163

FISHER, R. B. & GARDNER, M. L. G. (1974) A kinetic approach to the study of absorption of solutes by isolated perfused small intestine. *J. Physiol. (Lond.) 241*, 211-234

FOLIN, O. (1905) A theory of protein metabolism. *Am. J. Physiol. 13*, 117-138

GARDNER, M. L. G. (1975) Absorption of amino acids and peptides from a complex mixture in the isolated small intestine of the rat. *J. Physiol. (Lond.) 253*, 233-256

HEMMINGS, C., HEMMINGS, W. A. & PATEY, A. L. (1976*a*) The fate of oral doses of α-gliadin in suckling and adult rats. *Int. Res. Commun. Syst. Med. Sci. 4*, 38-39

HEMMINGS, C., HEMMINGS, W. A. & PATEY, A. L. (1976*b*) The fate of oral doses of beta- and gamma-gliadin in adult rats *Int. Res. Commun. Syst. Med. Sci. 4*, 152

HEMMINGS, C., HEMMINGS, W. A. & WILLIAMS, E. W. (1976*c*) Immunological assay for antigen in tissues following feeding of labelled α-gliadin to suckling and adult rats. *Int. Res. Commun. Syst. Med. Sci. 4*, 458

HEMMINGS, C., HEMMINGS, W. A., PATEY, A. L. & WOOD, C. (1977) The ingestion of dietary protein as large molecular weight degradation products in adult rats. *Proc. R. Soc. Lond. B Biol. Sci.* in press.

HEMMINGS, W. A. (1975*a*) Degradation products in gut and carcase tissue after injecting labelled ferritin into the ileum of suckling rats. *Int. Res. Commun. Syst. Med. Sci. 3*, 216

HEMMINGS, W. A. (1975*b*) Transport of protein across the small intestine of adult rats following injection of ferritin to the intestinal lumen. *Int. Res. Commun. Syst. Med. Sci. 3*, 262

HEMMINGS, W. A. (1975*c*) Transport of immunoglobulins across the gut of the adult rat. *Int. Res. Commun. Syst. Med. Sci. 3*, 382

HEMMINGS, W. A. (1976) The absorption of haemoglobin from oral doses by adult rats. *Int. Res. Commun. Sci. Med. Sci. 4*, 393-394

HEMMINGS, W. A. (1977) The absorption of large breakdown products of dietary proteins into the body tissues including brain, in *Current Research in Schizophrenia* (vol. 1: *The Biological Basis of Schizophrenia*), Reach, London, in press

HEMMINGS, W. A. (1977) The entry into the brain of large molecules derived from dietary protein. In preparation

HEMMINGS, W. A. & WILLIAMS, E. W. (1975) Transport of ferritin across the ileum of suckling rats. *Int. Res. Commun. Syst. Med. Sci. 3*, 215

HIGGINS, C. F. & PAYNE, J. W. (1977*a*) Peptide transport by germinating barley embryos. *Planta (Berl.) 134*, 205-206

HIGGINS, C. F. & PAYNE, J. W. (1977*b*) Characteristics of active dipeptide transport by germinating barley embryos: effects of pH and metabolic inhibitors. *Planta* (*Berl.*) in press

JONES, R. E. (1977) Deiodination of labelled proteins during transmission across the gut of suckling rats. *Proc. R. Soc. Lond. B Biol. Sci.* in press

LIDSTRÖM, F. & WRETLIND, K. A. J. (1952) Effect of dialysed casein hydrolysate. The effect of intravenous administration of a dialyzed enzymatic casein hydrolysate (Aminosol) on the serum concentration and on the urinary excretion of amino acids, peptides and nitrogen. *Scand. J. Clin. Lab. Invest. 4*, 167-178

MELLANDER, O. & HEDENSTEDT, S. (1957) Intravenös tillförsel av homolog mjölk. Några djurförsök. *Nord. Med. 57*, 148

NEWEY, H. & SMYTH, D. H. (1959) The intestinal absorption of some dipeptides. *J. Physiol. (Lond.) 145*, 48-56

NEWEY, H. & SMYTH, D. H. (1960) Intracellular hydrolysis of dipeptides during intestinal absorption. *J. Physiol. (Lond.) 152*, 367-380

RHODES, J. B., EICHHOLZ, A. & CRANE, R. K. (1967) Studies on the organisation of the brush border in intestinal epithelial cells. IV. Aminopeptidase activity in microvillus membranes of hamster intestinal brush borders. *Biochim. Biophys. Acta 135*, 959-965

SLEISENGER, M. H., PELLING, D., BURSTON, D. & MATTHEWS, D. M. (1977) Amino acid concentrations in portal venous plasma during absorption from the small intestine of the guinea-pig of an amino acid mixture simulating casein and a partial enzymic hydrolysate of casein. *Clin. Sci. Mol. Med. 52*, 259-267

SMYTH, D. H. & TAYLOR, C. B. (1957) Transfer of water and solutes by an *in vitro* intestinal preparation. *J. Physiol. (Lond.) 139*, 632-648

VAN SLYKE, D. D. & MEYER, G. M. (1913-14) The fate of protein digestion products in the body. 3. The absorption of amino acids from the blood by the tissues. *J. Biol. Chem. 16*, 197-212

A brush-border-bound peptidase and amino acid transport

HANS WACKER and GIORGIO SEMENZA

Laboratorium für Biochemie der Eidgenössischen Technischen Hochschule, Zürich

Abstract Aminopeptidase when in its integral form interacts with lecithin to form a stable lipoprotein complex. The reconstituted system is a single-bilayer vesicle about 30 nm in diameter. The hydrophobic part of the amphipathic hydrolase is incorporated in the lecithin bilayer with the portion carrying enzymic activity oriented towards the external aqueous phase. This part can be detached by proteases.

In recent years Matthews and his associates (e.g. Matthews 1975; Matthews & Adibi 1976) have written some exceptionally clear and comprehensive reviews on the subject of intestinal absorption of peptides. We shall therefore refer in this paper only to those reports of immediate relevance to our own observations.

From the work summarized and examined critically by Professor Matthews the conclusion is inescapable that in addition to the mechanism classically accepted, namely hydrolysis followed by transport of the free amino acids by sodium-dependent membrane carriers, other ways of handling the peptides arising from the digestion of proteins also exist, and may even be of overriding importance. In most cases, only *free* amino acids appear in the portal blood, but peptides may cross the membrane of the intestinal brush border intact; they are hydrolysed intracellularly, or during transport itself, or—more rarely—not at all. The nature of the peptide, the animal species and the conditions applied (in particular incubation time, peptide concentration, etc.) may be decisive for the way a given peptide is handled. In view of the bewildering complexity and variety of the ways in which peptides can be handled the observations and conclusions reported here are intended to apply only to the peptides and amino acids actually used in these experiments. Any generalizations made are to be regarded as part of a working hypothesis.

Some 90% of the total aminopeptidase activity of the small intestine is located within the intestinal cell. Thus, except for peptides which are unusually

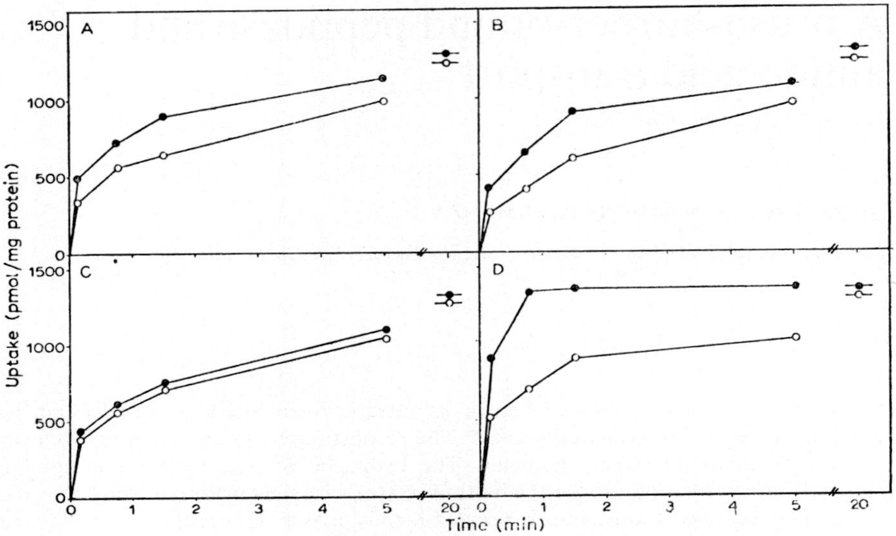

FIG. 1. Dipeptide and amino acid uptake by brush border membrane. Additions to incubation medium: NaSCN (100 mM, ●) or KSCN (100 mM, ○) and 1 mM (A) [1-^{14}C]glycyl-L-leucine, (B) glycyl-L-[U-^{14}C]leucine, (C) [2-^{3}H]glycine, (D) L-[4,5-^{3}H]leucine. (From Sigrist-Nelson 1975.)

resistant to hydrolysis, it may be difficult, when working with intact tissue, to decide whether a given peptide is absorbed during transport or afterwards. In the last few years preparations of essentially pure vesicles from brush border membranes have become available. Sigrist-Nelson (1975) has studied the uptake of glycylleucine, labelled at either a glycine or a leucine moiety, into such vesicles. She observed that the two moieties of the peptides were taken up at about the same rate (Fig. 1A, B) and were equally affected when potassium was substituted for sodium. As to the uptake of *free* amino acids, that of leucine was affected by sodium much more than that of glycine.

At very short incubation times (10–50 s), the radioactivity taken up was in the form of the intact dipeptide (Fig. 2). As incubation proceeded, more and more radioactivity was present as free amino acid, the peptide accounting for only 5% of the radioactivity at 5 min incubation. This peptidase activity was clearly associated with the brush border membrane itself and was not due to a cytosolic contaminant, as shown by the identical degree of purification of leucyl-nitroanilide-hydrolase activity and of sucrase (sucrose α-glucohydrolase, EC 3.2.1.48) (a typical marker of brush border membrane) during the preparation of the brush border membrane.

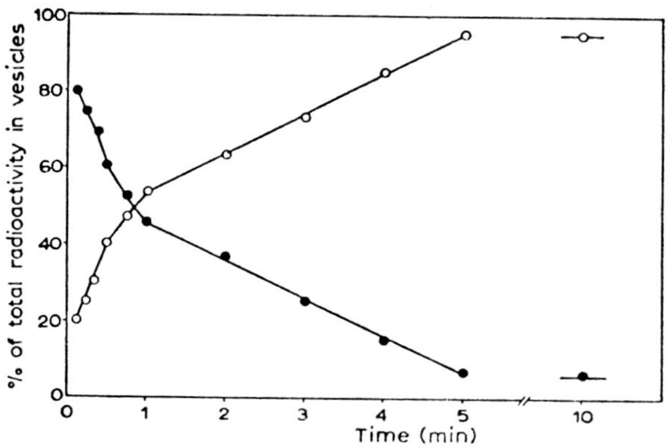

FIG. 2. Contents of vesicles incubated with [1-^{14}C]glycyl-L-leucine. Radioactivity appearing as peptide, ●, or glycine, ○. Additions to incubation medium: NaSCN (100 mM), [1-^{14}C]-glycyl-L-leucine (1 mM). (From Sigrist-Nelson 1975.)

Neither glycine nor leucine inhibited glycylleucine uptake, and neither glycylleucine nor leucylglycine inhibited the uptake of free glycine or free leucine (Table 1). On the other hand, a number of dipeptides inhibited the uptake of each of the two moieties of glycylleucine. In each experimental condition, the uptake of the glycine moiety was inhibited to the same extent as that of the leucine moiety (Table 2).

Thus, the results of Dr Sigrist-Nelson conclusively showed that in vesicles from brush borders of the rat small intestine the two moieties of glycylleucine were taken up by a system (or systems) independent of the system(s) for free glycine or free leucine. The identical extents of activation by Na^+ and of in-

TABLE 1

Influence of amino acids on dipeptide transport (from Sigrist-Nelson 1975)

Test inhibitors	*Peptide uptake pmol/mg protein*			
	[1-^{14}*C*]*Glycyl*-L-*leucine*		*Glycyl*-L-[U-^{14}*C*]*leucine*	
Mannitol	299±23/557±54	(0/0)	309±28/540±59	(0/0)
Glycine	291±29/575±58	(3±1/3±2)	290±27/540±50	(6±1/2±1)
L-Leucine	310±32/570±56	(4±2/2±1)	295±27/553±52	(5±1/2±1)

Additions to medium: NaSCN (100 mM), test inhibitor (40 mM), [1-^{14}C]glycyl-L-leucine (1 mM) or glycyl-L-[U-^{14}C]leucine (1 mM). Time of incubation was 10 s/45 s. The figures are means ± S.E. of five experiments. Percentage of inhibition indicated in parentheses.

TABLE 2

Influence of other dipeptides on glycyl-L-leucine transport (from Sigrist-Nelson 1975)

Test inhibitor	*Dipeptide uptake pmol/mg protein*			
	[1-^{14}C]Glycyl-L-leucine		*Glycyl-L-[^{14}C]leucine*	
Mannitol	302 ± 28	(0)	294 ± 25	(0)
L-Leucylglycine	149 ± 12	(51 ± 4)	145 ± 13	(51 ± 6)
L-Alanyl-L-leucine	139 ± 14	(54 ± 6)	147 ± 10	(50 ± 5)
Glycyl-L-methionine	187 ± 12	(38 ± 4)	171 ± 16	(42 ± 5)
Glycyl-L-alanine	208 ± 18	(31 ± 5)	212 ± 12	(28 ± 3)
D-Leucylglycine	245 ± 16	(19 ± 2)	250 ± 18	(15 ± 2)

Additions to incubation medium: NaSCN (100 mM), [1-^{14}C]glycyl-L-leucine (1 mM) or glycyl-L-[U-^{14}C]leucine (1 mM) and test inhibitors (40 mM). Length of incubation was 10 s. The figures are means ± S.E. of six experiments. Percentage of inhibition is given in parentheses.

hibition by various dipeptides further indicated that the two moieties of the peptide were handled by the same agency (or agencies). What is this agency (or agencies) and what is its (or their) transport mechanism?

It is difficult to give an unequivocal interpretation of the observation that most of the peptides were present intact after very short incubation times, whereas they were completely or almost completely hydrolysed after incubation for 5 min. It could be that, as the incubation proceeded, the contribution of the mechanism 'hydrolysis followed by uptake' increased; or it could be that the peptide was taken up intact and was hydrolysed subsequently. (A peptidase of cytosolic origin different from that in the brush border hydrolysing leucyl-4-nitroanilide might have been responsible.) On the other hand, the substrate might have access to the brush border aminopeptidase from *both* sides of the membrane, or it may have been transported back to the outer side of the membrane. It should also be kept in mind that perhaps some 90% of the vesicles in the preparation used by Sigrist-Nelson were leaky. From her results, therefore, it is not possible to distinguish between the following mechanisms for dipeptide transport (nor were the experiments conceived for this purpose):

(1) The peptide was transported first and hydrolysed later;

(2) The peptide was hydrolysed while being transported;

(3) The amino acids stemming from the peptide were liberated near to or transferred to a transport agency which was not as accessible to free amino acids as to peptides.

The elegant work of Louvard, Semeriva, Maroux and Desnuelle (Maroux & Louvard 1975; Louvard *et al.* 1975, 1976) has conclusively demonstrated that

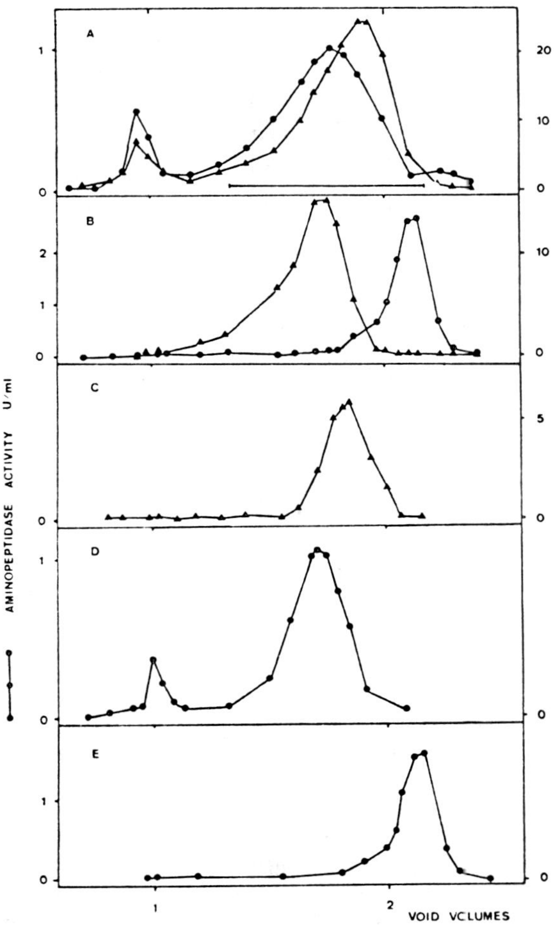

FIG. 3. Gel filtration on Sepharose 4B of the lipid–protein mixture after removal of cholate. The column (2.0 × 33 cm) was equilibrated and eluted with buffer, at a flow rate of 4 ml/h. Samples of about 1 mg protein and 10 mg lecithin in 1 ml of buffer were applied. Enzyme recovery was 85–90%. (A) Detergent form of aminopeptidase plus lecithin after removal of cholate by gel filtration from Sephadex G-50. (B) Trypsin-solubilized aminopeptidase submitted to an identical treatment. (C) Lecithin liposomes, without protein, prepared by the same procedure as in (A). (D) Detergent form of aminopeptidase. (E) Trypsin-solubilized aminopeptidase. (From Wacker *et al.* 1976*b*.)

brush border intestinal aminopeptidase spans the brush border membrane and that most of the molecule, including the hydrolytic site, protrudes out of the lipid bilayer. The hydrophobic peptide anchoring the protein in the bilayer amounts to not more than a few per cent of the total molecule. An unexpected

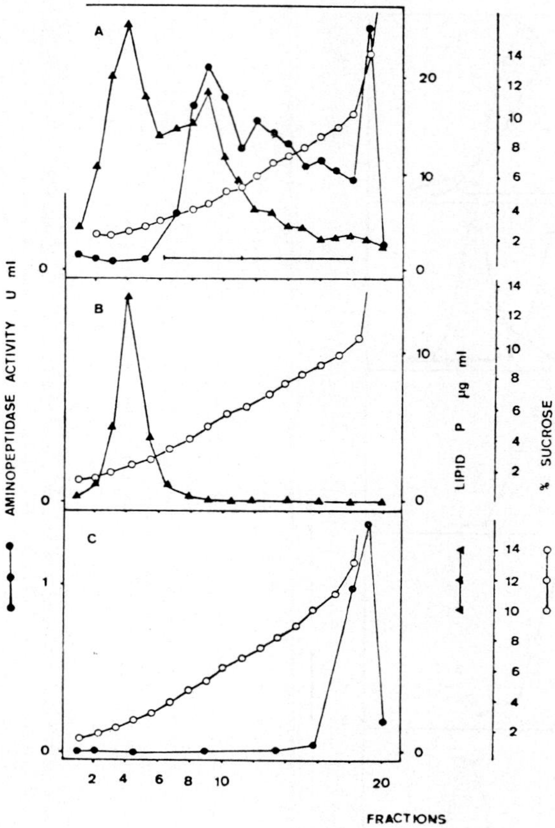

FIG. 4. Gradient centrifugation under quasi-equilibrium conditions. A linear sucrose gradient (0–12% sucrose) was made in a 5 ml cellulose nitrate centrifuge tube. The sample, in 0.15-0.25 ml buffer, was layered on top of the gradient. After centrifugation for 20 h at 25 000 rev/min (58 000 *g*) in a SW 50.1 rotor of a Beckman Spinco 50 B centrifuge the gradient was displaced upwards with a 40% sucrose solution which was pumped through a steel needle in the bottom of the tube. Fractions of 0.25 ml were collected. Sucrose concentration was read with an Abbe hand refractometer (Atago Optical Works Co., Ltd). (A) Sample from the retarded peak of Fig. 3A containing about 1 mg lipid and 0.1 mg protein. (B) Pure egg lecithin liposomes, about 1 mg, prepared by the cholate-Sephadex method. (C) Detergent form of aminopeptidase without lipids added, about 0.1 mg. (From Wacker *et al.* 1976*b*.)

observation of these workers was that it is located near the amino end, rather than near the carboxylate end, of the polypeptide chain. This is very interesting, also in relation to the biosynthesis of this membrane glycoprotein.

We decided to investigate a possible role of a brush-border-bound aminopeptidase in amino acid and peptide transport by isolating the enzyme and incorporating it into a lipid bilayer. Rather than aminopeptidase from the small intestine we chose to work with kidney brush border aminopeptidase because the latter enzyme can be obtained more easily intact, i.e. not partially attacked by trypsin (Wacker *et al.* 1976*a*). By so doing we assumed of course a fundamental similarity between the brush border of proximal renal tubuli and the brush border of the small intestine. Although little is yet known about the handling of peptides by renal proximal tubuli (Adibi, this volume, pp. 265-280), we thought that the close analogy between the two tissues in enzymic equipment and transport properties made our assumption reasonably safe.

The enzyme was prepared from hog kidney by either detergent solubilization or trypsin solubilization. Protein-containing liposomes were obtained by a procedure which we have already described elsewhere, with a few modifications (Wacker *et al.* 1976*b*). Essentially, egg lecithin was suspended in cholate buffer and mixed with the enzyme solution. Cholate was removed by gel filtration on Sephadex G-50; the peak containing the protein and the lipid was fractionated by gel filtration on Sepharose 4-B, followed by gradient centrifugation under quasi-equilibrium conditions in a sucrose gradient. In this way populations of monolamellar liposomes containing one, two or more molecules of aminopeptidase per liposome were obtained. The trypsin-solubilized form of the aminopeptidase failed to be incorporated into liposomes (Figs. 3 and 4).

Trypsin treatment of aminopeptidase-containing liposomes detached the aminopeptidase activity from the liposomes without destroying their bilayer structure. Lytic concentrations of Triton X-100 and cholate did not produce any increase in enzymic activity, a finding which also indicated that the substrate had access to the hydrolytic site of the enzyme in aminopeptidase-containing liposomes.

Preliminary uptake experiments with these aminopeptidase-carrying liposomes indicate that they can discriminate between L- and D-alanine. Further studies are in progress.

ACKNOWLEDGEMENTS

Our work was supported in part by the Swiss National Science Foundation, Berne, and by Nestlé Alimentana, Vevey. Mrs Ch. Storelli-Joss provided excellent technical assistance. Our sincere thanks are due to all of them.

References

ADIBI, S. A. (1977) Clearance of dipeptides from plasma: role of kidney and intestine, in this volume, pp. 265-280.

LOUVARD, D., MAROUX, S. & DESNUELLE, P. (1975) Topological studies on the hydrolases bound to the intestinal brush border membrane. II. Interactions of free and bound aminopeptidase with a specific antibody. *Biochim. Biophys. Acta 389*, 389-400

LOUVARD, D., SEMERIVA, M. & MAROUX, S. (1976) The brush-border intestinal aminopeptidase, a transmembrane protein as probed by macromolecular photolabelling. *J. Mol. Biol. 106*, 1023-1035

MAROUX, S. & LOUVARD, D. (1975) On the hydrophobic part of aminopeptidase and maltases which bind the enzyme to the intestinal brush border membrane. *Biochim. Biophys. Acta 419*, 189-195

MATTHEWS, D. M. (1975) Intestinal absorption of peptides. *Physiol. Rev. 55*, 537-608

MATTHEWS, D. M. & ADIBI, S. A. (1976) Peptide absorption. *Gastroenterology 71*, 151-161

SIGRIST-NELSON, K. (1975) Dipeptide transport in isolated intestinal brush border membrane. *Biochim. Biophys. Acta 394*, 220-226

WACKER, H., LEHKY, P., VANDERHAEGHE, F. & STEIN, E. A. (1976*a*) On the subunit structure of particulate aminopeptidase from pig kidney. *Biochim. Biophys. Acta 429*, 546-554

WACKER, H., MÜLLER, F. & SEMENZA, G. (1976*b*) Incorporation of hydrophobic aminopeptidase from hog kidney into egg lecithin liposomes: number and orientation of aminopeptidase molecules in the lecithin vesicles. *Febs (Fed. Eur. Biochem. Soc.) Lett. 68*, 145-152

Discussion

Kenny: Two questions: first, did the rate of transport correlate quantitatively with the hydrolytic activity of the aminopeptidase? Secondly, if you study the uptake of several amino acids, is there any correlation between the specificity of transport and that known for the aminopeptidase in hydrolysing peptides?

Semenza: It is very difficult to get initial rates of uptake. In our preliminary experiments, uptake corresponded to about 0.1 or 0.05% or even less of the total substrate present. On the second point, we have worked only with alanine, for which the two radioactively labelled stereoisomers are available. We hope to answer that question indirectly by studying a possible inhibition by other amino acids.

Norén: Did you use aminopeptidase M (EC 3.4.11.2) prepared from the kidney by solubilization with trypsin?

Semenza: Trypsin-solubilized aminopeptidase does not interact with the lipids. We use Triton X-100-solubilized aminopeptidase in these experiments.

Josefsson: I really doubt the whole story. I don't think aminopeptidase entered the liposome.

Semenza: Well, that is your problem. The aminopeptidase is not *inside*, but it is *anchored to* the liposome.

Josefsson: Then what are you studying?

Semenza: I would be very cautious and say that what we have seen is uptake.

Josefsson: What do you mean by that?

Semenza: Just that it is taken up by those liposomes which carry the *hydrophobic* aminopeptidase. As I pointed out in my talk, trypsin digestion of these liposomes destroys this capability to take up considerably more L-alanine than D-alanine.

Protein-free liposomes do not discriminate between the two isomers and aminopeptidase alone binds approximately only 1 mol L-alanine per enzyme molecule, and no D-alanine, under our conditions. Finally, a simple mixture of protein-free liposomes + aminopeptidase takes up much less L-alanine than those liposomes in which aminopeptidase is anchored to the lipids.

Josefsson: What is the difference between a mixture of aminopeptidase and liposomes and what you call association?

Semenza: When the monolamellar liposomes are made the way we have described (Wacker *et al.* 1976) and I have mentioned, they are physically associated with aminopeptidase, that is they are bound together. For this binding to occur aminopeptidase must have a hydrophobic part. (The bilayer structure in the eventual aminopeptidase-liposome is not severely disturbed by the association of aminopeptidase with the liposomes. We can show this by the nuclear magnetic resonance technique using shift reagents.) When lytic concentrations of either cholate or Triton are used there is no increase in aminopeptidase activity, as I have said, which indicates that the substrate has free access to the hydrolytic site from the outside. Trypsin treatment of those liposomes which carry aminopeptidase releases the hydrolytic activity from them completely. Thus, there can be no doubt that the monolamellar liposomes, as prepared by the cholate procedure in the presence of hydrophobic aminopeptidase (Wacker *et al.* 1976), carry the enzyme firmly but asymmetrically bound, with the catalytic site exposed to the surrounding medium.

Kim: I think it is a very important system but, as you have pointed out, it is a very complicated one. You were talking about intercalation of the hydrophobic enzyme into the lipid membrane. What is the orientation of the enzyme molecule in the liposomes? Is it possible that some enzymes are externally oriented in the liposomes while some are internally oriented? If that is the case, it might account for the relative inefficiency of peptide transport by your liposome system.

Semenza: The liposomes have a diameter of 28-30 nm so it would be difficult to think of the aminopeptidase molecule being wrapped by the lipid bilayer. In addition there should be an increase in aminopeptidase activity in these liposomes after detergent is added, and this is not the case (Wacker *et al.* 1976).

Kim: The reason for such an inefficient system for amino acid transport might be that 50% of the molecules are oriented inwards and 50% outwards.

Semenza: Then there should be an increase in activity if detergent is added, which is not the case.

Parsons: I think Professor Kim was wondering whether, when you add protein to the liposome, the protein would stick the membrane end on in a random fashion. Fifty per cent of the molecules would have their true 'outside' end at the outer surface of the liposome and the other 50% (the transport protein molecules) would have their true 'outside' end on the inside of the liposomes. Only half of the transport protein in the whole system would be used to take peptide into the liposomes.

Kim: That is known to happen with Na-ATPase.

Semenza: Do you mean that the liposome would just be a kind of belt around the molecule?

Kim: According to the Singer-Nicolson model of the membrane, membrane protein with hydrophobic and hydrophilic regions is intercalated into the membrane lipids through the hydrophobic region. From your description of the way you prepared the liposomes it seems as if some molecules are facing outwards and some inwards.

Semenza: If there were some hydrolytic sites on the inside of the lipid barrier, they would not be as easily accessible to the substrate, in this case a dipeptide. When Triton is added there should be an increase in hydrolytic activities, which there isn't.

Josefsson: It is impossible, from all physicochemical points of view, to put such a macromolecule inside the liposome.

Semenza: I don't believe that either.

Josefsson: The size of a single layer liposome is less than 20 nm.

Semenza: It is exclusively the hydrophobic part which goes into the liposome. When we treat these liposomes with trypsin a hydrophobic peptide remains associated with the liposomes, which still retain their bilayer structure.

Maroux: Aminopeptidase has a good affinity for free L-amino acids and perhaps your results could be explained by a binding of these amino acids to the aminopeptidase on the surface of the liposome, which would induce a higher local concentration.

Semenza: That was our first guess and it is still possible. But aminopeptidase can only bind one or perhaps two amino acids, and thus only an insignificant increase in concentration could be obtained. I am not saying that aminopeptidase acts as a membrane carrier. I suspect rather that the hydrophobic part can disturb the lipid structure locally. If there is a binding site there, it may provide for the specificity. I don't think the hydrophobic part alone can explain the specific increase in permeability, because trypsin treatment, which leaves the hydrophobic peptide associated with the liposomes, destroys their specific uptake of L-alanine.

Maroux: Have you studied the effect of sodium?

Semenza: Not yet.

Adibi: Whenever I have presented work on a peptide I have been asked how it resembles disaccharide. Is it possible that there is a carrier for dipeptides and none for disaccharides?

Semenza: I don't like trying to explain everything the same way, and you may be right.

Kim: But, if I recall correctly, you reported a selective transport of a disaccharide using liposomes with disaccharidases inserted into them.

Semenza: Yes, there is an increase in permeability but we don't know anything about the mechanism.

Radhakrishnan: In the vesicle system you used something like 1 mM labelled peptides and 40 mM unlabelled peptides, and you get about 50% less inhibition than we observed with monkey intestinal strips, even though we had lower ratios (labelled, 0.5 mM; unlabelled, 10 mM). Is there any basic difference to explain these results?

Semenza: I do not know.

Fottrell: How does the hydrophobic region of the aminopeptidase compare with the hydrophobic regions of other brush border enzymes?

Semenza: Dr Maroux is more qualified than I am to talk on the hydrophobic part of the aminopeptidase. Our work on sucrase indicates that there is a hydrophobic part in the brush border membrane. Interestingly enough, it seems that it is only the isomaltase unit which has a hydrophobic part associated with the membrane. As Dr Maroux has shown (Maroux & Louvard 1976), an area towards the amino end is on the inside of the membrane, and in sucrase it may also be so (J. Brunner & G. Semenza, unpublished results). This is a very interesting problem from a biosynthetic point of view.

Maroux: We have purified the hydrophobic part of aminopeptidases from pig kidney and intestine and of maltases from intestine. The comparison of their amino acid composition reveals close homology between the two aminopeptidases but no similarity between aminopeptidases and maltases. I shall develop this point later (p. 191).

Norén: Professor Semenza, you have purified a Triton-solubilized form of the sucrase–isomaltase complex. Have you incorporated this complex in liposomes and then studied the transport of dipeptides?

Semenza: The permeability of the liposomes for D- and L-glucose or for sodium is not increased by the incorporation (or binding) of aminopeptidase into the liposome bilayer.

Silk: Sigrist-Nelson's data showed that when isolated brush border membrane vesicles were incubated for increasing time periods with radioactive dipeptide, increasing proportions of radioactivity within the vesicles were in the form of free amino acids (Sigrist-Nelson 1975). Can you be quite certain that these vesicles were not 'outside-in'?

Semenza: The vesicles were prepared by the EDTA method (Hopfer *et al.* 1973). The osmotic space of these vesicles indicates that perhaps only 10% of the vesicles are actually closed. Treatment with Triton does not lead to an increase in their sucrase activity or leucylnaphthylamidase activity. In the electron microscope one can see 'bugs' only on the outside of the vesicles and these probably represent sucrase and other digestive enzymes. There are no data on the exposure of possible markers on the inside.

Silk: The difference in inhibition of free amino acids on uptake of peptide-bound amino residues when isolated brush border membrane vesicles are used (Sigrist-Nelson 1975) and when strips of monkey intestine are used, as in Professor Radhakrishnan's experiments, is interesting. Could one reason for the discrepancy in results be that one advantage of the vesicles is that they have no cytoplasmic peptidases?

Semenza: I cannot rule out that some cytosolic peptidases may be present in these vesicles.

Kenny: Have you looked at vesicles prepared without the use of EDTA?

Semenza: Not yet.

Baker: The rate of disappearance of the dipeptide within the vesicle seems to be much slower than within the intact cell, which suggests that the amount of enzyme activity is less.

Semenza: Let us assume for the sake of the argument that the peptides are taken up intact in these vesicles and that the aminopeptidase activity only looks to the outside. How do we explain why the amount of intact dipeptide present in the vesicles, both percentagewise and in abolute terms, nevertheless decreases with time of incubation?

Baker: The dipeptide would be hydrolysed inside the vesicle, wouldn't it?

Semenza: Yes, but how? What is hydrolysing the peptide if all the cytosolic peptidase has been washed out?

Baker: But have you washed it out? I don't believe you have.

Semenza: That is what I am saying: that some cytosolic dipeptidase may still be associated with these brush border vesicles.

Nicholson: Did you check whether any cytosolic marker enzymes such as lactate dehydrogenase or L-prolylglycine hydrolase were present in the vesicles?

Semenza: No detectable lactate dehydrogenase activity was associated with these vesicles (Sigrist-Nelson 1975).

Nicholson: Have you therefore any evidence that you have cleaned up the vesicles?

Semenza: Nobody knows what absorption properties a hypothetical cytosolic peptidase may have. They may be different from those of lactate dehydrogenase.

Norén: I don't think cytosol dipeptidases could be associated with the vesicles if these were washed before being used.

References

HOPFER, U., NELSON, K., PERROTTO, J. & ISSELBACHER, K. (1973) Glucose transport in isolated brush border membrane from rat small intestine. *J. Biol. Chem. 248*, 25-32

MAROUX, S. & LOUVARD, D. (1976) On the hydrophobic part of aminopeptidase and maltases which bind the enzyme to the intestinal brush border membrane. *Biochim. Biophys. Acta 419*, 189-195

SIGRIST-NELSON, K. (1975) Dipeptide transport in isolated intestinal brush border membrane. *Biochim. Biophys. Acta 394*, 220-226

WACKER, H., MÜLLER, F. & SEMENZA, G. (1976) Incorporation of hydrophobic aminopeptidase from hog kidney into egg lecithin liposomes: number and orientation of aminopeptidase molecules in the lecithin vesicles. *Febs (Fed. Eur. Biochem. Soc.) Lett. 68*, 145-152

On the function of the γ-glutamyl cycle in the transport of amino acids and peptides

ALTON MEISTER, SURESH S. TATE and GREGORY A. THOMPSON

Department of Biochemistry, Cornell University Medical College, New York

Abstract Recent work in this laboratory has been concerned with the enzymes that catalyse the synthesis and degradation of glutathione via the γ-glutamyl cycle. The synthesis of glutathione from glutamate, cysteine and glycine requires two ATP-dependent steps, catalysed by γ-glutamyl-cysteine synthetase and glutathione synthetase, respectively. The first step in the amino acid-dependent or peptide-dependent breakdown of glutathione is catalysed by γ-glutamyl-transpeptidase, a membrane-bound enzyme found in a variety of epithelial and lymphoid cells. γ-Glutamyltranspeptidase catalyses transfer of the γ-glutamyl moiety of glutathione to an amino acid or peptide acceptor to form the corresponding γ-glutamyl amino acid or peptide derivative and cysteinyl-glycine. Cysteinyl-glycine is hydrolysed to cysteine and glycine. The conversion of γ-glutamyl amino acids to 5-oxoproline and the corresponding amino acids is catalysed by γ-glutamylcyclotransferase. The enzyme 5-oxoprolinase catalyses the cleavage of 5-oxoproline to glutamate, a reaction that is coupled with the cleavage of ATP to ADP. These reactions constitute a metabolic pathway, the γ-glutamyl cycle, that accounts for the synthesis and degradation of glutathione. Studies on this cycle have (*a*) shown that 5-oxo-L-proline is a quantitatively significant metabolite, (*b*) led to purification and study of 5-oxoprolinase, an enzyme that catalyses a unique peptide bond cleavage that requires energy, (*c*) explained the aetiology and symptomatology of the disease 5-oxoprolinuria, (*d*) elucidated a physiologically significant mechanism that controls glutathione synthesis, and (*e*) focused new attention on the γ-glutamyl moiety of glutathione and on γ-glutamyltranspeptidase.

The evidence that the γ-glutamyl cycle functions *in vivo* is reviewed. The hypothesis that it functions as one of the systems that mediates amino acid and peptide transport is considered in the context of the available data, including recent studies on the interaction of γ-glutamyltranspeptidase with various dipeptides and neutral L-amino acids such as cystine and glutamine.

It seems appropriate to consider glutathione at this symposium because this tripeptide is present in virtually all cells, usually in concentrations much higher than those of other peptides. The synthesis and breakdown of glutathione take

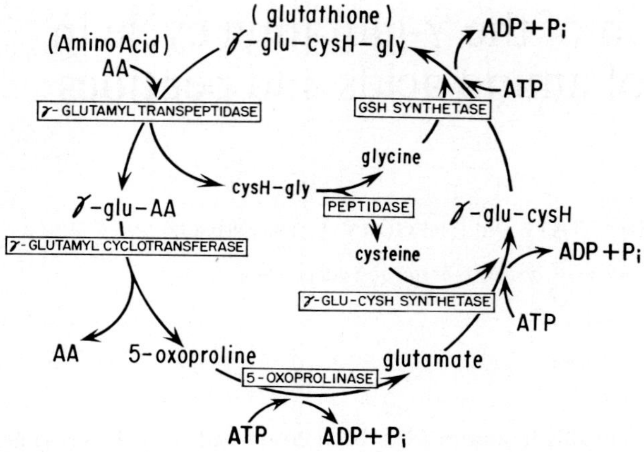

FIG. 1.

place by the reactions of the γ-glutamyl cycle (Fig. 1), a metabolic pathway that was elucidated mainly through investigations on the enzymology of glutathione. (The γ-glutamyl cycle and the biochemistry of glutathione have been reviewed recently [Meister & Tate 1976; Meister 1975]; these sources cite the relevant literature more comprehensively than is possible here.) Although studies on the γ-glutamyl cycle have tended to focus on the γ-glutamyl peptide bond of glutathione, there seem to be some significant relationships between the γ-glutamyl moiety of glutathione and the other characteristic structural feature of this tripeptide, its thiol group. Glutathione is the most abundant intracellular thiol and also the most abundant intracellular γ-glutamyl compound, except perhaps for glutamine which, in contrast to glutathione, occurs extracellularly as well as within cells.

THE γ-GLUTAMYL CYCLE

The first step in the breakdown of glutathione is catalysed by γ-glutamyltranspeptidase. Hanes *et al.* (1952) showed that this enzyme can catalyse the transfer of the γ-glutamyl moiety of glutathione to an amino acid or peptide acceptor to form the corresponding γ-glutamyl compound and cysteinyl-glycine (reaction 1). The enzyme also catalyses the slow hydrolysis of the γ-glutamyl bond of glutathione. A number of amino acids and peptides can serve as acceptors of the γ-glutamyl group. Much work has been done on the distribution, cellular disposition, specificity, kinetics, purification and characterization of γ-glutamyltranspeptidase. (See Tate 1975 and Meister *et al.* 1976 for recent reviews.)

(1) Glutathione + amino acid (or peptide) ⇌ γ-glutamyl-amino acid (or γ-glutamyl peptide) + cysteinyl-glycine

The cysteinyl-glycine formed in the transpeptidation reaction of glutathione is cleaved by dipeptidase to cysteine and glycine. The γ-glutamyl amino acids formed by transpeptidation are converted to 5-oxoproline (synonyms: pyroglutamate, pyrrolidone carboxylate) and the corresponding amino acids by the action of γ-glutamylcyclotransferase (reaction 2); this reaction was first observed by Connell & Hanes (1956). Woodward & Reinhart (1942) had earlier found that 5-oxoproline was formed when kidney preparations were incubated with glutathione. More recently, highly purified preparations of γ-glutamylcyclotransferase have been obtained from several tissues (Orlowski *et al.* 1969; Adamson *et al.* 1971; Orlowski & Meister 1973).

(2) L-γ-Glutamyl-L-amino acid → 5-oxo-L-proline + L-amino acid

The presence of high activities of γ-glutamyltranspeptidase and γ-glutamylcyclotransferase in the kidney seemed of interest since this tissue is known to contain a relatively high steady-state concentration of glutathione. The ability of kidney preparations to synthesize glutathione was therefore examined, and it was found that the kidney and other tissues have substantial amounts of the enzymes required for the synthesis of glutathione (Snoke & Bloch 1954; Meister 1974*a*). These enzymes, γ-glutamyl-cysteine synthetase and glutathione synthetase, catalyse the two ATP-dependent reactions shown in reactions (3) and (4), respectively.

(3) L-Glutamate + L-cysteine + ATP → L-γ-glutamylcysteine + ADP + P_i

(4) L-γ-Glutamylcysteine + glycine + ATP → glutathione + ADP + P_i

These studies thus showed that the kidney and other tissues have the enzymes that catalyse the degradation of glutathione and also the enzymes that catalyse glutathione synthesis. Since 5-oxoproline, a product of the degradation of glutathione by the transpeptidase-cyclotransferase pathway, does not accumulate to an appreciable extent in the kidney, liver, blood or urine, the question arose of whether 5-oxoproline might be metabolized. Indeed, this problem had been considered previously, but the earlier studies on the metabolism of 5-oxoproline in mammals had given somewhat conflicting results (Van der Werf & Meister 1975). Studies in our laboratory in which uniformly labelled 5-oxo-L-[^{14}C]-proline was administered to mice showed that the label appeared promptly in the respiratory carbon dioxide (Richman 1969; Van der Werf 1970; Orlowski & Meister 1970). Other studies showed that 5-oxoproline is rapidly metabolized by kidney slices and that the major product formed is glutamate

(Ramakrishna *et al.* 1970; Orlowski & Meister 1970; Van der Werf *et al.* 1971). When crude kidney preparations were incubated with labelled 5-oxoproline, magnesium ions, ATP and an ATP-generating system, both labelled glutamate and glutamine were formed. When this type of study was carried out in the presence of L-methionine sulphoximine, an irreversible inhibitor of glutamine synthetase (Ronzio & Meister 1968; Meister & Tate 1976), only L-glutamate was formed, and with more highly purified preparations 5-oxoproline was converted to L-glutamate in the presence or absence of methionine sulphoximine. Subsequently, a very highly purified preparation of the enzyme 5-oxoprolinase was obtained from rat kidney, and shown to catalyse reaction (5). We now know that this enzyme is widely distributed in mammalian tissues and that it also occurs in bacteria (Van der Werf & Meister 1974) and yeast (Mooz & Wigglesworth 1976).

(5) 5-Oxo-L-proline + ATP + $2H_2O$ → L-glutamate + ADP + P_i

The discovery of the reaction catalysed by 5-oxoprolinase indicates that there is a link between the reactions involved in glutathione degradation and those that catalyse its synthesis; it is therefore possible to visualize a cycle that requires glutathione, ATP and amino acids (Fig. 1).

EVIDENCE FOR FUNCTION OF THE γ-GLUTAMYL CYCLE IN VIVO

There is now good evidence that the reactions of the γ-glutamyl cycle take place *in vivo*. Thus, the turnover of glutathione was studied in mouse liver and kidney in experiments in which labelled L-glutamate was given to mice. The turnover of glutathione was found to be much more rapid in kidney than in liver, a result which parallels the relative activities of the enzymes of the γ-glutamyl cycle in these tissues. When 5-oxo-L-[^{14}C]proline was given to mice, the first-order rate constants for glutathione synthesis from 5-oxoproline in liver and kidney were found to be similar to those observed for glutathione synthesis from L-glutamate (Sekura & Meister 1974). Another experimental result that supports the function of the cycle *in vivo* is the finding of labelled 5-oxo-L-proline in kidney and liver after labelled L-glutamate was given to rats (Sekura *et al.* 1976). Additional evidence came from experiments in which animals were treated with L-2-imidazolidone-4-carboxylate, a competitive inhibitor of 5-oxoprolinase (Van der Werf *et al.* 1973, 1974). After mice were treated with the inhibitor, their ability to convert 5-oxoproline to respiratory carbon dioxide was markedly decreased and they excreted some of the administered 5-oxoproline in their urine. Mice injected with the inhibitor accumulated 5-oxoproline in several tissues including the liver and kidney.

When mice were given the inhibitor together with one of several L-amino acids, the accumulation of tissue 5-oxoproline and its urinary excretion were much greater than when the inhibitor was given alone. Such an increase in the accumulation of 5-oxoproline after administration of amino acid can be explained in terms of the γ-glutamyl cycle. Thus, it would be expected that the increase in concentrations of tissue amino acids produced by administering amino acids would be accompanied by an increased utilization of glutathione in transpeptidation, and thus by increased formation of 5-oxoproline. In the presence of a 5-oxoprolinase inhibitor, 5-oxoproline would accumulate. Further evidence consistent with the view that the γ-glutamyl cycle functions *in vivo* has come from studies on patients with the inborn error of glutathione metabolism, 5-oxoprolinuria. In this disease there is a modified γ-glutamyl cycle associated with a marked deficiency of glutathione synthetase activity.

ON THE PHYSIOLOGICAL ROLE OF THE γ-GLUTAMYL CYCLE

The studies reviewed above on the enzymology of glutathione and the γ-glutamyl cycle have led to several new approaches and ideas. (1) It is now evident that 5-oxo-L-proline is not an artifact, as often thought earlier, but that it is a quantitatively significant metabolite. The steady-state concentrations of 5-oxo-L-proline in tissues such as liver and kidney are about 0.05 mM. (2) The enzyme 5-oxoprolinase is itself of special interest in that it couples the energy-requiring cleavage of the internal peptide bond of 5-oxo-L-proline with the exergonic cleavage of ATP (Van der Werf *et al.* 1975; Griffith & Meister 1976). This seems to be the only known instance in which cleavage of a peptide bond requires energy. The mechanism of the coupling is being studied in our laboratory. (3) Elucidation of the γ-glutamyl cycle has led to an understanding of 5-oxoprolinuria. Patients with this disorder excrete huge amounts of 5-oxoproline in their urine and their blood concentration of 5-oxoproline is markedly elevated. Their altered metabolism and some of their symptomatology can now be explained in terms of the function of the γ-glutamyl cycle (Wellner *et al.* 1974; Meister 1974*b*). (4) Studies of 5-oxoprolinuria and experiments on γ-glutamyl-cysteine synthetase have elucidated a physiologically significant mechanism that controls glutathione synthesis. Thus, glutathione inhibits the first step in its biosynthesis. Glutathione binds non-covalently to γ-glutamyl-cysteine synthetase, attaching to the glutamate and cysteine binding sites of this enzyme. The available data indicate that γ-glutamyl-cysteine synthetase normally functions at considerably less than its maximal activity because of feedback inhibition by glutathione (Richman & Meister 1975). (5) Studies of the γ-glutamyl cycle have focused new attention on the formation and role of the γ-glutamyl moiety of glutathione and have also led to increased interest in the membrane-bound enzyme, γ-glutamyltranspeptidase. Thus, it has become evident that the enzyme is widely distributed in mammalian tissues. It is present in the membranes of many epithelial cells, including those of the renal tubule, jejunal villi, choroid plexus, ciliary body, visual receptor cells, retinal epithelium and cerebral astrocytes or their capillaries, in locations known to be involved in transport and secretory phenomena (Meister & Tate 1976). In recent studies, it was discovered that the enzyme is present in the membranes of certain lymphoid cells, thus indicating that the enzyme is more widely distributed than was previously recognized (Novogrodsky *et al.* 1976). Studies of the catalytic behaviour of the enzyme, including work on its kinetics, specificity and structure are currently being pursued; some of the results will be considered here. (6) Studies of the γ-glutamyl cycle have led to the hypothesis that this metabolic pathway represents one of the systems that mediate the transport of certain amino acids and peptides in mammalian tissues. A function of this system in transport is clearly more difficult to prove than its occurrence as a metabolic pathway, but the idea that the γ-glutamyl cycle functions in such transport is attractive and consistent with the available findings. The mechanism of group translocation proposed here implies that a γ-glutamyl amino acid or γ-glutamyl dipeptide is formed on or within the membrane, and that the amino acid or dipeptide is transported as a γ-glutamyl derivative, after which the γ-glutamyl moiety is removed by cyclization or hydrolysis.

In relation to the physiological function of the γ-glutamyl cycle, it seems notable that (1) the degradation of glutathione catalysed by membrane-bound γ-glutamyltranspeptidase is markedly stimulated by certain amino acids and peptides and that these are not metabolized in the cycle but are regenerated unchanged. (2) The cycle has enzymes that effectively resynthesize glutathione. (3) Of the several enzymes of the cycle, only γ-glutamyltranspeptidase is membrane-bound. (4) The breakdown of glutathione yields cysteine, glycine and 5-oxoproline. Glutamate may be formed directly by hydrolysis of γ-glutamyl amino acids. The cysteinyl-glycine formed in transpeptidation is evidently cleaved rapidly and does not accumulate. The possibility that the transpeptidase functions in the regulation of the intracellular concentration of glutathione must be considered; however, since there is an effective feedback mechanism for the control of glutathione biosynthesis, this appears unlikely. The transpeptidase is a membrane protein, and one must thus consider the likelihood that it plays a role in the transport of compounds into or out of the cell. It might also be a structural component of the membrane or serve as part of a mechanism to protect the cell or the cell membrane. It might function in the synthesis of cell membrane. One must also consider the idea that the transpeptidase functions as a receptor site for an external compound whose binding to the membrane-bound enzyme might trigger a signal which in some way would stimulate intracellular metabolic activity. In considering these possibilities, we cannot exclude the additional possibility that the transpeptidase might react with substrates not yet tested and thus that the enzyme has other functions that are not obviously related to its known catalytic activities.

Studies on amino acid transport have led to the proposal that this process involves a series of steps including the binding of amino acids to sites on the cell membrane, carrier-mediated translocation, intracellular release of the amino acid from the carrier, and the reactivation of the carrier in an energy-requiring process (Pardee 1968; Heinz 1972). The hypothesis that the γ-glutamyl cycle functions in transport was stimulated by the fact that the cycle has features previously postulated to be needed for amino acid transport. Thus, membrane-bound γ-glutamyltranspeptidase might mediate amino acid translocation. The enzyme, according to this hypothesis, would interact with extracellular amino acid and intracellular glutathione (or possibly another γ-glutamyl donor derived from glutathione) by transpeptidation via an intermediate γ-glutamyl enzyme. A mechanism for such translocation was previously suggested (Meister 1973). According to this mechanism the amino acid enters the cell as a peptide, i.e. the corresponding γ-glutamyl amino acid. (This postulate seems not very different from ideas expressed at this symposium about dipeptide transport systems.) As discussed previously, the interaction of the incoming amino acids and of glutathione (or other γ-glutamyl donor) with the membrane-bound transpeptidase may be associated with conformational changes in the enzyme. Possibly the enzyme is oriented so as to provide a channel through which the peptide can pass. It is clearly of importance to learn more about the structure and orientation of the transpeptidase in relation to the membrane. The release of the amino acid from its γ-glutamyl carrier within the cell would be catalysed by γ-glutamylcyclotransferase. Resynthesis of the precursor of the γ-glutamyl carrier, i.e. glutathione, is catalysed by three ATP-requiring steps in the cycle. This hypothesis seems consistent with the observations that the transpeptidase catalyses amino acid-stimulated and peptide-stimulated glutathione breakdown, and that administration of amino acids to animals leads to increased glutathione breakdown without depressing glutathione concentrations, presumably because of rapid resynthesis. Several modifications of the γ-glutamyl cycle in which less ATP is required have been considered (Meister & Tate 1976).

RECENT STUDIES ON γ-GLUTAMYLTRANSPEPTIDASE

γ-Glutamyltranspeptidase has been purified from a number of tissues, and several highly purified preparations of the enzyme are now available (Meister *et al.* 1976; Tate 1975). Transpeptidase preparations exhibiting molecular weights greater than about 200 000 (heavy form) have been obtained from kidney by methods involving treatment with deoxycholate and *n*-butanol. Preparations of the enzyme exhibiting a much lower molecular weight (68 000–90 000) (light form) have been obtained by procedures involving the use of proteases such as bromelain. A highly active purified preparation of the enzyme, which has a molecular weight of about 70 000, was obtained from rat kidney by a procedure involving extraction with Lubrol WX, ammonium sulphate and acetone fractionation, followed by digestion with bromelain and chromatography on DEAE-cellulose and Sephadex G-100 (Tate & Meister 1975). The light form of the enzyme exhibits a much higher specific activity than the heavy form. The available data suggest that the heavy form of the enzyme consists of a complex of proteins that may be organized structurally and functionally within the renal brush border. The heavy form of the enzyme exhibits dipeptidase activity and may also contain other enzymic activities. The light form of rat kidney transpeptidase was separated into twelve different enzymically active fractions by isoelectric focusing; each of these fractions could be dissociated with sodium dodecylsulphate in the presence of 2-mercaptoethanol into two non-identical subunits of molecular weights 46 000 and 23 000 (Tate & Meister 1976). Similar subunits can be obtained by treatment of the heavy enzyme with sodium dodecylsulphate, suggesting that the two subunits of the light enzyme are not produced by the action of bromelain and that they are present in the heavy form of the enzyme.

The specificity of partially and very highly purified γ-glutamyltranspeptidase preparations from various sources is remarkably similar. The isozymes separated by isoelectric focusing from the light form of rat kidney γ-glutamyltranspeptidase also exhibit similar substrate specificity. There are some species differences in acceptor specificity, but transpeptidase preparations obtained from different tissues of a given species exhibit similar relative activities with various acceptor amino acids and peptides. In general, the best amino acid acceptors are neutral amino acids, e.g. the L-isomers of glutamine, methionine, asparagine, serine and alanine (Tate & Meister 1974). Isoleucine, leucine, valine, threonine, aspartate and histidine are relatively less active, whereas L-proline, various D-amino acids and α-substituted amino acids are inactive as acceptors of the γ-glutamyl moiety. The relatively insoluble amino acids, L-tyrosine and L-cystine, could be studied only at concentrations of about 0.5

TABLE 1

Acitvity of γ-glutamyl transpeptidase toward various peptides*

Acceptor	*Relative activity*
Glycine	21
Gly-Gly	[100]
Gly-L-Ala	65
Gly-D-Ala	8
Gly-L-Asn	14
Gly-L-Asp	11
Gly-L-Thr	11
Gly-L-Val	10
Gly-L-Pro	19
Gly-D-Phe	7
Gly-Gly-Gly	9
L-alanine	24
L-Ala-Gly	32
L-Ala-L-Ala	9
L-methionine	31
L-Met-Gly	24
L-Met-L-Ala	7
L-Met-L-Ser	38
L-Met-L-Thr	16
L-Met-L-Met	28
L-Met-L-Phe	17
L-Met-L-Leu	12
L-Met-L-Pro	12
L-Met-L-Glu	19
L-Met-L-Ala-L-Ser	6
L-glutamate	22
L-Glu-Gly	23
L-Glu-L-α-aminobutyrate	27
L-glutamine	33
L-Glu-L-Gln	42
L-α-aminobutyrate	22
L-α-aminobutyryl-Gly	47

* Adapted from Tate & Meister (1974). The activities, expressed relative to that found with Gly-Gly [100], were determined as described (Tate & Meister 1974), with 2.5 mM-L-γ-glutamyl-*p*-nitroanilide and 20 mM acceptor; the control value (no acceptor) was 9.

mM; although L-tyrosine is only slightly active, L-cystine is an excellent acceptor; at a concentration of 0.5 mM, L-cystine is somewhat more active than L-glutamine and much more active than L-cysteine. Several amino acids related in structure to cystine (e.g. lanthionine, cystathionine, α,ε-diaminopimelate) are also very active acceptors (Thompson & Meister 1975).

The high acceptor activity of glycylglycine was noted by Hanes *et al.* (1952) in early studies of γ-glutamyltranspeptidase. Subsequently, it was found that certain other dipeptides are also very active acceptors of the γ-glutamyl moiety. Thus, in studies with highly purified rat kidney γ-glutamyltranspeptidase it was found that glycylglycine and glycyl-L-alanine were more active than glycine (Table 1) (Tate & Meister 1974). Methionylserine was more active than methionine, but other methionyl peptides were less active. L-Alanylglycine was more active than L-alanine, while L-alanyl-L-alanine was less active. In the five instances studied, aminoacylglycine dipeptides were found to be more active than the corresponding amino-terminal free amino acids. The corresponding aminoacyl-L-alanine dipeptides were less active than the aminoacylglycine dipeptides. It must be noted that the values obtained for the relative activities of various amino acids and peptide acceptors depend to some extent on the conditions employed in the assay. For this and other reasons, it seems difficult to assess the physiological significance of a given acceptor substrate solely on the basis of activity values. Thus, although glycylglycine is one of the most active acceptors, indeed more active than glutamine and methionine, there seems to be very little of this dipeptide in blood plasma and tissues. On the other hand, studies of the interaction of various substrates with the enzyme can lead to useful information about its catalytic properties and to deductions concerning the nature of its active centre.

Studies of the kinetics and specificity of γ-glutamyltranspeptidase are complicated by the fact that the enzyme can catalyse three types of reactions:

(6) γ-Glu-X + acceptor $\rightleftharpoons$ γ-Glu-acceptor + HX

(7) γ-Glu-X + H_2O $\rightarrow$ glutamate + HX

(8) γ-Glu-X + γ-Glu-X $\rightleftharpoons$ γ-Glu-γ-Glu-X + HX

Reaction (6) involves the transfer of the γ-glutamyl moiety of glutathione (and of other γ-glutamyl compounds) to an acceptor. The enzyme also catalyses the hydrolysis of γ-glutamyl compounds (reaction 7), as well as an autotranspeptidation reaction in which the γ-glutamyl compound itself serves as an acceptor of the γ-glutamyl moiety (reaction 8). In many studies of γ-glutamyltranspeptidase, L-γ-glutamyl-*p*-nitroanilide has been used as a substrate and the reaction has been followed by measuring the release of free *p*-nitroaniline (Orlowski & Meister 1963). However, this procedure does not permit separate determination of the individual rates of the several reactions, although the contributions of reactions (7) and (8) to the overall rate of *p*-nitroaniline formation may be reduced by using high concentrations of γ-glutamyl *p*-nitroanilide and the acceptor substrate.

In recent studies in our laboratory a direct spectrophotometric assay was designed for the separate determination of reaction (6) and reaction (7) by using conditions under which the L-γ-glutamyl donor does not bind to the acceptor site of the enzyme, and also by making use of a D-γ-glutamyl donor substrate (Thompson & Meister 1976). This has been possible because the enzyme exhibits a very high affinity for both of the optical isomers of γ-glutamyl *p*-nitroanilide. Thus, L-γ-glutamyl *p*-nitroanilide has a much higher affinity as a γ-glutamyl donor than it does as an acceptor, and D-γ-glutamyl *p*-nitroanilide, which also exhibits high affinity as a γ-glutamyl donor, is inactive as an acceptor of the D-γ-glutamyl moiety. Our studies have shown that the affinity of the enzyme for the L- and D-isomers of γ-glutamyl *p*-nitroanilide is of the same order of magnitude as that of the natural substrate, glutathione. It was also found that the V_{max} values for hydrolysis are about the same for both isomers of γ-glutamyl *p*-nitroanilide. When the enzyme is incubated with very low concentrations of L-γ-glutamyl *p*-nitroanilide this γ-glutamyl donor does not compete with water as an acceptor of the γ-glutamyl moiety, and it was thus readily established that the K_m value for hydrolysis is about 5 μM; on the other hand, the K_m value for L-γ-glutamyl *p*-nitroanilide as an acceptor of the γ-glutamyl moiety is much higher, about 2 mM. Since D-γ-glutamyl *p*-nitroanilide is a good γ-glutamyl donor substrate (K_m, 31 μM), it can serve usefully in studies of the acceptor specificity of the enzyme. D-γ-Glutamyl *p*-nitroanilide, like other D-amino acids, is not active as a γ-glutamyl acceptor. Since transpeptidation occurs between D-γ-glutamyl *p*-nitroanilide and L-amino acids, this system can be used for studies of transpeptidation without the complication of autotranspeptidation. The validity of this approach is indicated by studies performed with L-cystine and relatively low concentrations of L- and D-γ-glutamyl *p*-nitroanilide; the same K_m value (30 μM) was obtained for L-cystine with both L- and D-γ-glutamyl donors.

The use of D-γ-glutamyl *p*-nitroanilide as a γ-glutamyl donor substrate for the transpeptidase has greatly facilitated studies of the acceptor substrate specificity of the enzyme, and the results thus far obtained lead to some tentative conclusions about the nature of the acceptor site of γ-glutamyltranspeptidase. In Table 2, the values are given as a function of V_{max} divided by K_m. The various assumptions employed and the calculations, which are similar to those used by Folk & Chung (1973) in studies of transglutaminase, will be given in a forthcoming publication (Thompson & Meister 1977). Earlier studies showed that several neutral amino acids are very active as acceptors and that glycylglycine and certain similar peptides are more active than the most active free amino acids. The possibility that the activity of these peptides reflects their ability to bind to the cysteinyl-glycine site of the enzyme was considered previously, but it has remained unclear whether free amino acid acceptors bind to a part of the cysteinyl-glycine site or whether they bind to a separate site on the enzyme. In

TABLE 2

γ-Glutamyltranspeptidase activity determined with D-γ-glutamyl *p*-nitroanilide and various peptide acceptors

Acceptor	*Relative value**
L-Met-Gly	210
L-Gln-Gly	182
L-Ala-Gly	154
L-$(Cys)_2$)-Gly_2	134
Gly-Gly	[100]
L-Ser-Gly	99
L-CySH-Gly	81
L-Lys-Gly	57
L-Phe-Gly	28
L-Leu-Gly	19
L-Val-Gly	14
L-Pro-Gly	0
Gly-L-Ala	83
Gly-L-Pro	17
Gly-L-Leu	7
Gly-L-Lys	6
Gly-L-Met	5
Gly-L-Thr	3
Gly-L-Tyr	3

* These values are expressed relative to that obtained with Gly-Gly [100], and are a function of the ratio of V_{max} to K_m (Thompson & Meister 1977). The assays were carried out with D-γ-glutamyl-*p*-nitroanilide (0.02-0.1 mM) and various acceptors (0.25–10 mM).

recent studies in our laboratory, a number of dipeptides containing glycine were examined with highly purified γ-glutamyltranspeptidase.

The results indicate that dipeptides of the X-Gly type are almost always substantially more active than dipeptides of the Gly-X type. This suggests that the most active dipeptides (X-Gly) bind to the enzyme site which normally accepts the cysteinyl-glycine moiety of glutathione. That glycyl-L-alanine is very active is also consistent with this conclusion when one considers the close structural relationship between alanine and glycine. The findings are also in accord with another interesting conclusion. Thus, the relative activities observed with peptides of the X-Gly type fall in an order which is almost parallel to that found in earlier studies for the corresponding free amino acids (Tate & Meister 1974). For example, Met-Gly and Gln-Gly are the most active in this series of dipeptides, whereas Leu-Gly and Val-Gly are much less active. It may also be noted that Pro-Gly is not a substrate, nor is free proline a substrate. A reasonable interpretation of these findings would be that the free amino acids bind to the site which binds the cysteinyl moiety of cysteinyl-glycine, and that the presence of a C-terminal glycine residue in the dipeptide assists substantially in binding the dipeptide to the enzyme. It is also of interest that $(\dot{C}ys)_2$-Gly_2 is substantially more active as an acceptor substrate than is CysH-Gly; this finding seems to parallel the activities observed with L-cystine and L-cysteine as acceptor substrates, the former being much more active.

It is of particular interest that L-cystine is about four times more active than the most active X-Gly dipeptide listed in Table 2; this value was obtained under comparable conditions. This observation, and earlier findings on L-cystine, suggest that the high affinity and reactivity of γ-glutamyltranspeptidase with this amino acid may be of physiological significance. As noted elsewhere (Thompson & Meister 1975), γ-glutamylcystine is a substrate of γ-glutamyl-cyclotransferase, and there would thus appear to be a mechanism for utilization of this γ-glutamyl compound via the γ-glutamyl cycle. It is also possible, as previously suggested, that γ-glutamylcystine undergoes reduction to yield one molecule each of cysteine and γ-glutamylcysteine; the latter compound is a substrate of γ-glutamylcyclotransferase and also of glutathione synthetase. The available data indicate that γ-glutamyl dipeptides are not substrates of γ-glutamylcyclotransferase, suggesting that if such compounds are formed under physiological conditions, there is an alternative mechanism for release of the dipeptide from the γ-glutamyl moiety. One possibility is that γ-glutamyl dipeptides undergo an additional transpeptidation reaction leading to release of the dipeptide. The finding that administration of glycylglycine to rats if followed by markedly decreased kidney concentrations of glutathione suggests that transpeptidation of this dipeptide can occur *in vivo*; a similar effect does not occur after administration of equivalent amounts of glycine (Palekar *et al.* 1975). If the activity values thus far observed with various dipeptides reflect any physiological meaning, it would seem that the preferred substrates, at least as judged by *in vitro* studies, are dipeptides containing C-terminal glycine residues. The present findings suggest that the activity of the transpeptidase towards dipeptides is relatively restricted, but additional studies on other peptides would be desirable.

The broad amino acid specificity of γ-glutamyltranspeptidase is consistent with the idea that the γ-glutamyl cycle might mediate transport of the 'high capacity–low specificity' type (Young & Freedman 1971). The most active amino acid acceptors include such neutral amino acids as L-cystine, L-glutamine and L-methionine. L-Cystine is an active acceptor of the γ-glutamyl moiety at very low concentrations which are close to the concentration of L-cystine in mammalian blood plasma. It seems of interest that glutathione, the most prevalent intracellular form, is much more active as a γ-glutamyl donor than glutathione disulphide, while L-cystine, the predominant extracellular form, is a better γ-glutamyl acceptor substrate than L-cysteine. Thus, in certain respects the transpeptidase seems to be adapted to its proposed physiological role; it is highly active towards glutamine, the most abundant amino acid of blood plasma, and also towards L-cystine, present in the blood plasma in very low concentrations. It must be noted, however, that the function of γ-glutamyl-transpeptidase within the membrane may not be reflected exactly by the activities observed in studies in which the enzyme is dislodged from the membrane, solubilized and studied in aqueous solution. Under such circumstances we may be seeing reactions that only partially resemble the physiological function of the membrane-bound enzyme. These considerations emphasize again the importance of further work on the functional and structural characterization of this enzyme and on the γ-glutamyl cycle.

References

ADAMSON, E. D., SZEWCZUK, A. & CONNELL, G. E. (1971) Purification and properties of γ-glutamyl cyclotransferase from pig liver. *Can. J. Biochem. 49*, 218-226

CONNELL, G. E. & HANES, C. S. (1956) Enzymic formation of pyrrolidone carboxylic acid from γ-glutamyl peptides. *Nature (Lond.) 177*, 377-378

FOLK, J. E. & CHUNG, S. I. (1973) Molecular and catalytic properties of transglutaminases. *Adv. Enzymol. Relat. Areas Mol. Biol. 38*, 109-192

GRIFFITH, O. W. & MEISTER, A. (1976) Dependence of energy coupling on nucleotide base structure in the reaction catalyzed by 5-oxo-L-prolinase. *Biochem. Biophys. Res. Commun. 70*, 759-765

HANES, C. S., HIRD, F. J. R. & ISHERWOOD, F. A. (1952) Enzymic transpeptidation reactions involving γ-glutamyl peptides and α-aminoacyl peptides. *Biochem. J. 51*, 25-35

HEINZ, E. (1972) Transport of amino acids by animal cells, in *Metabolic Pathways* (Hokin, L. F., ed.), vol. 6, pp. 455-501, Academic Press, New York

MEISTER, A. (1973) On the enzymology of amino acid transport. *Science (Wash. D.C.) 180*, 33-39

MEISTER, A. (1974*a*) Glutathione synthesis, in *The Enzymes*, 3rd edn. (Boyer, P. D., ed.), vol. 10, pp. 671-697, Academic Press, New York

MEISTER, A. (1974*b*) The γ-glutamyl cycle; diseases associated with specific enzymatic deficiencies. *Ann. Intern. Med. 81*, 247-253

MEISTER, A. (1975) Biochemistry of glutathione, in *Metabolism of Sulfur Compounds, Metabolic Pathways*, 3rd edn., (Greenberg, D. M., ed.), vol. 7, pp. 101-188, Academic Press, New York

MEISTER, A. & TATE, S. S. (1976) Glutathione and related γ-glutamyl compounds; biosynthesis and utilization. *Annu. Rev. Biochem. 45*, 559-604

MEISTER, A., TATE, S. S. & ROSS, L. L. (1976) Membrane bound γ-glutamyl transpeptidase, in *The Enzymes of Biological Membranes* (Martinosi, A., ed.), vol. 3, pp. 315-347, Plenum, New York

MOOZ, E. D. & WIGGLESWORTH, L. (1976) Evidence for the γ-glutamyl cycle in yeast. *Biochem. Biophys. Res. Commun. 68*, 1066-1072

Novogrodsky, A., Tate, S. S. & Meister, A. (1976) γ-Glutamyl transpeptidase, a lymphoid cell-surface marker; relationship to blastogenesis, differentiation, and neoplasia. *Proc. Natl. Acad. Sci. U.S.A. 73*, 2414-2418

Orlowski, M. & Meister, A. (1963) γ-Glutamyl-*p*-nitroanilide: a new convenient substrate for determination and study of L- and D-γ-glutamyltranspeptidase activities. *Biochim. Biophys. Acta 73*, 679-681

Orlowski, M. & Meister, A. (1970) The γ-glutamyl cycle: a possible transport system for amino acids. *Proc. Natl. Acad. Sci. U.S.A. 67*, 1248-1255

Orlowski, M. & Meister, A. (1973) γ-Glutamyl cyclotransferase; distribution, isozymic forms, and specificity. *J. Biol. Chem. 248*, 2836-2844

Orlowski, M., Richman, P. G. & Meister, A. (1969) Isolation and properties of γ-L-glutamylcyclotransferase from human brain. *Biochemistry 8* 1048-1055

Palekar, A. G., Tate, S. S. & Meister, A. (1975) Decrease in glutathione levels of kidney and liver after injection of methionine sulfoximine into rats. *Biochem. Biophys. Res. Commun. 62*, 651-657

Pardee, A. B. (1968) Membrane transport proteins. *Science (Wash. D C.) 162*, 632-637

Ramakrishna, M., Krishnaswamy, P. R. & Rao, D. R. (1970) Metabolism of pyrrolidone-carboxylic acid in the rat. *Biochem. J. 118*, 895-897

Richman, P. G. (1969) *γ-Glutamylcysteine Synthetase and Related Studies on the Metabolism of the γ-Glutamyl Group.* Doctoral Dissertation, Cornell University Graduate School of Medical Sciences, New York, N.Y.

Richman, P. & Meister, A. (1975) Regulation of γ-glutamyl-cysteine synthetase by non-allosteric feedback inhibition by glutathione. *J. Biol. Chem. 250*, 1422-1426

Ronzio, R. & Meister, A. (1968) Phosphorylation of methionine sulfoximine by glutamine synthetase. *Proc. Natl. Acad. Sci. U.S.A. 59*, 164-170

Sekura, R. & Meister, A. (1974) Glutathione turnover in the kidney; considerations relating to the γ-glutamyl cycle and the transport of amino acids. *Proc. Natl. Acad. Sci. U.S.A. 71*, 2969-2972

Sekura, R., Van der Werf, P. & Meister, A. (1976) Mechanism and significance of the mammalian pathway for elimination of D-glutamate; inhibition of glutathione synthesis by D-glutamate. *Biochem. Biophys. Res. Commun. 71*, 11-18

Snoke, J. E. & Bloch, K. (1954) in *Glutathione* (Colowick, S. *et al.*, eds.), pp. 129-141, Academic Press, New York

Tate, S. S. (1975) γ-Glutamyl transpeptidase: properties in relation to its proposed physiological role, in *Isozymes: Physiology and Function* (Markert, C. L., ed.), vol. 2, pp. 743-765, Academic Press, New York

Tate, S. S. & Meister, A. (1974) Interaction of γ-glutamyl transpeptidase with amino acids, dipeptides, and derivatives and analogs of glutathione. *J. Biol. Chem. 249*, 7593-7602

Tate, S. S. & Meister, A. (1975) Identity of maleate-stimulated 'glutaminase' with γ-glutamyl transpeptidase in rat kidney. *J. Biol. Chem. 250*, 4619-4624

Tate, S. S. & Meister, A. (1976) Subunit structure and isozymic forms of γ-glutamyl transpeptidase. *Proc. Natl. Acad. Sci. U.S.A. 73*, 2599-2603

Thompson, G. A. & Meister, A. (1975) Utilization of L-cystine by the γ-glutamyl transpeptidase-γ-glutamyl cyclotransferase pathway. *Proc. Natl. Acad. Sci. U.S.A. 72*, 1985-1988

Thompson, G. A. & Meister, A. (1976) Hydrolysis and transfer reactions catalyzed by γ-glutamyl transpeptidase; evidence for separate substrate sites and for high affinity of L-cystine. *Biochem. Biophys. Res. Commun. 71*, 32-36

Thompson, G. & Meister, A. (1977) Specificity and kinetics of γ-glutamyl transpeptidase, in preparation

Van der Werf, P. (1970) *Metabolism of 5-Oxo-L-Proline; Conversion to L-Glutamate in an ATP-Dependent Reaction Catalyzed by 5-Oxoprolinase.* Doctoral Dissertation, Cornell University Gradutate School of Medical Sciences, New York, N.Y.

VAN DER WERF, P. & MEISTER, A. (1974) Isolation of 5-oxoprolinase from a prokaryote. *Biochem. Biophys. Res. Commun. 56*, 90-96

VAN DER WERF, P. & MEISTER, A. (1975) The metabolic formation and utilization of 5-oxo-L-proline (L-pyroglutamate, L-pyrrolidone carboxylate). *Adv. Enzymol. Relat. Areas Mol. Biol. 43*, 519-556

VAN DER WERF, P., ORLOWSKI, M. & MEISTER, A. (1971) Enzymatic conversion of 5-oxo-L-proline (L-pyrrolidone carboxylate) to L-glutamate coupled with ATP cleavage to ADP: a reaction in the γ-glutamyl cycle. *Proc. Natl. Acad. Sci. U.S.A. 68*, 2982-2985

VAN DER WERF, P., STEPHANI, R. A., ORLOWSKI, M. & MEISTER, A. (1973) Inhibition of 5-oxoprolinase by 2-imidazolidone-4-carboxylic acid. *Proc. Natl. Acad. Sci. U.S.A. 70*, 759-761

VAN DER WERF, P., STEPHANI, R. A. & MEISTER, A. (1974) Accumulation of 5-oxoproline in mouse tissues after inhibition of 5-oxoprolinase and administration of amino acids; evidence for function of the γ-glutamyl cycle. *Proc. Natl. Acad. Sci. U.S.A. 71*, 1026-1029

VAN DER WERF, P., GRIFFITH, O. & MEISTER, A. (1975) 5-Oxo-L-prolinase (L-pyroglutamate hydrolase); purification and catalytic properties. *J. Biol. Chem. 250*, 6686-6692

WELLNER, V. P., SEKURA, R., MEISTER, A. & LARSSON, A. (1974) Glutathione synthetase deficiency, an inborn error of metabolism involving the γ-glutamyl cycle in patients with 5-oxoprolinuria (pyroglutamic aciduria). *Proc. Natl. Acad. Sci. U.S.A. 71*, 2505-1509

WOODWARD, G. E. & REINHART, F. E. (1942) The effect of pH on the formation of pyrrolidone-carboxylic acid and glutamic acid during enzymatic hydrolysis of glutathione by rat kidney extract. *J. Biol. Chem. 145*, 471-480

YOUNG, J. A. & FREEDMAN, B. S. (1971) Amino acid transport. *Clin. Chem. 17*, 245-266

Discussion

Matthews: Both Gly-Gly and Gly-Sar are transported by the intestine and behave as if they share the same transport system. How does one interpret this? Does it suggest that the γ-glutamyl cycle is not after all important in peptide transport?

Meister: The possible function of the γ-glutamyl cycle in dipeptide transport needs further study. The interaction of the transpeptidase with dipeptides seems to reflect binding to the cysteinylglycine site on the enzyme. Nevertheless, it is possible that certain dipeptides are transported by the cycle. The experiment you mention might be difficult to interpret because multiple transport systems are probably involved. Glycylglycine might be transported by the cycle. Palekar *et al.* (1975) found markedly reduced glutathione concentrations in the kidney after administration of glycylglycine to rats. This seems to reflect transpeptidation between glutathione and glycylglycine.

A further point is that γ-glutamyl-glycylglycine (and probably other γ-glutamyl dipeptides) are not substrates of γ-glutamyl cyclotransferase. The further metabolism of such γ-glutamyl dipeptides might involve an additional transpeptidation reaction with an amino acid. This suggests that transpeptidation of dipeptides would add at least one more step.

Rubino: The carrier system for free neutral amino acids has been extensively investigated in terms of affinities for various substrates. Do the available data on the affinities of various amino acids for the carrier system fit with the affinities of various amino acids for γ-glutamyl transpeptidase?

Meister: The transpeptidase has a specificity that might fit reasonably well with its being a carrier of neutral amino acids. Although there have been, as you say, extensive studies on transport of neutral amino acids, there has been to my knowledge no chemical characterization of the carrier involved.

Adibi: Although carrier systems for amino acids have not yet been isolated, there is substantial biochemical and physiological evidence for their existence. Garvey *et al.* (1976) found that the relative specificity of the γ-glutamyl transpeptidase reaction with eight essential amino acids as substrates correlated well with the rate of intestinal absorption of these amino acids as described by us several years ago (Adibi & Gray 1967). In view of your data, would you accept that the γ-glutamyl transpeptidase might be involved with transport of amino acids and not physiologically involved with dipeptides?

Meister: Yes, this is possible.

Adibi: The correlation of specificity of γ-glutamyl transpeptidase is interesting. Have you had similar results to those of Garvey *et al.*?

Meister: Yes. The transport of amino acids across the rabbit ileum (Hajjar & Curran 1970) correlates fairly well with the specificity of the transpeptidase.

Rubino: Is γ-glutamyl transpeptidase dependent on sodium ion?

Meister: No. None of the enzymes of the γ-glutamyl cycle is specifically affected by Na^+.

Semenza: It is quite possible that various systems for transporting amino acids do exist and that one or more of them are Na-dependent. Such biologically important materials as amino acids may well require the presence of alternative transport mechanisms to secure survival.

Rubino: What happens to γ-glutamyl transpeptidase in Hartnup's disease, where the transport system for free neutral amino acids is impaired and the absorption of these amino acids is therefore defective?

Meister: Through the kindness of Dr Vivian Shih and Dr Dick Hoefnagel, we obtained skin biopsies from patients with Hartnup disease, and Dr Betty S. Danes cultured these cells. In our laboratory we found no evidence for deficient amino acid transport or deficient transpeptidase activity in these cultured skin fibroblasts.

Milne: With the inhibitor 2-imidazolidone-4-carboxylate or in the natural disease of 5-oxoprolinuria was there any indication of reduced amino acid transport? The obvious thing would be an amino aciduria. Can you produce this with your inhibitors or does it occur in 5-oxoprolinuria?

Meister: There is no amino aciduria in patients with 5-oxoprolinuria. One would not expect that the cycle would stop in this disease, because γ-glutamylcysteine is formed and can serve as a γ-glutamyl donor. Administration of the competitive inhibitor of 5-oxoprolinase, 2-imidazolidone-4-carboxylate (ICA) would not be expected to stop the cycle either. This inhibitor would prevent conversion of 5-oxoproline to glutamate. However, glutamate can be formed by a number of biochemical pathways. For example, in the kidney it can be formed by hydrolysis of glutamine, by transamination of α-oxoglutarate, or by reductive amination of α-oxoglutarate. Our observations indicate that administration of ICA does not produce amino aciduria, but this result is not unexpected. There is also a human inborn error of metabolism involving a marked deficiency of γ-glutamylcysteine synthetase. These patients are deficient in making γ-glutamylcysteine and therefore also glutathione. The two such patients now known have been found to have amino aciduria (Richards *et al.* 1974).

Silk: You said that γ-glutamyl transpeptidase is membrane-bound but that glutathione only exists within the cytoplasmic compartment of the cell. If this cycle plays a part in the translocation of the amino acid or peptide, how does the cycle link up from the mucosal side of the membrane to within the cell?

Meister: γ-Glutamyl transpeptidase is associated with the membranes of certain epithelial and lymphoid cells. One can demonstrate that it is located on the membrane in such a manner as to be accessible to externally supplied substrates. For example, in lymphoid cells, some of which are very well equipped with γ-glutamyl transpeptidase, one can show enzyme activity with cell suspensions by using a γ-glutamyl substrate that does not enter the cell (Novogrodsky *et al.* 1976). Similar considerations apply to the proximal renal tubular cells (Meister & Tate 1976). It is clear that the glutathione occurs intracellularly. Glutathione has been histochemically localized in several tissues (Ashgar *et al.* 1975). In these studies, the glutathione staining was in the proximal tubules in the cortex of the kidney. The problem is a crucial one. How does the enzyme on the membrane, or in the membrane, interact with glutathione, which is clearly *inside* the cell? One experimental result that seems relevant to this question is the finding that administration of glycylglycine to rats leads to prompt and substantial decrease in the glutathione content of the kidney (Palekar *et al.* 1975). There must be a mechanism by which glutathione can interact with the transpeptidase. Glutathione may move through the membrane and possibly interact with transpeptidase within the membrane. In the absence of the transpeptidase, glutathione might simply leak out of the cell. It is clear that we need to know much more about this.

Kenny: Is the phosphate-independent glutaminase the same enzyme as γ-glutamyl transpeptidase?

Meister: Phosphate-independent maleate-stimulated glutaminase has been shown to be identical with γ-glutamyl transpeptidase in rat kidney (Tate & Meister 1974*a*, 1975).

Kenny: Do you think that the maleate effect (in stimulating hydrolysis rather than transpeptidation) has a physiological role or is it just a biochemical quirk?

Meister: γ-Glutamyl transpeptidase interacts to a very slight extent with glutamine to form glutamate, ammonia, and some γ-glutamyl glutamine. The activity of the enzyme towards glutamine is about 1–2% of that with glutathione. In the presence of maleate, glutamine interacts more effectively with the enzyme and there is a considerable increase in its hydrolysis and also some increase in γ-glutamyl glutamine formation. At present, it seems unlikely that the interaction of γ-glutamyl transpeptidase with glutamine plays a significant role in the formation of ammonia by the kidney. It is conceivable that a compound might be formed in acidosis that would have an effect similar to that of maleate; this then might stimulate the production of ammonia from glutamine. As far as I am aware such a natural activator has not yet been found. On the basis of available published information, it would seem that most of the ammonia production by the kidney can be explained in terms of the action of phosphate-dependent glutaminase with perhaps some contribution by glutamate dehydrogenase. The mechanism of the effect of maleate on γ-glutamyl transpeptidase is still not clear. In addition to its apparent action in increasing the affinity of the enzyme for glutamine, it seems to uncouple transpeptidation in such a manner as to favour hydrolysis over transpeptidation. Maleate does convert the enzyme into a glutathionase. When maleate is given to animals, they exhibit a syndrome which resembles the Fanconi syndrome in man, and this is associated with amino aciduria. They also excrete large amounts of phosphate, glucose and perhaps other compounds. It would appear that this effect of maleate is rather non-specific and it seems difficult to ascribe all of these effects to an interaction between maleate and γ-glutamyl transpeptidase.

Parsons: The folates contain γ-glutamyl peptides. Have you considered whether γ-glutamyl transpeptidase has a role in folate metabolism?

Meister: Yes. The enzyme does not act on glutamyl derivatives in which the α-amino group is substituted.

Norén: γ-Glutamyl hydrazones of α-keto acids are inhibitors of the γ-glutamyl transpeptidase. Have you done any studies of amino acid transport in which such γ-glutamyl hydrazones were added to the system?

Meister: The γ-glutamyl hydrazones of α-keto acids were first prepared in our laboratory in the course of studies on glutamine transaminase (Cooper & Meister 1973). They were then examined as possible inhibitors of other enzymes and it was found that this class of compounds inhibits the transpeptidase (Tate & Meister 1974*b*). These compounds also inhibit γ-glutamylcysteine synthetase (Richman 1969). Despite the evident lack of specificity shown by this inhibitor, it was tested in an intact, perfused rat kidney system developed by Dr Thomas Maack (Maack *et al.* 1974). When the γ-glutamyl hydrazone of pyruvate was introduced into this system, there was a drastic reduction in the glomerular filtration rate which was not accompanied by a significant change in the fractional amino acid reabsorption. In view of the marked decrease in glomerular filtration rate, we have considered the results as being inconclusive.

Norén: Have you studied the intestine?

Meister: Yes. Villus tip cells and crypt cells of rat jejunal mucosa were separated and the activities of the enzymes of the γ-glutamyl cycle and the glutathione content were studied. The villus tip cells had much higher γ-glutamyl transpeptidase activities than the crypt cells (Cornell & Meister 1976). Our findings have been confirmed by Dr Thomas Q. Garvey in studies carried out in Dr Kurt Isselbacher's laboratory.

References

Adibi, S. A. & Gray, S. J. (1967) Intestinal absorption of essential amino acids in man. *Gastroenterology 52*, 837-845

Ashgar, K., Reddy, B. G. & Krishna, G. (1975) Histochemical localization of glutathione in tissues. *J. Histochem. Cytochem. 23*, 774-779

Cooper, A. J. L. & Meister, A. (1973) Activity of liver glutamine transaminase toward L-γ-glutamyl hydrazones of α-keto acids. *J. Biol. Chem. 248*, 8489-8498

Cornell, J. S. & Meister, A. (1976) Glutathione and γ-glutamyl cycle enzymes in crypt and villus tip cells of rare jejunal mucosa. *Proc. Natl. Acad. Sci. U.S.A. 73*, 420-422

Garvey, T. Q., Hyman, P. E. & Isselbacher, K. J. (1976) γ-Glutamyl transpeptidase of rat intestine: localization and possible role in amino acid transport. *Gastroenterology 71*, 778-785

Hajjar, J. J. & Curran, P. F. (1970) Characteristics of the amino acid transport system in the mucosal border of the rabbit ileum. *J. Gen. Physiol. 56*, 673-691

Maack, T., Johnson, V., Tate, S. S. & Meister, A. (1974) Effects of amino acids on the function of the isolated perfused rat kidney. *Fed. Proc. 33*, 305

Meister, A. & Tate, S. S. (1976) Glutathione and related γ-glutamyl compounds: biosynthesis and utilization. *Annu. Rev. Biochem. 45*, 559-604

Novogrodsky, A., Tate, S. S. & Meister, A. (1976) γ-Glutamyl transpeptidase, a lymphoid cell-surface marker: relationship to blastogenesis, differentiation & neoprasia. *Proc. Natl. Acad. Sci. U.S.A. 73*, 2414-2418

Palekar, A. G., Tate, S. S. & Meister, A. (1975) Decrease in glutathione levels of kidney and liver after injection of methionine sulfoximine into rats. *Biochem. Biophys. Res. Commun. 62*, 651-657

Richards, F., Cooper, M. R., Pearce, L. A., Cowan, R. J. & Spurr, C. L. (1974) Familial spinocerebellar degeneration, hemolytic anemia, and glutathione deficiency. *Arch. Intern. Med. 134*, 534-537

Richman, P. G. (1969) *γ-Glutamylcysteine Synthetase & Related Studies on the Metabolism of the γ-Glutamyl Group*, Ph.D. Thesis, Graduate School of Medical Sciences, Cornell University Medical College, New York, N.Y.

TATE, S. S. & MEISTER, A. (1974*a*) Stimulation of the hydrolytic activity & decrease of the transpeptidase activity of γ-glutamyl transpeptidase by maleate: identity of a rat kidney maleate-stimulated glutaminase and γ-glutamyl transpeptidase. *Proc. Natl. Acad. Sci. U.S.A. 71*, 3329-3333

TATE, S. S. & MEISTER, A. (1974*b*) Interaction of γ-glutamyl transpeptidase with amino acids, dipeptides & derivatives & analogs of glutathione. *J. Biol. Chem. 249*, 7593-7602

TATE, S. S. & MEISTER, A. (1975) Identity of maleate-stimulated glutaminase with γ-glutamyl transpeptidase in rat kidney. *J. Biol. Chem. 250*, 4619-4627

General discussion I

Parsons: I think that we should start by having a brief discussion of nomenclature.

Smyth: I should like to raise again the question of whether the difference between surface hydrolysis and intracellular hydrolysis is a semantic one rather than a real one. I believe Ugolev and myself hold very similar views but express them in different ways. We define intracellular hydrolysis as a process which liberates amino acids in such a way that they do not behave as if they were set free in the intestinal lumen: Ugolev's concept is really the same. The amino acids set free by surface hydrolysis are trapped in some way so that the process could come into our definition of intracellular.

Matthews: With all due respect, I think that to use the term 'intracellular' for something that is not literally inside the cell can lead to confusion.

Smyth: I accept this entirely. When we introduced the term intracellular in relation to peptide hydrolysis, the problem was a new one, and we had to use a term which could be defined in a way which could be tested experimentally. We did not attempt to define it in cytological terms. In another paper we used the term cellular hydrolysis but I don't think that gets us much further. It is difficult to get away entirely from the semantic problem.

Parsons: If a molecule has traversed the plasma membrane and gone right through it, one could say that it is intracellular, which seems a satisfactory definition. When a molecule is in the act of traversing the membrane and is actually within the membrane, you would say it is intracellular, would you not, Professor Smyth? I suppose I would say it had a cellular location and dodge the issue of whether it was intracellular.

Smyth: Our definition of intracellular or extracellular is very simple. If the peptidase is extracellular the amino acids liberated behave as if they had been added to the lumen. If it is intracellular they do not. If someone could refine that definition it would be useful.

Semenza: Professor Matthews (1975) has suggested nomenclature which is really unequivocal and in my view we should try to adopt it (see this volume, p. 353 *et seq.*).

Kenny: For an enzymologist there is usually no difficulty in distinguishing between membrane-bound and cytosolic enzymes. The former need special methods, involving, for example, proteases or detergents, to release them in a soluble form.

Adibi: It is important to know an enzyme's location, and to know, for example, whether the active site is facing outward or facing the lumen.

Matthews: I would suggest that Newey & Smyth's definition of intracellular hydrolysis (Newey & Smyth 1960), which was useful at the time when it was produced, is now inconveniently broad. We should try and refine things a bit, as Professor Smyth said.

Parsons: We can now proceed with our general discussion of peptidases and peptide transport.

Rhodes: It is my understanding, from the work of several people, that the intestinal microvillus membrane hydrolases appear, in the electron microscope, to protrude from the membrane facing the lumen. Some have called these protrusions 'knobs' or 'balls'.

Kenny: The dogma is that the glycosylated domains of membrane glycoproteins face the lumen and not the cytoplasm. In the case of the kidney brush border, all of the known peptidases are glycoproteins.

Rhodes: The intestinal microvillus membrane enzymes also contain glycoproteins.

Kenny: I agree that this determines their orientation in the membrane: the carbohydrate chains face outwards.

Smyth: The crux of the matter is what happens to amino acids set free by hydrolysis. Ugolev has the idea that they are captured in some way, and that is our concept of intracellular: if they are captured they do not behave as if they were free in the lumen.

Adibi: What is the evidence that there is no enzymic activity in the matrix of membrane?

Rhodes: Morphologically, with papain or trypsin, one can split off these 'knobs'. The enzymic activity goes with the knobs.

Adibi: Does it disappear totally from the membrane fraction?

Rhodes: Not totally, but 80–90 % of it goes.

Maroux: Papain treatment easily solubilizes 100 % of aminopeptidase, sucrase and maltase activities. Solubilization of alkaline phosphatase and trehalase is more difficult.

Semenza: The case of trehalase is a bit complicated because it is almost completely inactivated by papain treatment. It is solubilized by trypsin (Dahlqvist 1960).

Maroux: We have not observed inactivation of trehalase in the core of the pig brush border.

Meister: It seems interesting to consider whether an enzyme which is completely embedded within a membrane that is supposedly free of water can exhibit hydrolase activity. Perhaps when the enzyme is removed from the membrane and forced to be in water it will exhibit hydrolase activity, but it may not do this *in vivo*. This raises the question of whether some of the peptidase activities discussed at this meeting, including the γ-glutamyl transpeptidase that I discussed, in fact exhibit hydrolase activity *in vivo*. Are some of the activities observed actually artifacts or perhaps partial reactions which may serve as clues to the real reactions which have not yet been studied? It might be of interest to see whether some of these enzymes are active in non-aqueous media *in vitro*.

Kim: Regarding the release of brush-border enzymes by papain or trypsin, it is quite well known that glycoproteins are in general somewhat resistant to proteolysis. This is thought to be due in part to the steric hindrance caused by the carbohydrate side chains of the glycoprotein molecules. Therefore when one considers the differential release of various brush border enzymes by proteases, one must also consider the possibility of qualitative or quantitative differences in carbohydrate side chains.

Maroux: Papain does not release all the hydrolases from the membrane at the same rate. The same difference in the rate of hydrolysis of the peptide bond linking the hydrophobic and hydrophilic parts of these hydrolases is observed after their extraction by Triton X-100. So the first observation on the action of papain on the membrane cannot be interpreted in terms of peptide bond accessibility, as is often the case.

Hemmings: Does the distinction between internal and external attachment to the cell include an appreciation of the role of the glycocalyx? This is responsible for much protein attachment to cells and has a sort of patrol activity. There is a gross ingestion of glycocalyx in many cell types, which is a primitive character.

Parsons: So the glycocalyx is a sort of scaffolding?

Hemmings: Yes. The glycocalyx is ingested by the microvilli and by my favourite yolk sac cell.

Parsons: Does this happen in the pinocytosing epithelial cells of the newborn intestine?

Hemmings: Yes.

De Laey: We must consider three different areas within the enterocyte. The first is the glycocalyx, which has different kinetics from that of the extracellular space. A lot of exchange-adsorption and hydrolysis takes place in that area. The second site is the membrane and the asymmetry between its two sides. All the hydrolases and disaccharides are on the luminal side, and we presume that a number of dipeptidases are located there too. The third site is inside the cytoplasm, which is a terminal barrier to the progression of all large molecules going through the enterocyte before they get into the blood-stream. Whether we call the last two sites or only the last site intracellular is but a matter of words.

Josefsson: It is too limited to talk only about intracellular and membrane peptidases. Intracellularly there are a lot of compartments—the reticular endothelial systems, lysosomes, mitochondria and so on. We know there are large numbers of unidentified peptidases which seem to be of great importance, or at least there are a lot of regulatory factors for the biosynthesis of proteins. Professor P. N. Campbell's group (personal communication) have isolated precursors of enzymes during their biosynthesis and there must be peptidases which split off the small peptide fragments before the native protein is released, for example albumin. In the next few years we can therefore expect that a lot of new peptidases will be characterized and identified. We know they cannot be in the cytosol, and we shall therefore be looking at the cells for other possible sites for peptidases.

Parsons: Peptidases are located in organelles, and the plasma membrane that forms the brush borders can be considered to be one of the organelles.

De Laey: From the point of view of digestion, what is the most important organelle?

Josefsson: I don't think we should regard the peptidases as being merely digestive enzymes. Even the dipeptidases and the membrane-bound peptidases may have other biological functions of which we are not yet aware. They have a lot of suggested functions but we don't really know which of them are physiological functions. We need to learn more about the general metabolism. Also, we are concentrating here on the enterocytes, but we already know that all the enzymes that function in the intestinal transport and hydrolysis of peptides exist in other cells. They are located in those cells in the same way as they are in enterocytes. So what are their functions in those other cells which are not taking care of dietary proteins?

Silk: Although we have been focusing a lot of attention on membrane-bound externally-oriented peptidases, these enzymes cannot play an exclusive role during the transport of peptides. If, like disaccharides, peptides are hydrolysed to constituent free amino acids before absorption takes place, then it would be difficult to explain results of *in vitro* and *in vivo* experiments which show a faster absorption of peptide-bound amino acid residues than of free amino acid residues. Also it would be impossible to explain the observations that in Hartnup disease and cystinuria a transport system exists for dipeptide-bound amino acids which is independent of the system used for transport of free amino acids.

Nicholson: I want to discuss the interpretation of studies where amino acids have been shown to affect dipeptide transport. Professor Josefsson raised the possibility that inhibition of peptide transport by amino acids was due to inhibition of the intracellular cytosolic peptidases. If Professor Rubino had found any intact intracellular glycyl-L-proline during his absorption experiments with this peptide in the presence of L-proline, this could have been taken as evidence for the inhibition of the intracellular peptidases. In fact he found no such evidence. In other words, the proline seemed to be inhibiting the transport of the dipeptide without affecting intracellular hydrolysis. It seems, therefore, that we can interpret data from these experiments, where an inhibition of the surface transport mechanism is assumed, without involving inhibition of the intracellular enzymes.

Josefsson: In *in vitro* systems such as Professor Rubino is working with, we know that cytosol-located dipeptidases are released rather rapidly. Within one hour all the dipeptidases have leaked out. Therefore it is possible that some of the proline dipeptidase has leaked out and has been sufficiently inhibited by the high concentration of proline to be undetectable in the assay system.

Nicholson: Professor Rubino said the assay was sensitive.

Josefsson: The amount of proline in the system was high, about 40 mM, so it must inhibit all the proline dipeptidase leaking out. This is a possible explanation.

Rubino: There was no free glycine or proline in the medium at the end of incubation with [^{14}C-glycine]glycylproline. If this were the case, the addition of a large excess of glycine to the medium should have diluted the labelled glycine, thus resulting in inhibition of influx.

Josefsson: But the glycine is not inhibiting the proline dipeptidase.

Rubino: A large excess of glycine does not inhibit the influx of the Gly residue of glycylproline in either adult or newborn rabbits.

Silk: It is important to remember that in the experiments we are discussing, a dynamic situation exists. Thus, it is still possible for there to be a Gly-Pro dipeptidase in the mucosal medium, and for the released free amino acids to be immediately recaptured.

Rubino: It seems to me that if there were release of labelled glycine during incubation with [^{14}C-glycine]glycylproline, the presence of cold glycine in the medium should have inhibited the rate of influx of [^{14}C-glycine]glycylproline, no matter how rapidly the released amino acids were captured by the brush border.

Josefsson: If there was no inhibition of glycylproline uptake with an additional dipeptide then I misunderstood you. I thought that you had such a significant inhibition of uptake of the dipeptide that the glycylproline transport system must be inhibited by the other dipeptide. Then you did the same experiments with high concentrations of proline and high concentrations of glycine in the medium. You had no effect with glycine but a small effect with proline. I thought it was possible to explain the overall results by saying that the high inhibition of the transport mechanism of uptake by the dipeptides was due to the transport system being blocked. But the amino acid effect could be explained by inhibition of the proline dipeptidase. We know that glycine is not inhibiting at all but that proline is a product inhibitor. Then I suggested that with a really high concentration of proline there could be either an influx of proline into the cells, where it would inhibit hydrolysis of glycylproline uptake, or a leakage of proline dipeptidase and then inhibition, so that the same effect would be seen. I don't think that amino acids have an effect on the dipeptide transport system and I am just suggesting this as an alternative possibility.

References

DAHLQVIST, A. (1960) Characterization of hog intestinal trehalase. *Acta Chem. Scand. 14*, 9-16

MATTHEWS, D. M. (1975) Intestinal absorption of peptides. *Physiol. Rev. 55*, 537-608

NEWEY, H. & SMYTH, D.H. (1960) Intracellular hydrolysis of dipeptides during intestinal absorption. *J. Physiol. (Lond.) 152*, 367-380

Intestinal mucosal hydrolysis of proteins and peptides

YOUNG S. KIM

Gastrointestinal Research Laboratory, Veterans Administration Hospital, San Francisco and Department of Medicine, University of California School of Medicine, San Francisco, California

Abstract The major products of intraluminal hydrolysis of dietary proteins appear to be small peptides and amino acids. Recent studies indicate that the distal part of the small intestine may play an important role in the digestion and absorption of dietary proteins. Intestinal mucosal cellular hydrolysis of peptides and proteins represents the terminal stage of digestion of dietary proteins and appears to be carried out predominantly by amino-oligopeptidases in brush border membranes and cytoplasm.

These enzymes in the two main subcellular loci are distinct since they exhibit different electrophoretic mobilities, physicochemical properties, substrate specificities and responses to starvation and dietary manipulation.

Two amino-oligopeptidases have been purified from the intestinal brush border of the rat. The enzymes are remarkably similar to each other in many respects. They have an apparent molecular weight of 280 000 and are composed of two subunits of equal molecular weight. Both enzymes are glycoproteins having similar chemical compositions, common antigenic properties, substrate specificities and kinetic properties.

Dietary protein is derived from animal and vegetable sources and makes up about 11–14% of the average caloric intake. In western diets this amounts to about 70–100 g protein per day. Proteins derived from endogenous sources, such as gastric, biliary, pancreatic and intestinal secretions consisting of secretory glycoproteins and digestive enzymes, also enter the intestinal lumen. In addition, small quantities of plasma proteins and desquamated mucosal epithelial cells contribute to the total of the protein which is digested. Since healthy adults excrete only 1–2 g of faecal nitrogen per day, the digestion and absorption of dietary and endogenous proteins appear to be very efficient. Many studies of these processes have been done in the past few decades.

INTRALUMINAL PROTEIN DIGESTION

Until recently it was generally assumed that proteins are rapidly degraded by

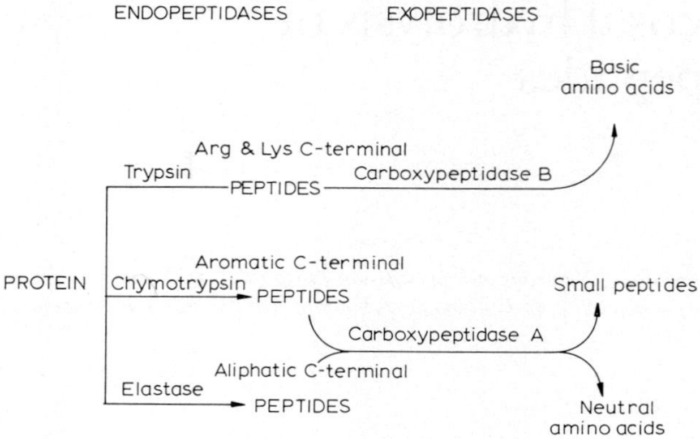

FIG. 1. Intraluminal digestion of dietary protein. (From Gray & Cooper 1971.)

the concerted action of gastric and pancreatic enzymes into their constituent amino acids which are then absorbed across the intestinal epithelium. Not surprisingly, therefore, attention was focused on the gastric and pancreatic proteolytic enzymes and the mechanisms of amino acid absorption.

As shown in Fig. 1 (Gray & Cooper 1971) pure pancreatic secretion from man and rat contains primarily endopeptidases and exopeptidases. Of the latter group of enzymes, carboxypeptidase activities are most active, although a trace of aminopeptidase activity can also be detected. Several investigators have recently indicated that the major products of the combined intraluminal hydrolysis of dietary protein by these enzymes are free amino acids and small peptides having two or six amino acid residues (Nixon & Mawer 1970*a*, *b*; Crampton *et al.* 1971; Adibi & Mercer 1973). There is little information, however, available on the precise characterization of these oligopeptides. For example, little is known about the relative amounts of peptides of different chain lengths or the type of amino acids occurring in the N-terminal or C-terminal residues of the peptides.

In addition, because of the variability in digestibility, amino acid composition and sequence of protein meals from different sources, the make-up of the peptides generated from each of these protein meals probably varies a great deal. Furthermore, the characteristics of the intraluminal hydrolysis probably will vary considerably, depending on the region of the small intestine where digestion is taking place.

In this regard it is of interest that, contrary to the earlier reports that protein digestion and absorption take place rapidly and almost entirely in the upper jejunum, a considerable amount of protein appears to be present in the distal

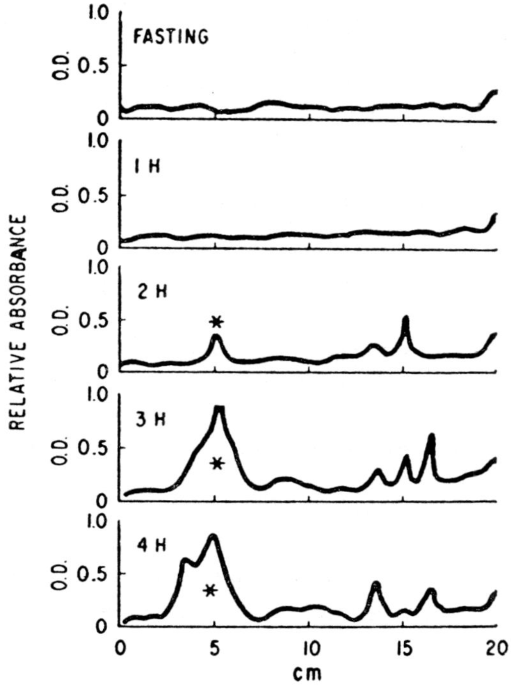

FIG. 2. Densitometric scans of protein-stained polyacrylamide gels of ileal aspirates obtained from normal subjects after a protein-rich meal. The protein peaks identified by an asterisk correspond to the peak of the fed protein (bovine serum albumin). Migration is towards the anode from right to left. (From Adibi & Mercer 1973.)

region of the small intestine even several hours after feeding of a test meal.

Fig. 2 shows results from a recent study by Adibi & Mercer (1973). The scanning profiles (protein stains, gel electrophoresis of ileal aspirates) were obtained from normal subjects at 1, 2, 3 and 4 h after oral ingestion of 50 g bovine serum albumin (BSA). The asterisk represents the electrophoretic mobility of intact BSA. Fed protein is clearly detectable in ileal fluids of man at 3 and 4 h after the ingestion of a meal. However, the quantitative determination of BSA appearing in the ileum has not yet been determined. We observed a similar result in a study in rats (Curtis *et al.* 1977).

Therefore, further information on the types and relative amounts of the products of intraluminal hydrolysis of different proteins in different parts of the small intestine is needed before we can fully understand the molecular mechanisms involved in the intestinal absorptive processes and in the final stage of digestion at the mucosal cell level.

Several families of peptidases and proteases are associated with mammalian cells. However, except for some peptidases, relatively little is known about the biological activities and the relative roles of these enzymes in the intestinal mucosal cells.

CELLULAR AND SUBCELLULAR DISTRIBUTIONS OF AMINO-OLIGO-PEPTIDASES OF SMALL INTESTINAL MUCOSA

The intestinal epithelium contains high levels of amino-oligopeptidase activity (Josefsson & Sjöström 1966; Peters 1970; Heizer *et al.* 1972) but the jejunal contents contain only traces of this activity. Since the amino nitrogen in the postprandial mesenteric and portal venous blood exists almost entirely in the form of free amino acids, it becomes apparent that the intestinal epithelium is the structure most likely to be associated with the terminal digestion of luminal peptides.

In the intestine, proliferative cells are localized in the crypt area. The cells formed there then migrate upwards into the villi. During this migration, these cells undergo differentiation which results in the development of mature absorptive epithelial cells. The activities of various enzymes also alter as the cells migrate from the crypts to the villus tips. For example, the activity of thymidine (EC 2.7.1.75) (which participates in DNA synthesis) was shown by a planar section technique to be confined primarily to crypt cells and to decline rapidly as the cells reach the crypt–villus cell junction (Imondi *et al.* 1969). The opposite is true for sucrase and lactase enzymes associated with the brush border membrane (Fortin-Magana *et al.* 1970). Using a similar technique, we examined the distribution of particulate and cytoplasmic peptidase activities (Kim & Kim 1977*a*). Their effect on phenylalanylglycine and leucyl-β-naphthylamide, which are predominantly hydrolysed by particulate enzymes, shows a sharp gradient from villus to crypt. A similar gradient is seen for other substrates such as prolylleucine, leucylalanine, glycylphenylalanine and glycylleucine, which are predominantly hydrolysed by the cytoplasmic enzymes. Thus, it appears that both brush border and cytoplasmic peptidases are concentrated in the mature epithelial cells of the villus.

The subcellular distribution of peptidases in the small intestinal mucosa is shown in Table 1 (Peters 1970; Kim *et al.* 1972, 1974). Most of the activity is located in two fractions, the brush border and the cytoplasm. The proportion of enzyme activity in these two fractions differs considerably, depending on the type and chain length of the substrates used. When dipeptides are used as substrates, less than 10% of total cellular activity is associated with the brush border fraction, while with tripeptides as much as 60% of the activity has been

TABLE 1

Subcellular distribution of peptidases in small intestinal mucosa. (Data from Peters 1970; Kim *et al.* 1972.)

	Guinea-pig					*Rat*		
Fraction	*Gly-Gly*	*Leu-Gly*	*Gly-Gly-Gly*	*Leu-Gly-Gly*	*Try-Gly-Gly*	*Leu-Ala*	*Leu-Leu-Leu*	*Ala-Gly-Gly-Gly*
	%					%		
Brush borders and nuclei	5	6	20	46	63	11	9	89
Mitochondrial fractions	2	1	7	5	3	3	1	4
Microsomes	1	1	9	5	4	5	2	7
Cytosol	92	91	64	44	30	81	88	0
Recovery (%)	98	90	87	76	87	98	107	97

found there. When two tetrapeptides were examined, nearly all the activity was in the brush border, with no detectable activity in the cytoplasm. To determine whether the same or different enzymes are active in the two fractions we examined three parameters of their properties: zymogram patterns, substrate specificities and thermal stabilities (Kim *et al.* 1972, 1974).

COMPARISON OF PROPERTIES OF AMINO-OLIGOPEPTIDASES OF SMALL INTESTINAL MUCOSA

Fig. 3 shows a typical zymogram demonstrating peptidase activity in various cell fractions when the dipeptide leucylleucine was used as substrate (Kim *et al.* 1972). Multiple zones of enzyme activity are seen in both the soluble and brush border fractions. The zymogram of either a crude or partially purified soluble fraction shows several bands. Without papain treatment of the brush border fraction virtually all the enzyme activity remained at the origin, but with papain treatment two major bands with electrophoretic mobilities quite distinct from those of the cytosol fraction were seen (Fig. 3, no. 5). There is little difference between this pattern and that of the untreated cytosol. Thus, the differences between the cytosol and brush border patterns were not due to papain treatment. We also used trypsin, chymotrypsin, Triton X-100 and sodium deoxycholate for solubilization and found the same zymogram pattern for the brush border fraction in each case.

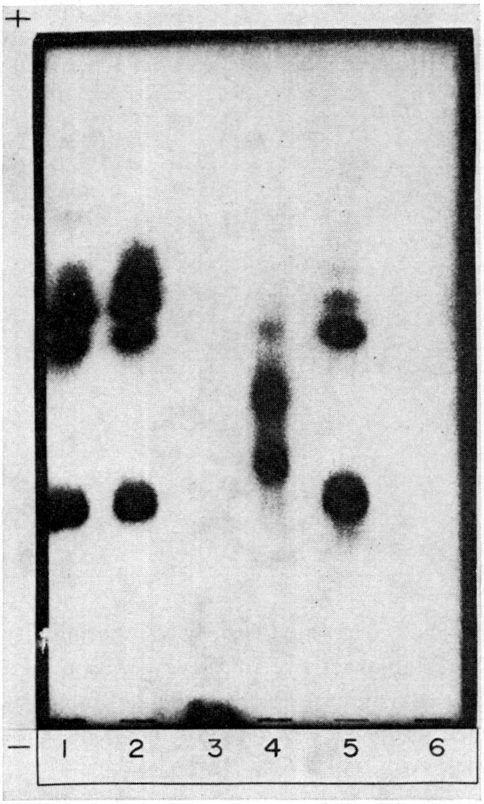

FIG. 3. Typical zymograms of peptide hydrolases from rat small intestinal mucosa with Leu-Leu as a substrate. The samples tested were: (1) soluble fraction; (2) soluble fraction partially purified on Bio-gel A 0.5 M column; (3) purified brush border membranes; (4) papain-treated purified brush border membranes; (5) papain-treated soluble fraction; and (6) papain alone. (From Kim *et al.* 1972.)

Zymograms obtained with different substrates from soluble and brush border fractions of the rat are shown in Fig. 4. A total of 20 dipeptides and six tripeptides were studied. Representative results are indicated in this figure. For the soluble fractions, the zymogram patterns obtained with most dipeptides resembled those obtained with leucylleucine or prolylleucine. All tripeptides showed a pattern similar to that seen with leucylleucylleucine. The main difference between the patterns obtained for dipeptides and tripeptides was the absence of one major and one minor band in the middle zone. In marked contrast to the zymograms obtained with soluble fractions, the brush border yielded uniform patterns with most substrates. However, very little if any

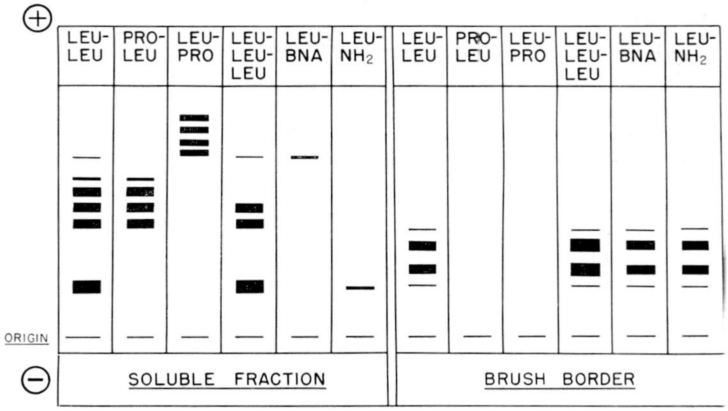

FIG. 4. A summary diagram of zymograms of peptidases from soluble and brush border membrane fraction of rat small intestinal mucosa with different substrates. (From Kim *et al.* 1972.)

brush border activity could be detected when dipeptides with proline at either the N-terminus or C-terminus were used.

Zymograms of cytosol fractions from human intestinal mucosa show multiple zones of enzyme activity (Kim & Kim 1977*b*). However, unlike the rat brush border, the human brush border shows only a single band. Although there is a band in the soluble fraction which has the same mobility as the band from the brush border, it does not represent the same enzyme since the cytosol enzyme did not utilize leucylleucylleucine, a substrate rapidly hydrolysed by the brush border enzyme. This indicates that there is an enzyme in the brush border with trileucine peptidase activity which is absent from the cytosol.

In order to examine the organ specificities of zymogram patterns, we prepared soluble and plasma membrane fractions from various organs of the rat. The patterns obtained from the soluble fractions of these organs were nearly identical (Kim *et al.* 1972). However, the patterns from papain-solubilized peptidase from plasma membranes of the small intestine, liver and kidney were different. Thus, the peptidase activities in plasma membrane vary depending on the organ from which they are derived.

We then compared the effects of partially purified cytoplasmic and brush border peptidases on peptide substrates of different chain lengths (Table 2) (Kim *et al.* 1974). Cytosol enzymes from rat intestine are capable of hydrolysing dipeptides and tripeptides. Longer peptides examined, with the exception of leucylglycylglycylglycine, were not hydrolysed even after prolonged incubation of up to 10 h. In contrast to the cytosol enzymes, brush border enzymes were

TABLE 2

Effects of rat intestinal peptidases on peptide substrates. (From Kim *et al.* 1974.)

Substrates	*Cytoplasmic enzymes*		*Brush border enzymes*	
	Alanine	*Glycine*	*Alanine*	*Glycine*
	nmol/10 min			
Ala-Gly	33.1	32.9	8.1	8.1
Ala-Gly-Gly	58.4	0.7	12.2	–
Ala-Gly-Gly-Gly	–	–	12.1	–
Ala-Gly-Gly-Gly-Gly	–	–	16.9	–
Gly-Gly		6.4		2.2
Gly-Gly-Gly		17.9		3.6
Gly-Gly-Gly-Gly		–		3.0
Leu-Trp-Met-Arg-Phe-Ala Phe-Val-Glu-Trp-Leu-Met-Asp-Thr	Not hydrolysed		Hydrolysed sequentially from N-terminus	

effective in hydrolysing peptides containing more than three amino acids, including a hexapeptide and an octapeptide.

Next we examined the heat stability of the cytosol and the brush border enzymes (Kim *et al.* 1974). Cytosol enzymes gave different results depending on the substrate used (Fig. 5). Enzyme activities in rat cytosol were relatively stable at 40 and 50 °C when alanylglycylglycine was the substrate, while inactivation occurred rapidly when alanylglycine was used. This suggests that the enzyme which hydrolyses alanylglycine is distinct from that which hydrolyses the tripeptides. In contrast, the brush border enzymes of both rat and human intestine showed considerable thermal stability even at 60 °C, irrespective of the chain length of the peptide substrate.

The properties of cytosol and brush border enzymes are summarized for comparison in Table 3. The percentage distribution of enzyme activity in the cytosol and brush border fractions varies, depending not only on the type of substrates but also on the chain length of the substrate. Zymograms show that cytosol enzymes have multiple bands, and that the band patterns are substrate-dependent and non-organ-specific. In contrast, the brush border enzymes show two major bands in rat and one major band in man. The band pattern does not depend on the substrate used and is organ-specific. Proline-containing dipeptides are hydrolysed by cytosol enzymes but are not readily hydrolysed by the brush border enzymes. Cytosol enzymes generally hydrolyse dipeptides and tripeptides only, whereas the brush border can utilize longer peptides as

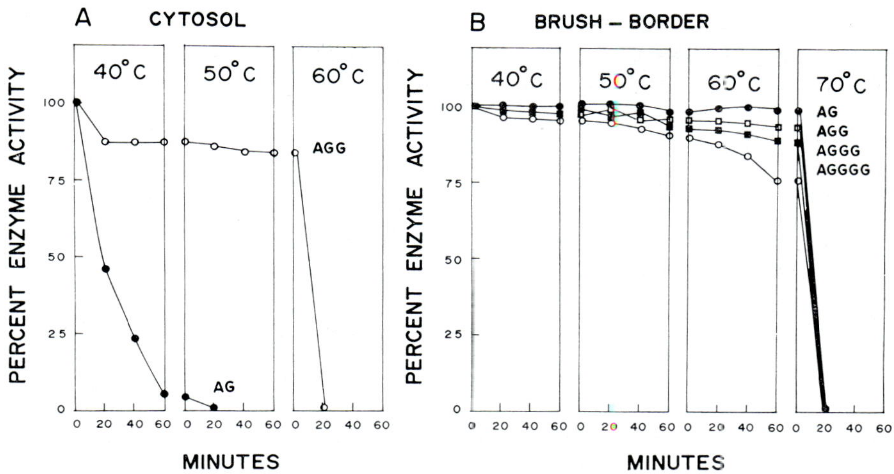

FIG. 5. Stability of peptidase activities of rat small intestinal mucosa for various peptide substrates at 40, 50, 60 and 70 °C, (A) cytosol enzymes and (B) brush-border enzymes. AG, Ala-Gly; AGG, Ala-Gly-Gly; AGGG, Ala-Gly-Gly-Gly; and AGGGG, Ala-Gly-Gly-Gly-Gly. (From Kim *et al.* 1974.)

well. Recently Dr Chung in our laboratory observed that a tetrapeptide, Leu-Gly-Gly-Gly, may be hydrolysed by both the cytosol and brush border fractions (Chung *et al.* 1977). Both enzymes appeared to be aminopeptidases. Cytosol enzymes are heat-labile while those of the brush border are relatively stable. The hydrolysis of some dipeptides by cytosol enzymes could be virtually abolished by *p*-hydroxymercuribenzoate, although this inhibitor had no effect on brush border enzymes. These data show that the peptidases from these two cellular fractions are distinct groups of enzymes. Since the brush border enzyme pattern is organ-specific while the cytosol is not, and there is a difference in the type of peptides hydrolysed, it is reasonable to assume that these two groups of enzymes subserve different cellular functions.

In order to gain some insight into the possible biological roles of peptidases of rat small intestinal mucosa, we did several experiments. These included dietary manipulation and intestinal resections to determine whether the responses of the enzymes in the two fractions differed.

In the first experiment five days of starvation markedly reduced specific enzyme activity as well as the enzyme activity per cell of the brush border, while the activities of the cytoplasmic fraction rose significantly, as did the activities of gluconeogenic enzymes and enzymes involved in transamination of amino acids (Kim *et al.* 1973). Non-organ specificity of intestinal cytosol

TABLE 3

Comparison of the properties of intestinal cytosol and brush border peptidases

	Cytosol	*Brush border*
Substrates (% of total cellular activity)		
Dipeptides	80-95	5-12
Tripeptides	30-90	10-60
Tetra- and higher peptides	Nil	90
Zymogram pattern	(a) Multiple bands (b) Heterogeneity of band pattern (c) Non-organ-specific pattern	(a) Two major bands (b) Homogeneity of band pattern (c) Organ-specific pattern
Proline-containing dipeptides	Hydrolysed	Not readily hydrolysed
Chain length of substrate	2, 3 (4)	2, 3, 4, 5, 6, 7, 8, -
End-terminal specificity	Amino peptidase	Amino peptidase
Terminal stability	Heat-labile	Relatively heat-stable
Effect of *p*-hydroxymercuri - benzoate	Inhibited	Not inhibited

peptidases had suggested some general cellular function in protein catabolism. These starvation data further suggested that during protein starvation a more rapid hydrolysis of intestinal intracellular protein might occur. These amino acids might then be transported to other more vital organs.

In the second study, the effect of one week's feeding of isocaloric 10% and 55% casein diets on the enzyme activities of the brush border and cytoplasmic fractions were compared (Nicholson *et al.* 1974). The effect of high protein feeding was much greater on the peptidase activity of the brush border membranes than on that of the cytosol. This 'adaptive' response of the brush border peptidase, which is analogous to disaccharidase adaptation, suggested to us the possible role of brush border peptidase in the digestion of dietary proteins.

In the last study, changes in the activities of peptidases in the two subcellular fractions of the remaining ileum were examined after proximal small intestinal resection (McCarthy & Kim 1973). The specific activities and the segmental total activities of the enzymes rose to the same extent in both fractions. This indicates that the remaining intestinal segment has the ability to compensate for the loss of peptidase activity resulting from the resection.

In the light of these studies, which suggest an important role for brush border peptidases in the terminal stages of digestion of dietary proteins, we have recently turned our attention to the purification and characterization of brush border peptidases.

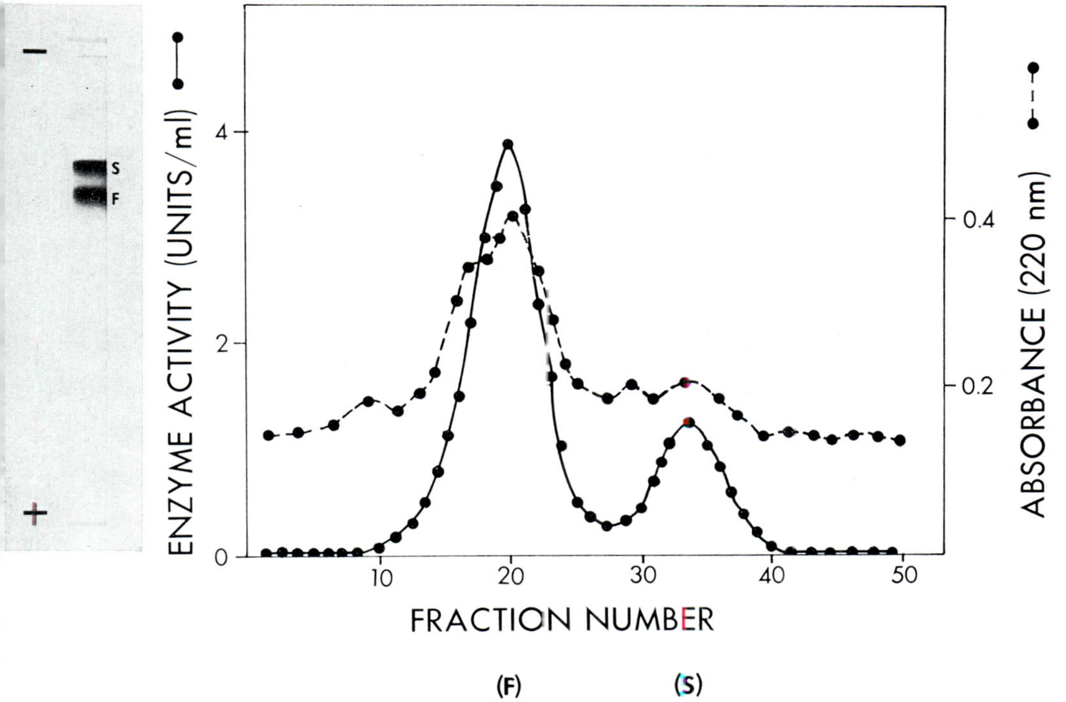

FIG. 6. Gel electrophoresis of partially purified peptidases (left) and the resolution of the enzymes into two fractions after preparative polyacrylamide gel electrophoresis. Gel electrophoresis shows the enzyme F and S after staining with Coomassie brilliant blue. (From Kim *et al.* 1976.)

PURIFICATION OF BRUSH BORDER PEPTIDASES

Our studies included the solubilization, isolation and characterization of peptidases of the brush border membrane from rat small intestine (Kim & Brophy 1976; Kim *et al.* 1976).

Triton X-100 was used to solubilize peptidases from the membranes. The enzymes were purified by $(NH_2)_2SO_4$ fractionation, followed by DEAE-Sephadex and gel filtration chromatography. At this stage 200-fold purification of the enzyme was achieved.

At the left in Fig. 6 is shown the gel electrophoretic pattern at this stage of purification. The gel shown here was stained for protein. To obtain quantities of the two components, preparative electrophoresis was carried out. The results of this procedure are shown on the right. The solid line represents enzyme activity while the broken line shows the protein profile. The first

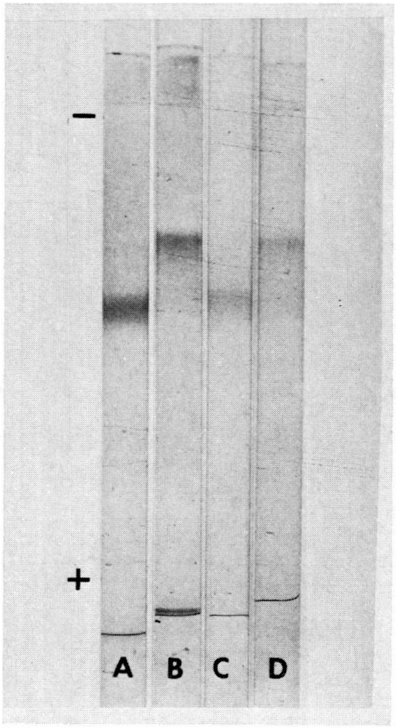

FIG. 7. Polyacrylamide gel electrophoresis of purified peptidases F and S. Peptidases F (A, C) and S (B, D) show a single band when stained for protein with Coomassie brilliant blue (A, B) or when stained for carbohydrate with PAS reagent (C, D). (From Kim & Brophy 1976.)

fraction, i.e. the faster-migrating fraction, is designated F, and the slower-eluting fraction is designated S. The two fractions obtained from preparative electrophoresis were purified further on DEAE-Sephadex, and the results they gave upon gel electrophoresis are presented in Fig. 7. Gels A and B were stained for protein and showed that the enzymes had been separated by these procedures. Gels C and D were stained for carbohydrate with periodic acid-Schiff (PAS) reagent. From the results it appears that both enzymes are glycoproteins.

We determined the amino acid and carbohydrate compositions of the two enzymes and found that their amino acid compositions are nearly identical. Carbohydrate analyses showed some differences and are presented in Table 4. Both peptidases contained the same sugars. The neutral sugars, mannose, galactose, fucose and glucose, accounted for most of the carbohydrates.

TABLE 4

Carbohydrate content of peptidases F and S (means of two analyses, with a molecular weight of 280 000 assumed for both enzymes on the basis of gel filtration experiments). (From Kim & Brophy 1976.)

Carbohydrate	*Peptidase F*	*Peptidase S*
	Mol carbohydrate residue/mol enzyme	
Mannose	56	81
Galactose	57	76
Fucose	30	34
Glucose	108	96
Glucosamine	97	142
Galactose	13	30
Sialic acid	9	6

Although *N*-acetylgalactosamine was present in both enzymes, *N*-acetylglucosamine was the predominant amino sugar. A small amount of sialic acid was also present. Of interest is the presence of glucose in these enzymes. This is not generally found in glycoproteins, and we are therefore investigating its mode of association with the enzyme.

It is not surprising to find that these intestinal brush border peptidases are glycoproteins, since many membrane-bound proteins which are externally oriented are glycoproteins. Kelly & Alpers (1973*a*) have reported that purified maltase (α-glucosidase; EC 3.2.1.20) and sucrase (β-fructofuranosidase; EC 3.2.1.26) from rat intestine contain 15–20% carbohydrate, a result similar to our findings for brush border peptidases from rat intestine. Intestinal alkaline phosphatase (EC 3.1.3.1) (Fosset *et al.* 1974) and glucoamylase (exo-1,4-α-glucosidase; EC 3.2.1.3) (Kelly & Alpers 1973*b*) are also glycoproteins.

The molecular weights of the enzymes were determined by two techniques. By gel filtration an apparent molecular weight of 280 000 was found for both enzymes (Fig. 8). When the enzymes were electrophoresed under dissociating conditions on SDS-polyacrylamide gels, a single band was obtained corresponding to 140 000 molecular weight. These data suggest that both brush border enzymes are composed of two subunits of equal molecular weight.

Next we examined the immunological properties of the two peptidases. The results of immunodiffusion studies are shown in Fig. 9. The centre wells contain the antisera to the fast and slow enzymes, while the peripheral wells contain slow enzyme, fast enzyme, rat small intestinal mucin and a concentrated cytoplasmic fraction. Both enzymes reacted with either antiserum and formed a fused immunoprecipitin line. These results indicate that the two enzymes are immunologically identical. No immunoprecipitin line was formed between

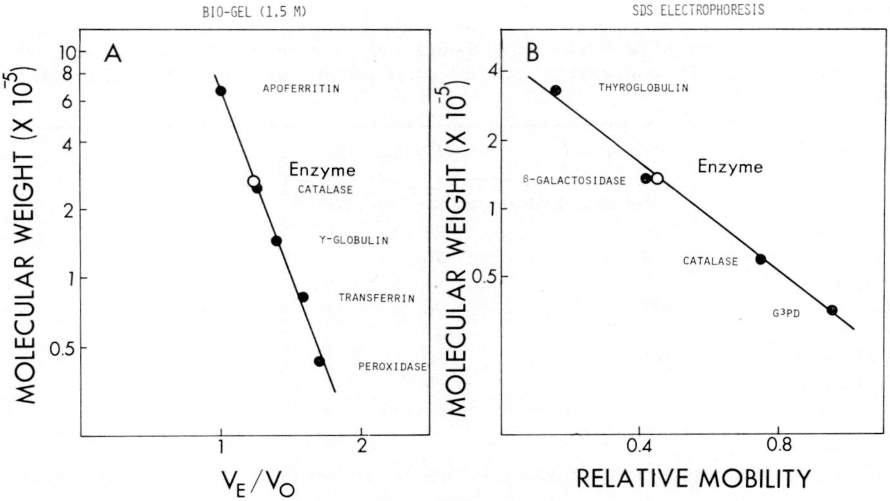

FIG. 8. Molecular weight determination of purified peptidases F and S by Bio-gel (1.5 M) chromatography and by SDS–polyacrylamide gel electrophoresis. o, peptidases F or S. (From Kim & Brophy 1976.)

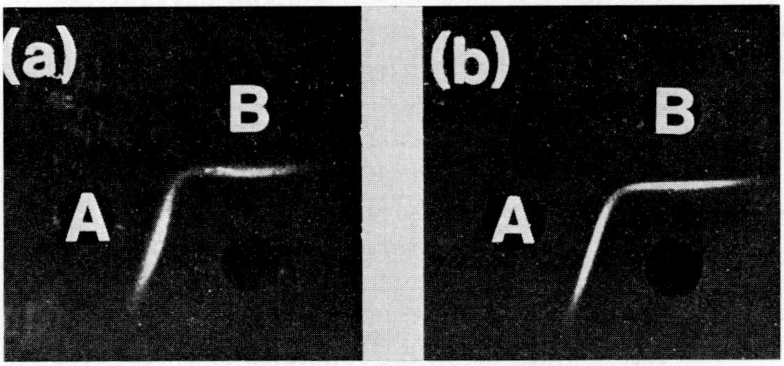

FIG. 9. Immunodiffusion of peptidase antisera against purified peptidases. A, peptidase F; B, peptidase S; (a) centre well, rabbit antipeptidase F; (b) centre well, rabbit antipeptidase S. (From Kim *et al.* 1976.)

these antisera and cytoplasmic fractions of intestinal cells or purified intestinal mucin. Purified brush border membrane vesicles were quantitatively precipitated by both antisera indicating that the two purified peptidases are indeed brush border enzymes.

We have also investigated the interactions of the peptidases with lectins which specifically bind different sugar moieties at the ends of oligosaccharide

TABLE 5

Effects of rabbit antipeptidase antisera and of phytohaemagglutinin (PHA) on rate of hydrolysis of various peptides by peptidases F and S (these concentrations of antiserum or lectin had previously been shown to result in a 50% precipitation of the enzymes when Phe-Gly was used as substrate for the experiment). (From Kim *et al.* 1976.)

Substrate	*Percentage inhibition of enzyme activity*			
	Peptidase F		*Peptidase S*	
	Anti-F	*PHA*	*Anti-S*	*PHA*
Phe-Gly	65	33	76	26
Phe-$(Gly)_2$	69	30	77	32
Gly-Phe	60	38	72	28
Leu-Gly	65	22	74	16
Leu-$(Gly)_2$	67	28	80	32

Anti F and Anti S: rabbit antipeptidase F and antipeptidase S antisera.

side chains of glycoproteins, causing the aggregation or precipitation of the four lectins tested (concanavalin A, wheat germ agglutinin, ricin and phytohaemagglutinin). Only phytohaemagglutinin gave a line of precipitation when tested against the two peptidases on immunodiffusion plates. Since the agglutinating ability of this lectin is inhibited by *N*-acetylgalactosamine, it is likely that the peptidase contains *N*-acetylgalactosamine at or near the terminal non-reducing end of a carbohydrate side chain. We also examined the effect of phytohaemagglutinin antisera on peptidase.

Table 5 shows the effects of concentrations of lectin activity or antiserum which had previously been shown to result in a 50% inhibition of the enzymes when phenylalanylglycine was used as substrate. For each agglutinin a similar inhibition of peptidase activity occurs, irrespective of the substrate used. These findings suggest that the binding sites for both lectin and antibody are near the active site of the enzyme.

The relative rates of hydrolysis of various peptides by the two enzymes were measured (Table 6) and were found to be similar. All peptides were examined at 5 mM concentration except the largest peptides. The amino acid composition of the substrates influenced the rates of hydrolysis greatly. It is clear from Table 6 that the dipeptides containing proline or D-amino acids at either the C-terminal or N-terminal position of the peptide are not hydrolysed.

Fig. 10 shows data obtained from hydrolysis of an octapeptide by purified enzymes. There was a gradual increase in all amino acids released from the substrate with time. The data suggest that sequential hydrolysis of the octapeptide from the N-terminus takes place with no evidence for endopeptidase

TABLE 6

Relative rates of hydrolysis of peptides by the peptidases. (From Kim *et al.* 1976.)

Activity related to L-*Phe-Gly* = *100*[a]	*Substrates*
200-270	L-Ala(Gly)$_2$, L-Phe(Gly)$_2$
150-200	L-Ala(Gly)$_3$
100-150	L-Ala(Gly)$_4$, L-Leu(Gly)$_2$, L-Leu-L-Met, L-Leu-L-Val
50-100	L-Ala-Gly, L-Leu-L-Ala, L-Leu-L-Leu
10-50	L-Leu-Gly, Gly-Gly, L-Phe-L-Glu, L-Ala-L-Glu, L-Glu-L-Ala, L-Arg-L-Phe, L-Met-L-Ser, L-Leu-BNA[b]
1-10	Gly-L-Phe, Gly-L-Leu, L-Phe-L-Asp, L-Asp-L-Phe, L-Lys-L-Phe, L-Ser-L-Phe, L-Ser-L-Met, L-Val-L-Leu
0	L-Pro-L-Leu, L-Pro-L-Phe, L-Leu-L-Pro, L-Phe-L-Pro, L-Met-L-Pro; D-Leu-D-Leu, L-Leu-D-Leu, D-Leu-L-Leu, D-Leu(Gly)$_2$; hippuryl-L-Arg; bradykinin,[c] insulin B chain, casein

[a] Values are initial rates of release of NH_2-terminal amino acids relative to the initial rate of hydrolysis of L-Phe-Gly.

[b] L-Leu-BNA: L-Leucyl-β-naphthylamide.

[c] Bradykinin: L-Arginyl-L-prolyl-L-prolyglycyl-L-phenylalanyl-L-seryl-L-prolyl-L-phenylalanyl-L-arginine.

or carboxypeptidase activities. This indicates that the enzymes are aminoexopeptidases.

We then did experiments to determine whether different peptides hydrolysed by these enzymes compete for the same active site. Double reciprocal plots of data from the hydrolysis of three peptides (leucylalanine, phenylalanylglycine and phenylalanylglycylglycine) in the absence and presence of increasing concentrations of alanylglycine are shown in Fig. 11. The plots show competitive inhibition, indicating that alanylglycine competes for the same active site as the tripeptide and the two dipeptides.

To summarize, we have isolated and purified two enzymes from the intestinal brush border of the rat. The enzymes are remarkably similar to each other in many respects. They are both glycoproteins and have similar chemical compositions. They have common antigenic properties, and these antigenic properties are distinct from those of cytoplasmic enzymes. Substrate specificities and kinetic properties of the two enzymes are similar. The isolation and characterization of these enzymes from the brush border membrane should eventually aid in elucidation of the details of protein absorption in the intestine.

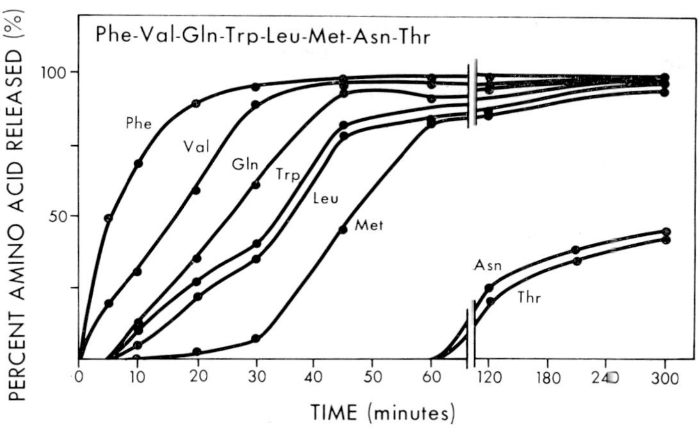

FIG. 10. Effect of purified peptidase F on hydrolysis of an octapeptide. The peptidase was incubated with the octapeptide at 37°C for various time periods. (From Kim *et al.* 1976.)

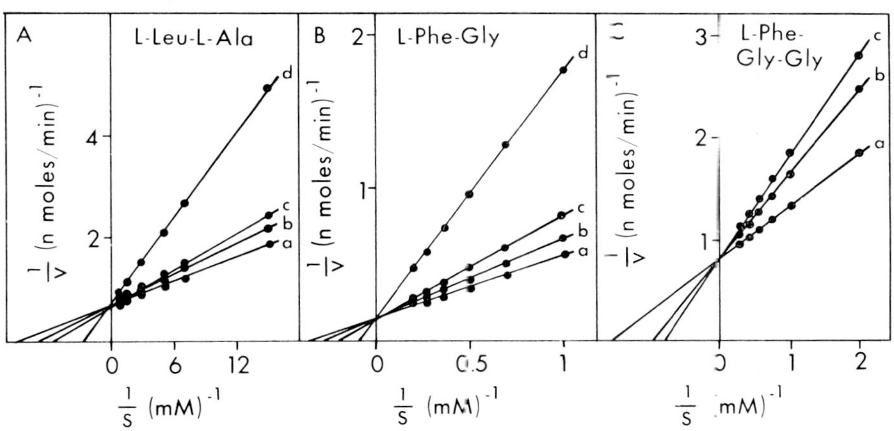

FIG. 11. Competition between Ala-Gly and Leu-Ala, Phe-Gly and Phe-Gly-Gly for peptidase F. Various concentrations of the last three peptides were assigned in the presence of fixed concentrations of Ala-Gly. a = 0 mM, b = 0.5 mM, c = 1.0 mM, d = 5.0 mM. (From Kim *et al.* 1976.)

DISCUSSION

The early observation of Newey & Smyth (1962) that dipeptides could be taken up intact from the intestinal lumen and hydrolysed by the intestinal mucosal cells was followed by the observation by Matthews and his co-workers (1968) and Adibi (1971) that glycine was absorbed faster when fed as peptides than when fed as free amino acids. Subsequently, Milne and his co-workers demonstrated that patients with Hartnup disease, who have a transport defect for neutral free amino acids, could absorb these normally when the amino acids were fed as peptides (Asatoor *et al.* 1970). Taken together, these and later experiments clearly showed that amino acids and dipeptides have separate transport mechanisms.

The precise biological roles of peptidases in the brush border and cytoplasmic fractions of small intestinal mucosal cells are not yet clear. However, from the biochemical and physiological data available (Kim *et al.* 1973; McCarthy & Kim 1973; Nicholson *et al.* 1974), it would appear that peptidases from both subcellular fractions play important roles in the terminal digestion of dietary proteins. Furthermore, the enzymes in the cytoplasmic fraction may play a more general role in cellular metabolism. For instance, during starvation or protein malnutrition these cytoplasmic enzymes may contribute to hydrolysis of intracellular protein to provide amino acids for vital metabolic functions (Kim *et al.* 1973).

A schematic representation of possible mechanisms for peptide hydrolysis and transport is shown in Fig. 12. All the available evidence points to the fact that mechanism 3, which invokes surface hydrolysis followed by absorption of liberated free amino acids by conventional amino acid carriers, is unlikely to be of major importance. Mechanism 2 is currently held to be the major route for dipeptide transport. In this scheme, unhydrolysed dipeptides are transported intact across the brush border membrane, bypassing the brush border peptidase, to be hydrolysed by cytoplasmic enzymes. Although this scheme probably explains the absorption data derived from experiments with prolyl peptides and other peptides that are poorly hydrolysed by brush border enzymes, it is difficult to imagine that dipeptides with high affinities for brush border peptidases would avoid hydrolysis by brush border enzymes and be absorbed solely by this mechanism. Crane and his associates, in studying sugar transport, have described an exclusive mechanism for the transport of disaccharide-bound monosaccharides (Malathi *et al.* 1973). In this model, disaccharide hydrolysis and transport are closely integrated. The existence in the intestinal brush border of peptidases with high affinity for many dipeptides raises the possibility of a similar relationship between dipeptide hydrolysis and transport. This model is depicted for a dipeptide in mechanism 1. In this scheme, membrane hydrolysis of a dipeptide is followed by uptake of amino acids by carrier mechanisms available only to amino acids liberated by the action of these enzymes.

It should be pointed out that none of these models is mutually exclusive. For longer peptides which are not hydrolysed by cytosol peptidases, it is clear

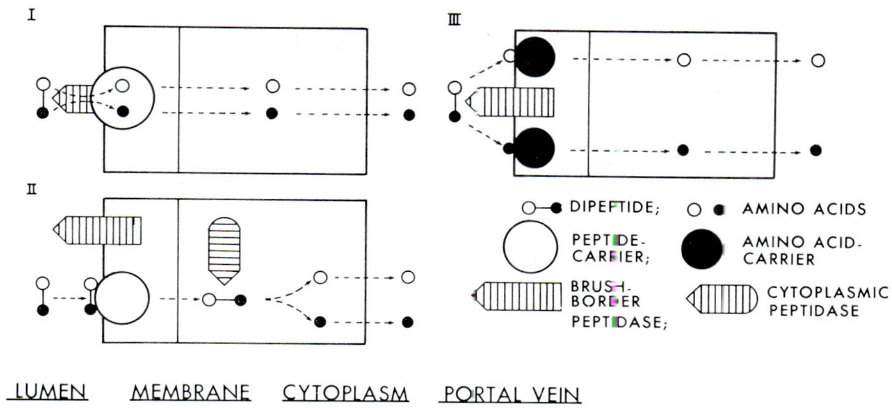

FIG. 12. Schematic representation of possible mechanisms for peptide hydrolysis and transport.

that the major, if not exclusive, site of hydrolysis is the brush border membrane.

An understanding of the structure of membrane-bound peptidases, the location within the membranes of these enzymes, and their interaction with peptide and amino acid carriers, should help to elucidate the mechanisms of peptide and amino acid transport.

ACKNOWLEDGEMENT

This work was supported by Public Health Service Grant AM 17938 from the National Institute of Health and Veterans Administration Research Grant.

References

ADIBI, S. A. (1971) Intestinal transport of dipeptides in man: relative importance of hydrolysis and intact absorption. *J. Clin. Invest. 50*, 2266-2275

ADIBI, S. A. & MERCER, D. W. (1973) Protein digestion in human intestine as reflected in luminal, mucosal and plasma amino acid concentrations after meals. *J. Clin. Invest. 52*, 1586-1594

ASATOOR, A. M., CHENG, B., EDWARDS, K. D. G., LANT, A. F., MATTHEWS, D. M., MILNE, M. D., NAVAB, F. & RICHARDS, A. J. (1970) Intestinal absorption of two dipeptides in Hartnup disease. *Gut 11*, 380-387

CHUNG, Y. C., SILK, D. B. A. & KIM, Y. S. (1977) Intestinal transport and hydrolysis of di-, tri- and tetrapeptides. To be published

CRAMPTON, R. F., GANGOLLI, S. D., MATTHEWS, D. M. & SIMPSON, P. (1971) Rates of absorption from tryptic hydrolysates of proteins and the corresponding acid hydrolysates or amino acid mixtures. *J. Physiol. (Lond.) 213*, 43P-44P

CURTIS, K. J., WHITEHEAD, J. S., PERDOMO, J. M., SILK, D. B. A. & KIM, Y. S. (1977) Protein digestion and absorption of the rat. *Am. J. Physiol.* in press

FORTIN-MAGANA, R., HURWITZ, R., HERBST, J. J. & KRETCHMER, N. (1970) Intestinal enzymes: indicators of proliferation and differentiation in the jejunum. *Science (Wash. D.C.) 167*, 1627-1628

FOSSET, M., CHAPPELET-TORDO, D. & LAZDUNSKI, M. (1974) Intestinal alkaline phosphatase. Physical properties of quarternary structure. *Biochemistry 13*, 1783-1788

GRAY, G. M. & COOPER, H. L. (1971) Protein digestion and absorption. *Gastroenterology 61*, 535-544

HEIZER, W. D., KERLEY, R. L. & ISSELBACHER, K. J. (1972) Intestinal peptide hydrolases differences between brush border and cytoplasmic enzymes. *Biochim. Biophys. Acta 264*, 450-461

IMONDI, A. R., BALIS, M. E. & LIPKIN, M. (1969) Changes in enzyme levels accompanying differentiation of intestinal epithelial cells. *Exp. Cell Res. 58*, 323-330

JOSEFSSON, L. & SJÖSTRÖM, H. (1966) Intestinal dipeptidases. IV. Studies on the release and subcellular distribution of intestinal dipeptidases of the mucosal cells of the pig. *Acta Physiol. Scand. 67*, 27-33

KELLY, J. J. & ALPERS, D. H. (1973*a*) Blood group antigenicity of purified human intestinal disaccharidases. *J. Biol. Chem. 248*, 8216-8221

KELLY, J. J. & ALPERS, D. H. (1973*b*) Properties of human intestinal glycoamylase. *Biochim. Biophys. Acta 315*, 113-120

KIM, Y. S. & BROPHY, E. J. (1976) Rat intestinal brush border membrane peptidases. I. Solubilization, purification and physicochemical properties of two different forms of the enzyme. *J. Biol. Chem. 251*, 3199-3205

KIM, Y. S. & KIM, Y. W. (1977*a*) Cellular localization of brush border and cytoplasmic peptide hydrolases. To be published

KIM, Y. S. & KIM, Y. W. (1977*b*) Zymogram studies of intestinal peptide hydrolases of man. To be published

KIM, Y. S., BIRTWHISTLE, N. & KIM, Y. W. (1972) Peptide hydrolases in the brush border and soluble fractions of small intestinal mucosa of rat and man. *J. Clin. Invest. 51*, 1419-1430

KIM, Y. S., MCCARTHY, D. M., LAND, W. & FONG, W. (1973) Alterations in the levels of peptide hydrolases and other enzymes in brush border and soluble fractions of rat small intestinal mucosa during starvation and refeeding. *Biochim. Biophys. Acta 321*, 262-273

KIM, Y. S., KIM, Y. W. & SLEISENGER, M. H. (1974) Specificities of peptide hydrolases in brush border and cytosol fractions of rat small intestine. *Biochim. Biophys. Acta 370*, 283-296

KIM, Y. S., BROPHY, E. J. & NICHOLSON, J. A. (1976) Rat intestinal brush border membrane peptidases. II. Enzymatic properties, immunochemistry and interactions with lectins of two different forms of the enzyme. *J. Biol. Chem. 251*, 3206-3212

MALATHI, P., RAMASWAMY, K., CASPARY, W. F. & CRANE, R. K. (1973) Studies on the transport of glucose from disaccharides by hamster small intestine *in vitro*. I. Evidence for a disaccharidase-related transport system. *Biochim. Biophys. Acta 307*, 613-626

MATTHEWS, D. M., CRAFT, I. L., GEDDES, D. M., WISE, I. J. & HYDE, C. W. (1968) Absorption of glycine and glycine peptides from the small intestine of the rat. *Clin. Sci. (Oxf.) 35*, 415-424

MCCARTHY, D. M. & KIM, Y. S. (1973) Changes in sucrase, enterokinase and peptide hydrolase after intestinal resection. The association of cellular hyperplasia and adaptation. *J. Clin. Invest. 52*, 942-951

NEWEY, H. & SMYTH, D. H. (1962) Cellular mechanisms in intestinal transport of amino acids. *J. Physiol. (Lond.) 164*, 527-551

NICHOLSON, J. A., MCCARTHY, D. M. & KIM, Y. S. (1974) The responses of intestinal brush border and cytosol peptide hydrolase activities to variation in dietary protein content. *J. Clin. Invest. 54*, 890-898

NIXON, S. E. & MAWER, G. E. (1970*a*) The digestion and absorption of protein in man. 1. The site of absorption. *Br. J. Nutr. 24*, 227-240

NIXON, S. E. & MAWER, G. E. (1970*b*) The digestion and absorption of protein in man. 2. The form in which digested protein is absorbed. *Br. J. Nutr. 24*, 241-258

PETERS, T. J. (1970) The subcellular localization of di- and tripeptide hydrolase activity in guinea-pig small intestine. *Biochem. J. 120*, 195-203

Discussion

Fottrell: Professor Kim showed a brush border peptidase activity of 5–10% with dipeptide substrates. We have purified soluble peptidases from the guinea-pig, where about 90% of the activity is in the cytosol fraction with about 5–10% of activity against dipeptides detectable in the brush border fraction (Donlon & Fottrell 1973; O'Cuinn & Fottrell 1975; Piggott & Fottrell 1975). We recently showed (Piggott *et al.* 1976) that during the preparation of brush borders from guinea-pig intestinal mucosa, cytoplasmic peptidases are bound to the brush borders.

Rhodes: There is other evidence that binding of cytoplasmic peptidases to microvillus membranes is unlikely. We isolated hamster microvillus membranes and looked for contamination. There are four observations against adsorption of the cytosol enzymes. First, the activity stays there after exposure to 1 M-Tris. Secondly, after deoxycholate treatment, the activity is still there (Rhodes *et al.* 1967). Thirdly, sodium chloride up to 1 M does not separate the peptidase activity. The activity is destroyed with close to 1 M-NaCl. Finally, we used Triton X-100 with centrifugation and column chromatography and we still couldn't separate the dipeptidase activity (J. B. Rhodes, A. Eichholz. & R. K. Crane, unpublished observations).

Fottrell: What substrate were you using?

Rhodes: We used four different dipeptides, leucylglycine, alanylglycine, glycylglycine and leucineamide. They were consistent in having the same distribution pattern.

Semenza: Is the cytosol dipeptidase a glycoprotein, Professor Fottrell?

Fottrell: We could not detect any carbohydrate in this enzyme.

Radhakrishnan: The same is true in our preparations.

Norén: We did immunological studies on the pig small intestine with antibodies raised against glycylleucine dipeptidase and proline dipeptidase obtained from the cytosol. These antibodies did not react with Triton X-100-solubilized material from brush border membranes.

Fottrell: I think there are species problems here. As we showed in our paper (Piggott *et al.* 1976), brush borders from rat intestine adsorb more peptidase from the cytosol than guinea-pig mucosa. I have had no experience with pig intestinal mucosa.

Josefsson: Confusion in this matter becomes worse when we talk without precise definitions. We talk about dipeptidase activity and dipeptidases, which are different things, and I think everybody will agree that we should use the right terms. There is no doubt that there is dipeptidase activity in the brush border. Professor Kim's enzyme has a high dipeptidase activity but it is not a dipeptidase. The true dipeptidases, which are responsible for most of the dipeptidase activity in the cellular fractionation experiments, are located in the cytosol, as I shall discuss later. However, as Professor Fottrell said, there may be a mixture of the different enzymes in the brush border preparation if this is not very carefully made. We must be careful with the terminology. A dipeptidase is a defined enzyme, while dipeptidase activity is an enzyme activity that splits dipeptides.

Parsons: So the distinction between a dipeptidase and a tripeptidase lies in whether the enzyme hydrolyses dipeptides and higher peptides? If it hydrolyses tripeptides it cannot be a true 'dipeptidase'.

Josefsson: No; the true dipeptidases so far purified have no activity whatsoever on peptides other than dipeptides. The brush border peptidases like Professor Kim's have a great variety of activities, including activity against dipeptides, tripeptides and larger peptides.

Kim: I suppose we should call it amino-oligopeptidase.

Josefsson: Yes, but we should recognize that their range of activities also includes dipeptidase activity.

Fottrell: There is no real conflict here, but I think there are species differences. One of our peptide hydrolases from guinea-pig (the α-peptide hydrolase) (Donlon & Fottrell 1973) hydrolyses both dipeptides and tripeptides. A second peptide hydrolase (the β-peptide hydrolase) prefers dipeptides but will also hydrolyse some tripeptides. We have used very sensitive criteria to establish the homogeneity of our enzyme preparations which seem to have broader substrate specificity than Professor Josefsson's enzyme. One of the points I wish to emphasize is that some of the brush border peptide hydrolase activity when tested with dipeptides may be due to adsorption of cytoplasmic enzymes.

Parsons: But if you have an enzyme in the brush border that is derived from, say, the pancreas, it ought to be possible to desorb it, to displace it. Whereas if it is an enzyme that is an integral part of the plasma membrane that constitutes the brush border, the activity in the membranes will not be capable of being washed away.

Fottrell: Yes, I don't disagree. Professor Kim's brush border enzyme activity against a dipeptide satisfies the criteria already established for brush border enzymes.

Rhodes: I agree with Professor Fottrell that we have to be very careful about what is 'no activity'. When I say 'no activity' I mean less than 1 nmol/(min mg protein).

Fottrell: I consider that when brush border dipeptide hydrolase activity against a given substrate is 5–10% of the total mucosal activity, the possibility of adsorption of cytosol peptide hydrolase activity should be investigated.

Radhakrishnan: There is certainly a lot of confusion about the terms 'activity' and the 'enzyme' and we should be very careful. In the cytosol, where there are specific enzymes, they may have overlapping specificities. Professor Fottrell's caution about the brush border enzymes is very timely, but Professor Kim's observations are also very interesting. I am not sure whether the enzyme isolated by Kim is the same as the leucine aminopeptidase that has been known for a long time. The substrate specificity would be somewhat similar. It could certainly act on dipeptides and on oligopeptides. It is very important that we should be clear whether we are talking in terms of activity or in terms of enzyme.

Mellander: Professor Kim, did you try dipeptides with phosphorylated serine?

Kim: No, we haven't.

Mellander: Did you use synthetic substrates only or did you test your purified enzymes under biological conditions, that is in peptide mixtures from food proteins?

Kim: No, we haven't done that.

Parsons: What was the effect of heavy metals on your brush border enzyme, Professor Kim?

Kim: It is rather confusing. In the presence of 0.5 to 1 mM-cobalt, manganese or magnesium, the activity of the brush border enzyme that we have isolated is inhibited. Zinc completely abolishes the enzyme activity. If we pretreat the enzyme with EGTA or phenanthroline and then add metal ions such as cobalt the enzyme activity is restored, but manganese and magnesium do not restore the enzyme activity.

Semenza: What about zinc?

Kim: Zinc does not restore it.

Rhodes: Dr Maroux has shown that in the hog membrane aminopeptidase there are two zinc molecules per aminopeptidase molecule.

Radhakrishnan: Was there any sialic acid in your enzymes, and would that make a difference?

Kim: Both fast- and slow-moving enzymes contain sialic acid. The fast-moving enzyme has slightly more sialic acid than the slow one. But when we treated the fast-moving enzyme with *Clostridia* or *Vibrio cholera* neuraminidase, we observed no change in electrophoretic mobility. However, the difference in the electrophoretic mobility of fast- and slow-moving enzymes may be due to other chemical groups on the molecule, such as the sulphate groups.

Radhakrishnan: If it is not terminal (or buried) you may not pick up a difference.

Kim: From what is generally known about the structure of glycoprotein, one can safely assume that sialic acid is located terminally. The question is whether the neuraminidases that we used can cleave the specific glycosidic bond between sialic acid and the penultimate sugar.

Adibi: What are the relative rates of hydrolysis of dipeptides and tripeptides by membrane-bound enzymes? And in view of all this confusion, is it fair to say that the activity against dipeptides is probably trivial and of little biological significance? These membrane-bound enzymes appear efficient against some tripeptides and higher peptides.

Kim: I don't think that there is any confusion. The brush border enzyme is quite active against some dipeptides and also against most tripeptides and higher peptides. In the cytoplasm, there are enzymes which predominantly have hydrolytic activity against dipeptides and some against tripeptides.

Adibi: Professor Radhakrishnan has shown that the master dipeptidase sits in the cytosol. The membrane fraction seems to be better suited for hydrolysis of some tripeptides and larger peptides, whereas hydrolysis of dipeptides by the membrane-bound enzymes seems to be non-existent or very small.

Parsons: You have a measure of the specific activity of this enzyme but do you know what fraction of the protein digestion can go through that pathway? You could calculate how many moles of amino nitrogen could be handled by this pathway per day in your rats, and you could compare that with your estimate of the amino-N input to the intestinal tract.

Kim: We have not done that.

Radhakrishnan: The molecular activity may be very low.

Silk: For many years now, enzymologists have commented on the fact that only 5–10% of the total mucosal peptidase activity against dipeptides resides in the brush border fraction of mucosal homogenates and a common conclusion drawn is that because such a small proportion of enzyme activity is present at the cell surface these surface enzymes play only an insignificant role in mucosal hydrolysis of luminal dipeptides. The fact remains, however, that 5–10% of total mucosal activity could well be sufficient, *in vivo*, to hydrolyse major components of luminal peptide presented to the mucosa for absorption. Indeed we have shown this to be the case in a recent study of comparisons of *in vivo* absorption rates of two dipeptides and *in vitro* rates of hydrolysis of the two dipeptides by brush border peptidases (Silk *et al.* 1976). It should be remembered, however, when interpreting the results of such studies, that protein digestion products may not come in contact *in vivo* with all the enzyme activity measured *in vitro*.

Josefsson: It is important to recognize that this 5–10% dipeptidase activity varying between the different substrates in the brush border preparation is a significant activity. There is some doubt that this activity has a role in the digestion process but it does not belong to the dipeptidases: it belongs to the membrane-bound peptidases. We have to recognize even more that we shouldn't regard the enterocytes as the only cells that contain enzymes for peptide digestion. All those membrane-bound peptidases, as well as the cytosol-located dipeptidases, are present in other tissue cells. That means that there is no proof that special enzymes are adopted for the processing of dietary proteins. This is a very important point to remember. All we know is that there are tremendous amounts of the enzymes in the intestinal wall. That is perhaps why those enzymes have the capacity for doing all the hydrolysing processes. In addition they have the capacity to take care of the hydrolysis of dietary proteins. I stress this point because when we know more about the enzyme molecules, we can use very specific means for studying them in the different tissues, for example by using immunological methods. We can recognize that the same molecules, or very closely related molecules, are present in kidney, liver, brain, muscle and so on. This applies to both the cytosol-located peptidases and the membrane-located peptidases. We must not be too narrow in our thinking.

Kenny: One is always bothered by the confusion of nomenclature surrounding 'leucine aminopeptidase' (LAP). The confusion dates from the period when histochemists first used chromogenic substrates like leucine 2-naphthylamide to locate LAP. Later Patterson *et al.* (1963) proved that the enzymes attacking the new substrates differed from LAP, for which L-leucineamide is the traditional substrate. The Enzyme Commission is quite clear: the cytosolic enzyme, LAP, is 3.4.11.1 and the membrane-bound enzyme (aminopeptidase M, i.e. microsomal) is 3.4.11.2. They are indeed different enzymes, immunologically distinct, with different substrate specificities and different responses to metal ions. Incidentally, one must be careful in interpreting metal ion effects: activation of an enzyme by, say, Mn^{2+} or Ca^{2+} does not mean that the native enzyme contains that metal ion.

Fottrell: Our findings using immunological techniques support what Professor Josefsson said. We found (O'Cuinn *et al.* 1975) that peptide hydrolases immunologically similar to intestinal peptide hydrolases are present in several tissues of the guinea-pig such as brain and kidney.

References

DONLON, J. & FOTTRELL, P. F. (1973) Purification and characterization of one of the forms of peptide hydrolases from guinea-pig small intestinal mucosa. *Biochim. Biophys. Acta 327*, 425-436

O'CUINN, G. & FOTTRELL, P. F. (1975) Purification and characterization of an aminoacyl proline hydrolase from guinea-pig intestinal mucosa. *Biochim. Biophys. Acta 391*, 388-395

O'CUINN, G., PIGGOTT, C. O. & FOTTRELL, P. F. (1975) Immunological relationship between three intestinal peptide hydrolases and similar enzymes in various tissues of the guinea-pig and of other animal species. *FEBS (Fed. Eur. Biochem. Soc.) Lett. 53*, 164-166

PATTERSON, E. K., HSIAO, S. H. & KEPPELL, A. (1963) Distinction between leucine aminopeptidase and enzymes that hydrolyze L-leucyl-β-naphthylamide. *J. Biol. Chem. 238*, 3611-3620

PIGGOTT, C. O. & FOTTRELL, P. F. (1975) Purification and characterization from guinea-pig intestinal mucosa of two peptide hydrolases which preferentially hydrolyse dipeptides. *Biochim. Biophys. Acta 391*, 403-409

PIGGOTT, C. O., O'CUINN, G. & FOTTRELL, P. F. (1976) Similarities between a dipeptide hydrolase from brush border and cytosol fractions of guinea-pig intestinal mucosa. *Biochem. J. 159*, 715-717

RHODES, J. B., EICHHOLZ, A. & CRANE, R. K. (1967) Studies on the organisation of the brush border in intestinal epithelial cells. IV. Aminopeptidase activity in microvillus membranes of hamster intestinal brush borders. *Biochim. Biophys. Acta 135*, 959-965

SILK, D. B. A., NICHOLSON, J. A. & KIM, Y. S. (1976) Relationships between mucosal hydrolysis and transport of two phenylalanine dipeptides. *Gut 17*, 870-877

Intestinal brush border peptidases

OVE NORÉN, HANS SJÖSTRÖM, BIRTE SVENSSON*, LARS JEPPESEN, MICHAEL STAUN and LARS JOSEFSSON

Department of Biochemistry C, University of Copenhagen, Copenhagen

Abstract A brush-border-specific antiserum was raised in rabbits, with Triton X-100-solubilized brush border proteins from pig intestine being used as antigens. The antiserum was used in immunoelectrophoretic studies of brush border proteins solubilized with Triton X-100. Five immunoprecipitates were obtained which corresponded to microsomal aminopeptidase (EC 3.4.11.2), aspartate aminopeptidase (EC 3.4.11.7), lactase (β-galactosidase, EC 3 2.1.23), maltase (exo-1,4-α-glucosidase, EC 3.2.1.3) and sucrase–isomaltase (sucrose α-glucohydrolase, EC 3.2.1.48). A faint immunoprecipitate was also found for the glycylprolyl dipeptidyl peptidase (EC 3.4.14.–). The brush border proteins were solubilized on a large scale from a brush border membrane preparation by the use of Triton X-100; the peptidases obtained were homogeneous in size and had hydrophobic properties. By chromatography on columns of concanavalin A-Sepharose, hydroxyapatite, Ultrogel AcA 34, DEAE-cellulose and immunosorbent, γ-glutamyl transpeptidase (γ-glutamyl transferase, EC 2.3.2.2) and microsomal aminopeptidase were each isolated in separate fractions. Glycylprolyl dipeptidyl peptidase and aspartate aminopeptidase were obtained in another fraction. Immunoelectrophoretic, inhibitor and chromatographic studies showed that the intestinal brush border peptidases are similar to the corresponding particulate peptidases obtained from other organs.

It can be regarded as an established fact that the intestinal protein digest disappears from the intestinal lumen both as free amino acids and as small peptides in physiologically significant amounts (Matthews 1975). The existence of separate uptake mechanisms for amino acids and small peptides has been demonstrated, but for additional insight into the absorptive mechanisms on a molecular level a detailed knowledge of the brush border proteins and peptidases is required.

In order to contribute to the elucidation of the relation of the intestinal

* *Present address:* Carlsberg Laboratory, Dept of Chemistry, Gammel Carlsbergvej 10, DK - 2500 Valby, Denmark

peptidases to peptide transport and hydrolysis we earlier purified and characterized two pig dipeptidases which are localized in the cytoplasm of the enterocyte (Josefsson *et al.* 1977).

The brush border proteins have been studied on an analytical scale (Critchley *et al.* 1975; Louvard *et al.* 1975; Maestracci *et al.* 1975), but some of them have also been purified on a large scale, proteolytic enzymes being used as solubilizing agents (Cogoli *et al.* 1972; Maroux *et al.* 1973). However, solubilization of proteins by proteolytic enzymes cannot be considered to yield native proteins suitable for structural studies, as limited proteolysis of the proteins to be solubilized often occurs. Non-ionic detergents have been used as an alternative and they have proved to be lenient solubilization agents, since most enzymes studied retained their biological activity during the solubilization procedure (Tzagoloff & Penefsky 1971). Recently brush border enzymes solubilized with Triton X-100 were reported as having been purified to homogeneity (Sigrist *et al.* 1975; Maroux & Louvard 1976; Kim & Brophy 1976).

ANALYTICAL STUDIES ON INTESTINAL BRUSH BORDER PROTEINS

Analytical methods hitherto used to study brush border proteins solubilized with non-ionic detergents suffer from several drawbacks which greatly hamper the study of the native forms of brush border proteins. Triton X-100-solubilized proteins have low electrophoretic mobility in polyacrylamide gels, implying poor resolution in such systems (Wacker *et al.* 1976; Louvard *et al.* 1975). Electrophoresis in polyacrylamide gels in the presence of sodium dodecylsulphate was reported to give good resolution when applied to brush border proteins (Critchley *et al.* 1975; Maestracci *et al.* 1975), but sodium dodecylsulphate is a strong denaturing agent and therefore the electrophoretic pattern obtained does not reflect the different brush border proteins in a native state.

Crossed immunoelectrophoresis in agarose gels in the presence of non-ionic detergents has been used in studies of erythrocyte membrane proteins (Bjerrum & Lundahl 1973). We have used this method in studies of brush border membrane proteins.

Preparation of brush border antigen. Brush borders were prepared from the pig small intestine by the method of Eichholz & Crane (1974). The brush border proteins were solubilized by incubating the brush borders in a 50 mM-Tris-HCl buffer, pH 8.0, containing Triton X-100 (1%) for 1 h at 4°C; the insoluble fraction was then centrifuged off and the supernatant (brush border antigen) was collected.

Preparation of microsomal antigen. A microsomal fraction was prepared from pig intestinal mucosa by homogenization, removal of large particles by centrifugation, and subsequent ultracentrifugation at 100 000 g for 1 h. The pellet obtained was suspended in 50 mM-Tris-HCl buffer, pH 8.0, containing 1% Triton X-100 and stirred for 1 h at 4 °C. The insoluble material was removed by centrifugation and the supernatant was collected (microsomal antigen).

Preparation of antibodies. The brush border antigen and microsomal antigen were each repeatedly injected into rabbits during a two-month period (Harboe & Ingild 1973). The antisera obtained from the rabbits were purified by ammonium sulphate fractionation and ion-exchange chromatography and then used in crossed immunoelectrophoresis.

Purity of antigens. The microsomal antigen was subjected to crossed immunoelectrophoresis against the corresponding purified antiserum. Ten precipitates were obtained. The brush border antigen produced only five precipitates in the same electrophoretic system, indicating that the brush border antigen is considerably purified by the preparation procedure. Crossed immunoelectrophoresis of the microsomal antigen against the brush border antiserum also resulted in the same five precipitates (Fig. 1), thus making it possible to study brush border antigens without the need to prepare brush borders.

Specificity of the brush border antiserum. The specificity of the antibodies obtained against the brush border antigen was tested by using the double-layer immunofluorescence technique for a section of pig small intestine. The purified antiserum against the brush border antigen was used as the first layer and swine anti-rabbit IgG conjugated with fluorescein–isothiocyanate was used as the second layer. Fluorescence was only conspicuous in the brush border region of the enterocytes, indicating the brush border specificity of the antiserum.

Identification of the immunoprecipitates. The five precipitates obtained by immunoelectrophoresis (Fig. 1) were identified by histochemical techniques (Uriel 1971), using the substrates listed in Table 1. From the hydrolysis pattern obtained it is concluded that precipitate 1 corresponds to the aminopeptidase earlier studied by Maroux *et al.* (1973) (microsomal aminopeptidase, EC 3.4.11.2). Precipitates 2 and 4 correspond to the maltase (EC 3.2.1.3) and the sucrase–isomaltase complex (sucrose α-glucohydrolase, EC 3.2.1.48),

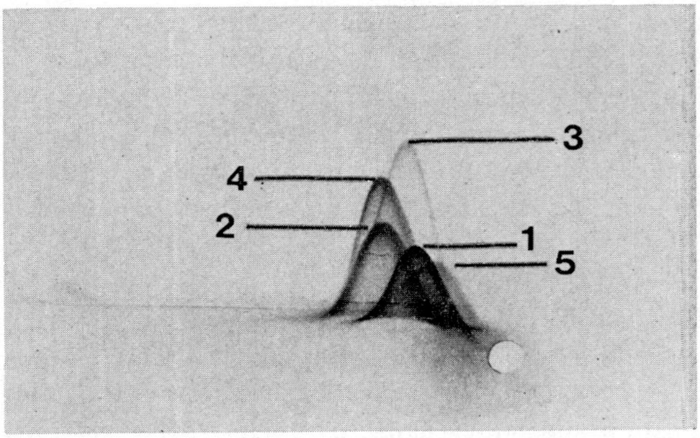

FIG. 1. Crossed immunoelectrophoresis of a microsomal antigen against brush border antiserum. The numbers on the precipitates refer to the text and to Table 1.

TABLE 1

Identification of immunoprecipitates

Substrate	*Hydrolysed by precipitate no.*
α-Asp-β-naphthylamide	5
Gly-Pro-β-naphthylamide	Distinct colour above the application well
γ-Glutamyl-β-naphthylamide	Not hydrolysed
Isomaltose	4
Lactose	3
Leu-β-naphthylamide	1
Lys-β-naphthylamide	1
Maltose	2 and 4
Naphthol AS-MX phosphoric acid	Not hydrolysed
Sucrose	4
Trehalose	Not hydrolysed

After crossed immunoelectrophoresis the immunoplates were incubated at 37 °C for 20 min with each substrate and combined with the appropriate colour-developing technique. Numbers refer to the immunoprecipitates in Fig. 1.

respectively. The intestinal lactase (β-galactosidase, EC 3.2.1.23) is represented by precipitate 3. An intestinal brush border aminopeptidase which releases acid amino acids from a peptide chain was reported by Andria *et al.* in 1976 (aspartate aminopeptidase, EC 3.4.11.7) and precipitate 5 represents this enzyme. Three of the substrates used were not hydrolysed by any precipitate, indicating that antibodies against alkaline phosphatase (EC 3.1.3.1), γ-glutamyl transpeptidase (γ-glutamyltransferase, EC 2.3.2.2) and trehalase (EC 3.2.1.28) were not raised during the preparation of the antibodies. Neither was glycyl-prolyl-β-naphthylamide hydrolysed by any of the five precipitates in Fig. 1, but a distinct staining appeared just above the application well. When a purified glycylprolyl dipeptidyl peptidase (EC 3.4.14.–) fraction was studied in the same system an immunoprecipitate was seen that hydrolysed glycyl-proline β-naphthylamide. These results may arise either because these enzymes are only present in low concentrations in the brush border antigen or because the antigens have poor antigenicity. The results obtained in this work demonstrate the usefulness of crossed immunoelectrophoresis in the study of brush border proteins solubilized with non-ionic detergent.

ISOLATION AND SEPARATION OF THE TRITON X-100-SOLUBILIZED FORMS OF THE INTESTINAL BRUSH BORDER HYDROLASES

With the aim of studying the native forms of the intestinal brush border peptidases we have started to purify microsomal aminopeptidase, aspartate aminopeptidase, glycylprolyl dipeptidyl peptidase and γ-glutamyl transpeptidase in their Triton X-100-solubilized forms from pig small intestine. Crossed immunoelectrophoresis was used as an analytical tool for following the progress of purification.

A technique adapted to preparative work on brush border membranes was reported by Sigrist *et al.* (1975). We have explored the usefulness of this technique for purifying brush border peptidases from pig intestine. Most of the activity of three of the peptidases studied is obtained in the membrane fraction (Table 2). The peptidases were solubilized by incubation in 1 % Triton X-100 for 1 h at 4 °C. The efficiency of the solubilization procedure was confirmed by ultracentrifugation (Table 2) and gel filtration (Fig. 2). The three peptidases are solubilized in high yields, with considerable purification (Table 2). The size homogeneity of the peptidases in the solubilized membrane fraction as tested by gel filtration is apparent (Fig. 2), as they are eluted from the gel filtration column as symmetrical peaks.

Crossed immunoelectrophoresis of the solubilized membrane fraction against the microsomal antiserum showed the same five precipitates as are demonstrated in Fig. 1. The hydrolases of the solubilized membrane fraction were isolated in separate fractions by different chromatographic systems (Fig. 3). All chromatographic procedures were undertaken in 0.1 % Triton X-100 to avoid aggregation of the peptidases.

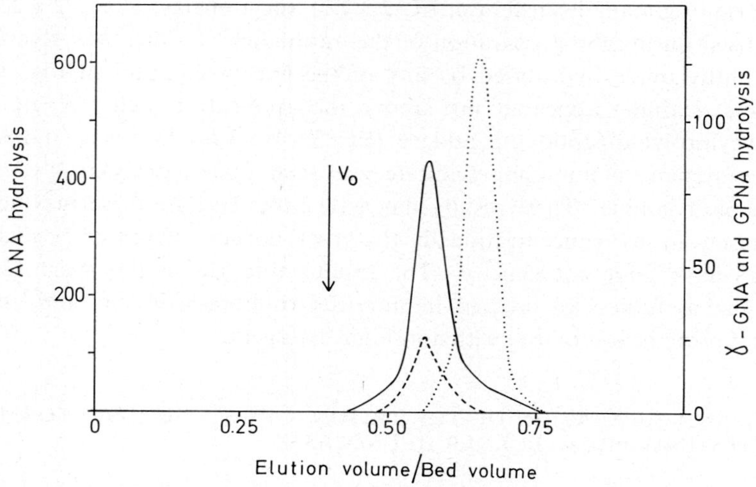

FIG. 2. Gel filtration of a solubilized membrane fraction on a column of Ultrogel AcA 34. —, alanyl-*p*-nitroanilide (ANA) hydrolysis;..., γ-glutamyl-3-carboxy-4-nitroanilide (GNA) hydrolysis; ---, glycylprolyl-β-naphthylamide (GPNA) hydrolysis. V_0: exclusion volume of the column.

TABLE 2

Distribution of three pig intestinal brush border peptidases in different fractions obtained with the brush border membrane preparation technique

Fraction	*ANA hydrolysis*		*γ-GNA hydrolysis*		*GPNA hydrolysis*	
	Recovery (%)	*Specific activity*	*Recovery (%)*	*Specific activity*	*Recovery (%)*	*Specific activity*
Homogenate	100	0.16	100	0.024	100	0.014
Supernatant after Ca^{2+} precipitation	80	0.31	73	0.044	67	0.024
Membrane fraction	57	1.2	38	0.12	36	0.07
Solubilized membrane fraction	23	1.4	38	0.45	36	0.15

Specific activity is expressed as μmol substrate hydrolysed at 37 °C/(min mg protein). Protein measurements as described by Wang & Smith (1975).

Assay conditions

ANA: 1 mM-L-alanyl-p-nitroanilide, 50 mM-Tris-HCl buffer, pH 7.3.

γ-GNA: 4 mM-L-γ-glutamyl-3-carboxy-4-nitroanilide, 0.1 M-glycyl-glycine, 0.1 M-Tris -HCl buffer, pH 8.25.

GPNA: 0.2 mM-glycyl-L-prolyl-β-naphthylamide, 50 mM-Tris-HCl, pH 8.0.

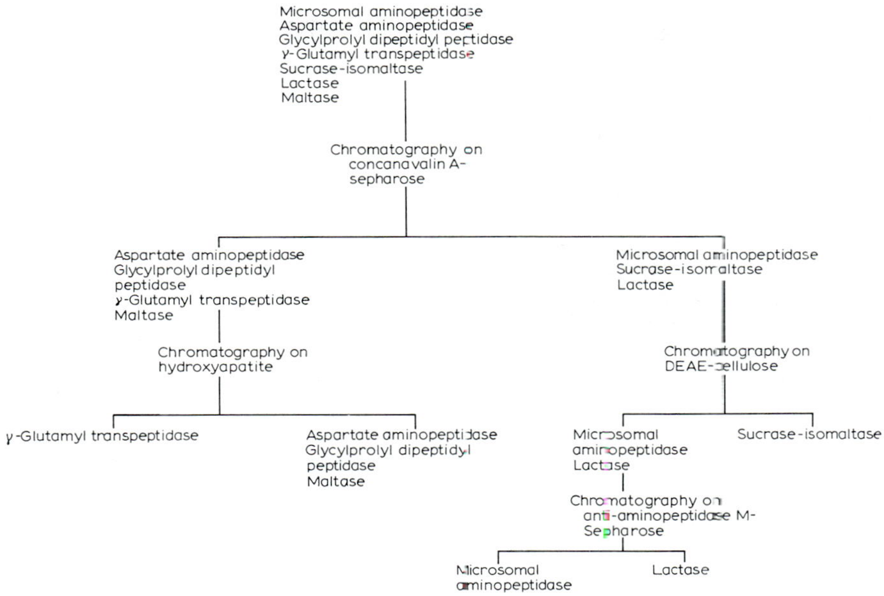

FIG. 3. Flow sheet for preparation of brush border hydrolases.

Chromatography on concanavalin A-Sepharose. The solubilized membrane fraction (Table 2) was chromatographed on concanavalin A-Sepharose equilibrated with 50 mM-Tris-HCl buffer, pH 8.0 at 4 °C. Almost none of the γ-glutamyl transpeptidase, glycylprocyl dipeptidyl peptidase, maltase and aspartate aminopeptidase adsorbed to the column. After the column was washed with buffer, microsomal aminopeptidase was eluted with 0.25 M-α-methyl mannoside at room temperature together with sucrase and lactase. The immunoelectrophoretic pattern of the eluted material is shown in Fig. 4.

Chromatography on hydroxyapatite. The non-adsorbed fraction obtained by concanavalin A-Sepharose chromatography containing the γ-glutamyl transpeptidase, glycylprolyl dipeptidyl peptidase and aspartate aminopeptidase, together with maltase, was dialysed against 5 mM-potassium phosphate buffer, pH 7.0, containing 0.2 M-potassium chloride. The solution was applied to a hydroxyapatite column equilibrated in the same buffer. Most of the γ-glutamyl transpeptidase did not adsorb to the column but could be washed out free from other known peptidase activities. Then the buffer was changed to 0.2 M-potassium phosphate, pH 7.0 containing 1% Triton X-100, and glycylprolyl dipeptidyl peptidase, aspartate aminopeptidase and maltase were eluted.

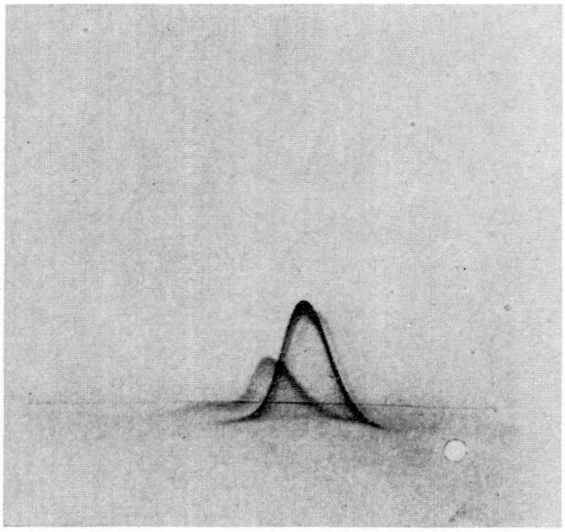

FIG. 4. Crossed immunoelectrophoresis of the fraction eluted with 0.25 M-α-methyl-mannoside in concanavalin A-Sepharose chromatography. Microsomal antiserum was used in immunoelectrophoresis.

Chromatography on DEAE-cellulose. The fraction eluted with 0.25 M-α-methyl mannoside from concanavalin A-Sepharose chromatography, and containing the microsomal aminopeptidase together with lactase and sucrase-isomaltase, was dialysed against 50 mM-Tris-HCl buffer, pH 9.0, and then passed through a column of DEAE-cellulose equilibrated in the same buffer. The column was eluted with a linear sodium chloride gradient in the buffer. The microsomal aminopeptidase was eluted together with the lactase at a lower sodium chloride concentration than the sucrase-isomaltase.

Chromatography on anti-aminopeptidase M-Sepharose. The microsomal aminopeptidase- and lactase-containing fraction obtained by DEAE-cellulose chromatography was subjected to chromatography on a column of anti-aminopeptidase M-Sepharose. The chromatographic material of the column was synthesized by coupling rabbit antibodies against aminopeptidase M to Sepharose 4B by the cyanogen bromide method (March *et al.* 1974). The lactase did not adsorb to the column but the microsomal aminopeptidase did. The latter was eluted by 0.1 M-sodium carbonate buffer, pH 10.8, containing 0.15 M-sodium chloride.

As described above, it is possible to obtain the intestinal γ-glutamyl transpeptidase and the microsomal aminopeptidase free of other brush border peptidases and glycosidases. The non-adsorbed fraction obtained by hydroxyapatite chromatography and containing the γ-glutamyl transpeptidase was free of immunoprecipitates as tested by crossed immunoelectrophoresis under the conditions stated in Fig. 1. The aminopeptidase preparation obtained after chromatography on anti-aminopeptidase M-Sepharose produced only one immunoprecipitate which corresponded to the aminopeptidase activity. The γ-glutamyl transpeptidase and microsomal aminopeptidase preparations must, however, be further analysed by other analytical methods before the purity of the preparations can be confirmed. So far we have not succeeded in separating glycylprolyl dipeptidyl peptidase, aspartate aminopeptidase and maltase from each other.

PROPERTIES OF THE TRITON X-100-SOLUBILIZED FORMS OF THE BRUSH BORDER PEPTIDASES

Hydrophobic properties. γ-Glutamyl transpeptidase, microsomal aminopeptidase and glycylprolyl dipeptidyl peptidase as obtained in the solubilized membrane preparation have been studied with respect to their solubility and hydrophobic properties. All three enzymes precipitated to 50% at 10% ammonium sulphate saturation and precipitation was complete at 37% saturation. The three peptidases bind strongly to alkyl-Sepharose and aminoalkyl-Sepharose columns, a property which underlines their hydrophobic nature. A further indication of their hydrophobic properties is the poor yield and asymmetrical peaks obtained in gel filtration experiments on Ultrogel AcA 34 when Triton X-100 is omitted from the elution buffer. The properties of the peptidases obtained from the solubilized membrane fraction therefore differ from the corresponding peptidases solubilized by proteolytic enzymes (Maroux *et al.* 1973; Kenny *et al.* 1976; Tate & Meister 1974). They also differ from purified microsomal aminopeptidase solubilized with Triton X-100 in respect of their solubility (Maroux & Louvard 1976; Kim & Brophy 1976). These differences may be explained by the different preparation procedures used.

Isoelectric focusing. The solubilized membrane fraction was subjected to isoelectric focusing in a pH gradient ranging from 3 to 10. The microsomal aminopeptidase appeared as a narrow symmetrical peak (Fig. 5) and the isoelectric point was determined to be 4.5. The glycylprolyl dipeptidyl peptidase was heterogeneous. A main component focused at pH 4.6 and a minor peak at pH 5.0. The γ-glutamyl transpeptidase appeared as a broad zone between pH 4.0 and 6.5.

Binding to lectins. The enterocyte brush border is rich in carbohydrate as indicated by its ability to react with lectins (Etzler & Branstrator 1974). The brush border enzymes so far purified have been reported to be glycoproteins

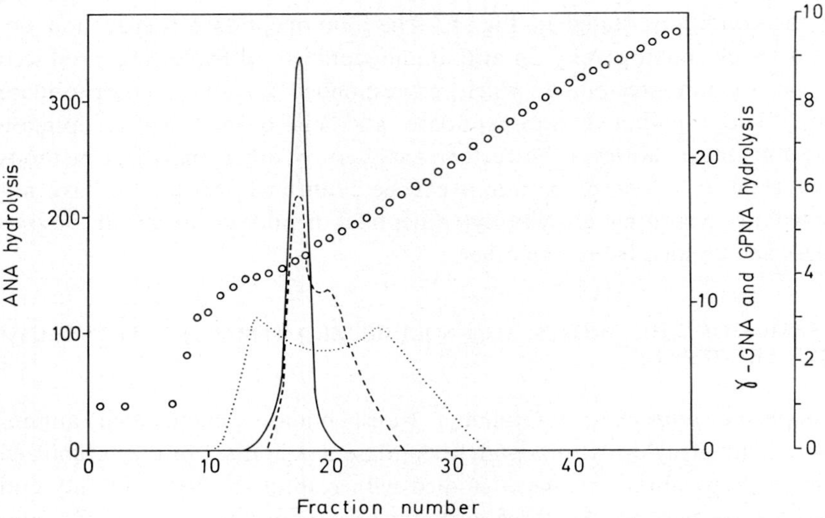

FIG. 5. Isoelectric focusing of a solubilized membrane fraction on an Ampholine 3-10 gradient. Enzyme activities are as indicated in Fig. 2. Circles indicate the pH.

(Cogoli *et al.* 1972; Maroux *et al.* 1973; Kim & Brophy 1976). The binding of the brush border peptidases to various lectins was therefore investigated (Table 3). The results obtained clearly indicate that all three peptidases studied are glycoproteins but that they differ in their carbohydrate moieties. It was also found that each peptidase may be heterogeneous in its carbohydrate composition, as only a fraction of the enzyme was bound to some of the lectins.

Influence of miscellaneous compounds. EDTA was found to inhibit microsomal aminopeptidase at a concentration of 1 mM. This finding is in agreement with that reported for the corresponding kidney enzyme (Lehky *et al.* 1973). γ-Glutamyl hydrazones of several α-keto acids (1 mM) inhibited γ-glutamyl transpeptidase, as did serine in combination with a borate buffer. This inhibition pattern has also been found for the kidney enzyme (Szewczuk & Connell 1965; Tate & Meister 1974). The serine–borate inhibition was not specific for γ-glutamyl transpeptidase as it also was found for microsomal aminopeptidase. When phenylmethyl sulphonylfluoride (2 mM) was added to the solubilized membrane fraction, glycylprolyl dipeptidyl peptidase was inactivated; this suggests that the enzyme is a serine peptidase, as also reported for the corresponding kidney enzyme (Kenny *et al.* 1976).

Immunological comparison. Aminopeptidase M prepared from pig kidney

TABLE 3

Binding of brush border peptidases to various lectins covalently bound to Sepharose

Lectin	*ANA hydrolysis*	*γ-GNA hydrolysis*	*GPNA hydrolysis*
Concanavalin A			
Not adsorbed	0.12	0.95	0.93
Eluted with 0.25 M-methyl mannoside	0.62	0.03	0.0
Soy bean			
Not adsorbed	0.14	0.20	0.58
Eluted with 1 M-galactose	0.85	0.53	0.19
Ricinus communis (mol. wt. 120 000)			
Not adsorbed	0.10	0.03	0.10
Eluted with 0.2 M-lactose	0.80	0.54	0.56
Ricinus communis (mol. wt. 60 000)			
Not adsorbed	0.05	0.02	0.06
Eluted with 0.2 M-lactose	0.87	0.30	0.20
Wheat germ			
Not adsorbed	0.57	0.53	0.42
Eluted with 0.25 M-*N*-acetylglucosamine	0.39	0.33	0.51

Columns were packed with the appropriate lectin–Sepharose material equilibrated in 50 mM-Tris-HCl, pH 8.0, containing 0.1% Triton X-100. A sample of the solubilized membrane fraction was applied. The columns were washed with the buffer and eluted with the same buffer containing the appropriate carbohydrate. Fractions were collected and analysed for peptidase activities. The figures given represent fractions of the applied peptidase activities. The peptidases were assayed as described in Table 2.

and a purified microsomal aminopeptidase preparation obtained from pig intestine were run together in a fused rocket immunoelectrophoresis (Svendsen 1973) against the antibodies prepared against the brush border protein solution. A distinct precipitate was conspicuous for each enzyme. The precipitates were fused, with only thin extensions under the point of fusion. This result indicates that there is a close structural relationship between intestinal microsomal aminopeptidase and kidney aminopeptidase M.

Heterogeneity of γ-glutamyl transpeptidase. A non-adsorbed fraction from concanavalin A-Sepharose chromatography was subjected to chromato-

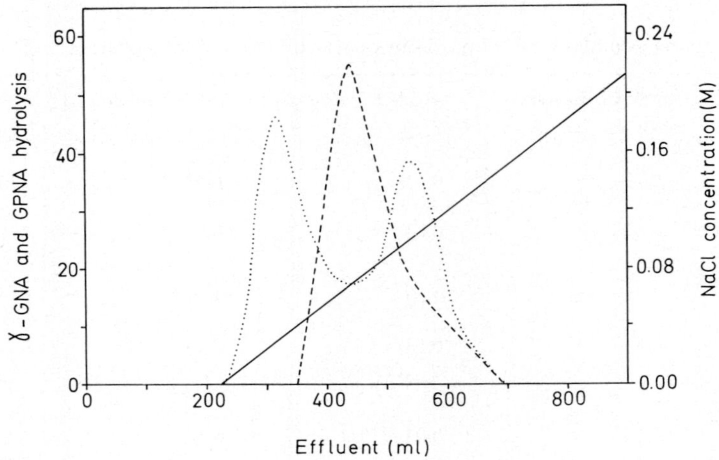

FIG. 6. A non-adsorbed fraction from concanavalin A-Sepharose chromatography chromatographed on a DEAE–cellulose column. Enzyme activities are as indicated in Fig. 2. The straight line shows the NaCl concentration.

graphy on a DEAE–cellulose column in 20 mM-Tris–HCl buffer, pH 8.0, containing 0.1% Triton X-100. The chromatography was developed with a linear sodium chloride gradient. Glycylprolyl dipeptidyl peptidase was eluted as a single peak but γ-glutamyl transpeptidase appeared as two partially separated peaks (Fig. 6). A similar chromatographic pattern was found for kidney γ-glutamyl transpeptidase (Tate & Meister 1974). When the applied material was pretreated with neuraminidase, glycylprolyl dipeptidyl peptidase was eluted at a lower salt concentration and the two γ-glutamyl transpeptidase peaks fused to a single peak, indicating that the two enzymes contain sialic acids and that the heterogeneity of the γ-glutamyl transpeptidase depends on differences in sialic acid content.

DISCUSSION

By analytical and preparative techniques we have demonstrated the existence of a microsomal and an aspartate aminopeptidase, a γ-glutamyl transpeptidase and a glycylprolyl dipeptidyl peptidase in the brush borders of the pig small intestine. These findings are in agreement with results obtained in other laboratories (Maroux *et al.* 1973; Andria *et al.* 1976; Cornell & Meister 1976; Kenny *et al.* 1976).

The specificities of the known peptidases in the gastrointestinal tract fit well with each other, making it possible for all types of known peptides to be hydrolysed to a form that can be effectively taken up by the enterocytes. The amino acids, especially the basic, aromatic and neutral ones, are released by the combined action of trypsins, pepsins, chymotrypsins and elastases, followed by the action of carboxypeptidases A and B, before absorption by the intestinal epithelium. Many dipeptides have been demonstrated to be rapidly absorbed from the intestinal lumen. So far only one dipeptidyl peptidase releasing such dipeptides from oligopeptides has been demonstrated in the small intestine. A similar kidney enzyme was reported to produce dipeptides composed of two neutral amino acids or dipeptides composed of aminoacyl-proline (Kenny *et al.* 1976). The dipeptide composed of two neutral amino acids can then be absorbed directly by the peptide transport system and subsequently hydrolysed by cytosol dipeptidases or hydrolysed by microsomal aminopeptidase to amino acids before transport. The aminoacyl-proline peptide must be transported directly into the enterocyte, because the only enzyme known to be capable of splitting the aminoacyl-proline bond is proline dipeptidase, which is localized in the cytoplasm (Norén *et al* 1977). Oligopeptides are probably simultaneously hydrolysed by dipeptidyl peptidase and microsomal and aspartate aminopeptidases, resulting in free amino acids and dipeptides which are then transported into the enterocyte. Which enzymes and which transport mechanisms are of greatest importance must depend on the compositon of the dietary and endogenous protein in the gastrointestinal tract.

It has been argued that the brush border peptidases are the enzymes of most importance in the final digestion of proteins, as these peptidases occur only in the small intestine. Our studies have shown that there are great similarities between the intestinal brush border peptidases and the corresponding particulate enzymes from other tissues, a finding which is also valid for cytosol dipeptidases. The conclusion is that the physiological role played by the peptidases depends more on localization in a particular organ than on differentiation of the various enzymes for a special function.

ACKNOWLEDGEMENTS

We are indebted to Mrs T. Buhl and Miss A. Mørch for their skilful technical assistance. Dr E. Dabelsteen and Dr M. Danielsen performed some of the immunological studies. We thank Dr P. Lehky for giving us the pig kidney aminopeptidase M.

References

ANDRIA, G., MARZI, A. & AURICCHIO, S. (1976) α-Glutamyl-β-naphthylamide hydrolase of rabbit small intestine. Localization in the brush border and separation from other brush border peptidases. *Biochim. Biophys. Acta 419*, 42-50

BJERRUM, O. & LUNDAHL, P. (1973) Detergent-containing gels for immunological studies of solubilized erythrocyte membrane components, in *A Manual of Quantitative Immunoelectrophoresis* (Axelsen, N. H., Krøll, J. & Weeke, B., eds.), pp. 139-143, Universitetsforlaget, Oslo

COGOLI, A., MOSIMANN, H., VOCK, C., VON BALTHAZAR, A. K. & SEMENZA, G. (1972) A simplified procedure for the isolation of the sucrase-isomaltase complex from rabbit intestine. *Eur. J. Biochem. 30*. 7-14

CORNELL, J. S. & MEISTER, A. (1976) Glutathione and γ-glutamyl cycle enzymes in crypt and villus tip cells of rat jejunal mucosa. *Proc. Natl. Acad. Sci. U.S.A. 73*, 420-422

CRITCHLEY, D. R., HOWELL, K. E. & EICHHOLZ, A. (1975) Solubilization of brush borders of hamster small intestine and fractionation of some of the components. *Biochim. Biophys. Acta 394*, 361-376

EICHHOLZ, A. & CRANE, R. K. (1974) Isolation of plasma membranes from intestinal brush borders, in *Methods in Enzymology*, vol. 31 (Fleischer, S. & Packer, L., eds.), pp. 123-134, Academic Press, New York

ETZLER, M. E. & BRANSTRATOR, M. L. (1974) Differential localization of cell surface and secretory components in rat intestinal epithelium by use of lectins. *J. Cell Biol. 62*, 329-343

HARBOE, N. & INGILD, A. (1973) Immunization, isolation of immunoglobulins, estimation of antibody titre, in *A Manual of Quantitative Immunoelectrophoresis* (Axelsen, N. H., Krøll, J. & Weeke, B., eds.), pp. 161-164, Universitetsforlaget, Oslo

JOSEFSSON, L., SJÖSTRÖM, H. & NORÉN, O. (1977) Intracellular hydrolysis of peptides, in this volume, pp. 199-206

KENNY, A. J., BOOTH, A. G., GEORGE, S. G., INGRAM, J., KERSHAW, D., WOOD, E. J. & YOUNG, A. R. (1976) Dipeptidyl peptidase IV, a kidney brush border serine peptidase. *Biochem. J. 155*, 169-182

KIM, Y. S. & BROPHY, E. J. (1976) Rat intestinal brush border membrane peptidases. 1. Solubilization, purification and physicochemical properties of two different forms of the enzymes. *J. Biol. Chem. 251*, 3199-3205

LEHKY, P., LISOWSKI, J., WOLF, D. P., WACKER, H. & STEIN, E. A. (1973) Pig kidney particulate aminopeptidase. A zinc metalloenzyme. *Biochim. Biophys. Acta 321*, 274-281

LOUVARD, D., MAROUX, S., VANNIER, C. & DESNUELLE, P. (1975) Topological studies on the hydrolases bound to the intestinal brush border membrane. 1. Solubilization by papain and Triton X-100. *Biochim. Biophys. Acta 375*, 236-248

MAESTRACCI, D., PREISER, H., HEDGES, T., SCHMITZ, J. & CRANE, R. K. (1975) Enzymes of the human intestinal brush border membrane. Identification after gel electrophoretic separation. *Biochim. Biophys. Acta 382*, 147-156

MARCH, S. C., PARIKH, I. & CUATRECASAS, P. (1974) A simplified method for cyanogen bromide activation of agarose for affinity chromatography. *Anal. Biochem. 60*, 149-152

MAROUX, S. & LOUVARD, D. (1976) On the hydrophobic part of aminopeptidase and maltases which bind the enzyme to the brush border membrane. *Biochim. Biophys. Acta 419*, 189-195

MAROUX, S., LOUVARD, D. & BARATTI, J. (1973) The aminopeptidase from hog intestinal brush border. *Biochim. Biophys. Acta 321*, 282-295

MATTHEWS, D. M. (1975) Intestinal absorption of peptides. *Physiol. Rev. 55*, 537-608

NORÉN, O., DABELSTEEN, E., SJÖSTRÖM, H. & JOSEFSSON, L. (1977) Histological localization of two dipeptidases in the pig small intestine and liver, using immunofluorescence. *Gastroenterology 72*, 87-92

SIGRIST, H., RONNER, P. & SEMENZA, G. (1975) A hydrophobic form of the small intestinal sucrase-isomaltase complex. *Biochim. Biophys. Acta 406*, 433-446

SVENDSEN, P. J. (1973) Fused rocket immunoelectrophoresis, in *A Manual of Quantitative Immunoelectrophoresis* (Axelsen, N. H., Krøll, J. & Weeke, B., eds.), pp. 69-70, Universitetsforlaget, Oslo

SZEWCZUK, A. & CONNELL, G. E. (1965) The reaction of iodoacetamide with the active center of γ-glutamyl transpeptidase. *Biochim. Biophys. Acta 105*, 352-367

TATE, S. & MEISTER, A. (1974) Interaction of γ-glutamyl transpeptidase with amino acids, dipeptides and derivatives and analogs of glutathione. *J. Biol. Chem. 249*, 7593-7602

TZAGOLOFF, A. & PENEFSKY, H. S. (1971) Extraction and purification of lipoprotein complexes from membranes, in *Methods in Enzymology*, vol. 22 (Jacoby, W. B., ed.), pp. 219-230, Academic Press, New York

URIEL, J. (1971) Colour reactions for the identification of antigen-antibody precipitates in gels, in *Methods in Immunology and Immunochemistry*, vol. 3 (Williams, C. A. & Chase, M. W., eds.), pp. 294-321, Academic Press, New York

WACKER, H., LEHKY, P., VANDERHAEGHE, F. & STEIN, E. A. (1976) On the subunit structure of particulate aminopeptidase from pig kidney. *Biochim. Biophys. Acta 429*, 546-554

WANG, C. S. and SMITH, R. L. (1975) Lowry determination of protein in the presence of Triton X-100. *Anal. Biochem. 63*, 414-417

Discussion

Maroux: With D. Louvard, Ch. Vannier and M. Semeriva, I have been working on the amphipathic structure of aminopeptidase from pig intestinal and renal brush border and its mode of integration into the membrane. Our aim was to answer the question: is the structure of the aminopeptidase and its position with respect to the membrane lipid bilayer compatible with its involvement in the transport of amino acids that arises from the hydrolysis of its substrates?

The brush border membrane spontaneously forms closed right-side-out vesicles very suitable for studying the mode of integration of surface proteins. An improved technique has been devised for the purification of large quantities of these vesicles (Louvard *et al.* 1973).

Almost all the surface proteins of these vesicles are hydrolases (aminopeptidase, maltases, sucrase, alkaline phosphatase, γ-glutamyl transferase) (Louvard *et al.* 1975*a*). They can be solubilized either by papain treatment or by neutral detergent without any loss of activity. Comparison of the molecular properties of the enzymes released by these two different treatments (Louvard *et al.* 1975*a*) suggests that, like other membrane proteins (Segrest *et al.* 1973; Spatz & Strittmatter 1971, 1973; Utermann & Simons 1974), the brush border hydrolases have an amphipathic structure. In other words, as shown in Fig. 1, they would be composed of a hydrophobic part ensuring their fixation to the lipid matrix and of a hydrophilic part bearing the hydrolytic activity. Triton X-100 would liberate the integral molecule which aggregates in the absence of detergent, whereas papain would liberate only the hydrophilic

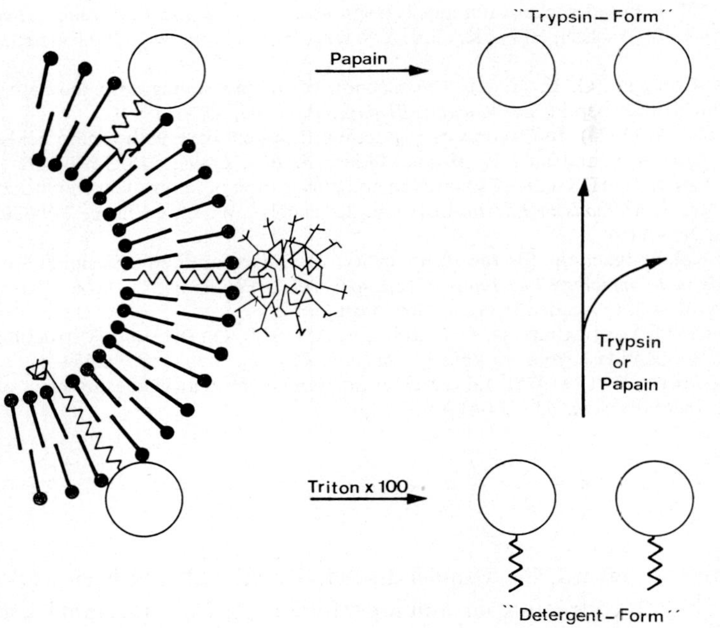

FIG. 1 (Maroux). Model proposed for the integration of intestinal hydrolases into the brush border membrane and their solubilization. (From Maroux *et al.* 1975.)

part which is soluble in aqueous medium in the absence of detergent. The detergent form always displays a shorter electrophoretic migration on polyacrylamide gel than the papain form (Louvard *et al.* 1975*a*). Papain, and in some cases trypsin, converts the detergent form into a form similar to the papain, or trypsin, form and would split the molecule into its two parts.

This scheme has been confirmed: the detergent forms of intestinal and renal brush border aminopeptidases and intestinal maltases have been purified and their hydrophobic and hydrophilic domains were separated and characterized from a trypsin hydrolysate (Maroux *et al.* 1973; Maroux & Louvard 1976; Vannier *et al.* 1976).

The hydrophilic parts (trypsin or papain forms) of the enzymes have exactly the same chromatographic behaviour in the absence of detergent as the entire molecule (detergent form) has in the presence of detergent, and the same high molecular weight (280 000 in the case of aminopeptidases). The hydrophobic parts have a low molecular weight (8 000/10 000 by gel electrophoresis on polyacrylamide gel in the presence of sodium dodecyl sulphate). Like lipids, they are extractable by the chloroform-methanol mixture of Folch. This

TABLE 1 (Maroux)

Amino acid composition of the hydrophobic peptides from renal and intestinal aminopeptidase and intestinal maltases

	Amino acid composition of the hydrophobic part of:		
	Renal aminopeptidase	*Intestinal aminopeptidase*	*Intestinal maltases* $(M_2 + M_3)$
Asx	6	3	5
Thr	4	3	6
Ser	7	5	6
Glx	6	5	6
Pro	2	2	3
Gly	9	9	4-5
Ala	11	11	5
½ Cys	0	0	0
Val	7	9	8
Met	–	0-1	1
Ile	9	9	4
Leu	10	10	16
Phe	3	3	6
Tyr	2	4	1-2
Lys	4-5	5	4
His	1	1	0-1
Arg	3	1	3
Total number* of residues	85	80	79-82
Total weight*	8970	8460	–
Polarity index*	37	35	35

* Tryptophan not included

hydrophobic character can be explained by their high content of apolar residues, as given in Table 1. The hydrophobic parts of the aminopeptidases from both origins have an identical composition in apolar residues, except for valine and tyrosine, but they differ completely from the hydrophobic parts of the maltases. This suggests that the hydrophobic structure is characteristic of hydrolases. This structure would be responsible for the specificity of the hydrolases in the transport process of the degradation products released by the hydrophilic part.

This hypothesis is supported by our finding that the intestinal aminopeptidase spans the membrane (Louvard *et al.* 1976*a*). Fig. 2 shows the experimental procedure we used to specifically label the hydrophobic part of the aminopeptidase, supposing that it spans the membrane.

(1) Using a preparation of brush borders as starting material (Hopfer *et al.*

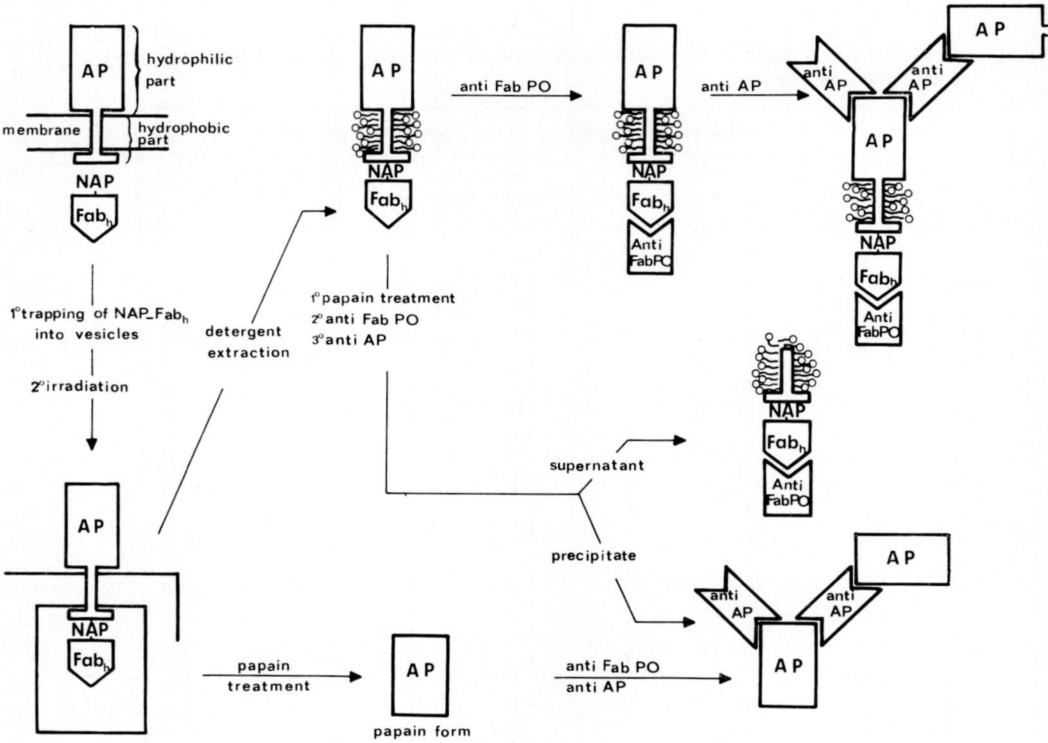

FIG. 2 (Maroux). Experimental procedure for labelling the part of aminopeptidase spanning the membrane. AP, aminopeptidase; Fab/$_H$, Fab fragment from human IgG, myeloma protein Zeg; NAP-, 2-nitro-4-azidophenyl group; NAP-Fab/$_H$, 2-nitro-4-azidophenyl group covalently attached to Fab/$_H$; anti-AP, IgG 'anti-papain form' of aminopeptidase; anti Fab-PO, sheep Fab anti-human IgG labelled with peroxidase. (From Louvard *et al.* 1976*a*.)

1973), closed right-side-out membrane vesicles can be obtained. It is then easy to trap into these vesicles a high molecular weight reagent added to the medium while vesiculization proceeds. The 4-fluoro-3-nitrophenyl azide was covalently attached to the Fab fragment of a human immunoglobulin to produce a photogenerated macromolecular reagent.

(2) After extensive washing, photolysis induces the cross-linking of the trapped reagent to accessible proteins.

(3) The membrane proteins are solubilized by detergent or papain.

(4) The human Fab bearing the photogenerated reagent is labelled by its monospecific antibody coupled to peroxidase.

(5) Aminopeptidase is then specifically precipitated by its antibody. The

extent of the aminopeptidase photolabelling is determined by peroxidase activity in the precipitate. The specific labelling of the hydrophobic part corresponds to the difference between the labelling of aminopeptidase after detergent and papain solubilization.

The results obtained clearly show that the trapped reagent cannot react with the hydrophilic part of the molecule, whereas the hydrophobic part is strongly labelled. Thus the hydrophobic part of aminopeptidase spans the membrane and is composed of two sections: one embedded within the lipid matrix of the membrane and one at the inner face of the membrane.

An original quantitative immunological technique has been used for evaluating the degree of homology existing between the hydrophilic parts of renal and intestinal aminopeptidases and their position with respect to the membrane (Louvard *et al.* 1975*b*, 1976*b*; Vannier *et al.* 1976). The results show that, as indicated in Fig. 1, in both cases the hydrophilic parts emerge almost entirely from the membrane: at least 80% of their surface is accessible to their specific antibodies.

The two enzymes have 50% of their antigenic determinants in common. They are located in the active site region and in the region masked by integration into the membrane.

In conclusion, the structure of the aminopeptidase and its mode of integration into the membrane are compatible with its involvement in the transport of amino acids resulting from hydrolysis. Two facts are particularly important and argue in favour of this hypothesis: (1) aminopeptidase is a transmembrane protein; (2) the hydrophobic part embedded in the lipid matrix, which is the region of the molecule that could be involved in the transport process, seems specific to the enzyme, as is also the case for the hydrolytic active site region in the hydrophilic part. But of course we now have to find direct proof showing that hydrolysis of peptides by aminopeptidase is of the 'intramembranous' type (see Matthews, this volume, pp. 5–14).

Semenza: Dr Norén, did you test for aryl-β-glucosidase activities such as phlorizin hydrolase (phloretin-glucosidase, EC 3.2.1.62) in your immunoelectrophoretogram studies? I would expect it to go along with lactase, which is bound to it, forming the so-called β-glycosidase complex (Schlegel-Haueter *et al.* 1972; Leese & Semenza 1973).

Norén: We didn't test that.

Semenza: When membrane proteins are solubilized with proteases one is always aware of the resulting chemical modification. I would like to give a similar word of caution about the use of detergents. It is perfectly true that anionic detergents such as Triton seem to alter the tertiary and quarternary structure of proteins less than sodium dodecyl sulphate does, but there is no fundamental difference between detergents. It is only a quantitative difference and one may well dissociate subunits even with Triton.

Kenny: I agree with Dr Norén about the immunological identity of dipeptidyl peptidase IV in both gut and kidney. Dr Maroux, are the two hidden antigenic sites part of the papain-released portion of the aminopeptidase?

Maroux: Yes.

Norén: You presented a sedimentation coefficient for the solubilized form of the aminopeptidase. Did you do this analysis in the presence of detergent?

Maroux: Yes.

Semenza: In the presence of appropriate concentrations of detergents we have almost the same sedimentation constant and Stokes' radius for the Triton-solubilized sucrase-isomaltase complex as for the papain-solubilized complex.

Kim: Dr Maroux, you said that the hydrophobic fragment has a molecular weight of about 9000. Does the calculation of the membrane thickness account for that molecular weight of the fragment?

Maroux: If we assume that the structure of this fragment is 100% α-helix, the length obtained is two or three times the thickness of the membrane.

References

HOPFER, U. H., NELSON, K., PERROTTO, J. & ISSELBACHER, K. J. (1973) Glucose transport in isolated brush border membrane from rat small intestine. *J. Biol. Chem. 248*, 25-32

LEESE, H. J. & SEMENZA, G. (1973) On the identity between the small intestinal enzymes phlorizin-hydrolase and glycosyl-ceramidase. *J. Biol. Chem. 248*, 8170-8173

LOUVARD, D., MAROUX, S., BARATTI, J., DESNUELLE, P. & MUTAFSCHIEV, S. (1973) On the preparation and some properties of closed membrane vesicles from hog duodenal and jejunal brush border. *Biochim. Biophys. Acta 291*, 747-763

LOUVARD, D., MAROUX, S., VANNIER, Ch. & DESNUELLE, P. (1975*a*) Topological studies on the hydrolases bound to the intestinal brush border membrane. I. Solubilization by papain and Triton X-100. *Biochim. Biophys. Acta 375*, 236-248

LOUVARD, D., MAROUX, S. & DESNUELLE, P. (1975*b*) Topological studies on the hydrolases bound to intestinal brush border membrane. II. Interaction of free and bound aminopeptidase with a specific antibody. *Biochim. Biophys. Acta 289*, 389-400

LOUVARD, D., SÉMÉRIVA, M. & MAROUX, S. (1976*a*) The brush border intestinal aminopeptidase, a transmembrane protein as probed by macromolecular photolabelling. *J. Mol. Biol. 106*, 1023-1035

LOUVARD, D., VANNIER, Ch., MAROUX, S., PAGÈS, J. M. & LAZDUNSKI, C. (1976*b*) Quantitative immunochemical technique for evaluation of the extent of integration of membrane proteins and of protein conformational changes and homologies. *Anal. Biochem. 76*, 83-94

MAROUX, S. & LOUVARD, D. (1976) On the hydrophobic part of aminopeptidase and maltases which bind the enzyme to the intestinal brush border membrane. *Biochim. Biophys. Acta 419*, 189-195

MAROUX, S., LOUVARD, D. & BARATTI, J. (1973) The aminopeptidase from hog intestinal brush border. *Biochim. Biophys. Acta 321*, 282-295

MAROUX, S., LOUVARD, D. & DESNUELLE, P. (1975) in *Biological Membranes. Neurochemistry*

(*Proc. 10 FEBS Meet., Paris, 1975*), vol. 41 (Montreuil, J. & Mandel, P., eds.), pp. 55-69, North-Holland, Amsterdam

SCHLEGEL-HAUETER, S., HORE, P., KERRY, K. R. & SEMENZA, G. (1972) The preparation of lactase and glucoamylase of rat small intestine. *Biochim. Biophys. Acta 258*, 506-519

SEGREST, J. P., KAHANE, I., JACKSON, R. L. & MARCHESI, V. T. (1973) Major glycoprotein of the human erythrocyte membrane: evidence for an amphipathic molecular structure. *Arch. Biochem. Biophys. 155*, 167-183

SPATZ, L. & STRITTMATTER, P. (1971) A form of cytochrome b_5 that contains an additional hydrophobic sequence of 40 amino acid residues. *Proc. Natl. Acad. Sci. U.S.A. 68*, 1042-1046

SPATZ, L. & STRITTMATTER, P. (1973) A form of reduced nicotinamide adenine dinucleotide cytochrome b_5 reductase containing both the catalytic site and additional hydrophobic membrane binding segment. *J. Biol. Chem. 248*, 793-799

UTERMANN, G. & SIMONS, K. (1974) Studies on the amphipathic nature of the membrane proteins in Semliki Forest virus. *J. Mol. Biol. 85*, 569-587

VANNIER, Ch., LOUVARD, D., MAROUX, S. & DESNUELLE, P. (1976) Structural and topological homology between porcine intestinal and renal brush border aminopeptidase. *Biochim. Biophys. Acta 455*, 185-199

Intracellular hydrolysis of peptides

LARS JOSEFSSON, HANS SJÖSTRÖM and OVE NORÉN

Department of Biochemistry C, University of Copenhagen, Copenhagen

Abstract Purification of the first dipeptidases, glycylleucine dipeptidase (EC 3.4.13.2) and proline dipeptidase (EC 3.4.13.9), from intestine, have shown them to be true dipeptidases, hydrolysing only dipeptides in their *laevo* form. Although they have quite different specificities they show great similarities in their chemical and physicochemical properties. Specific antibodies against the two dipeptidases have been raised and used in combination with the double-layer immunofluorescent staining technique to study their histological localization in the small intestine. The results conclusively demonstrated that both glycylleucine dipeptidase and proline dipeptidase are exclusively located in the cytosol of the enterocytes. This location fits well with the current idea of transport and hydrolysis of dipeptides in the intestine where the dipeptides are taken up by a specific transport mechanism and are hydrolysed intracellularly. Although it is tempting to add a specific role for the dipeptidases of the enterocytes in the final digestion of exogenous proteins, studies so far have shown their identity with the corresponding dipeptidases of other tissues. It is therefore suggested that their role in the digestion process may be based entirely on their abundance in the intestine.

The hydrolysis of peptides in animal tissues has been the subject of increasing interest for the last decade. Thus extensive studies of the digestion and absorption of the peptides originating from the dietary proteins and of the various peptidases of the enterocyte have given us significant new information about these processes (recently reviewed by Matthews & Payne 1975). Our knowledge about the intracellular proteolytic enzymes involved in the catabolism of the endogenous proteins of the cells has also widened during the last years, and lately special attention has been focused on their role in the regulation of intracellular enzyme concentrations (Katunuma 1975). A topic of current interest in intestinal enzymology is whether the enterocytes, as receivers and processors of the exogenous proteins, contain peptidases specially adapted for the digestion process which are different to the peptidases in other cells enrolled only in endogenous protein metabolism.

TABLE 1

Subcellular distribution of isolated and characterized peptidases in the enterocyte

Cytosol	Glycylleucine dipeptidase[1] (EC 3.4.13.2) Glycylglycine dipeptidase[2] (EC 3.4.13.1) Proline dipeptidase[1] (EC 3.4.13.9) Aminotripeptidase[2]
Lysosomes	Cathepsins A, B, C and D[3]
Brush border	Aspartate aminopeptidase [4] (EC 3.4.11.7) Glycylprolyl dipeptidylpeptidase[5] γ-Glutamyl transpeptidase[6] (EC 2.3.2.2) Microsomal aminopeptidase[7] (EC 3.4.11.2) Aminotripeptidase[2]

[1] Josefsson & Sjöström (1966)
[2] Peters (1970)
[3] Peters (1975)
[4] Andria *et al.* (1976)
[5] Kenny *et al.* (1976)
[6] Cornell & Meister (1976)
[7] Maroux *et al.* (1973)

The enterocyte is rich in peptidases; those which have been isolated and characterized are listed in Table 1, which also gives their subcellular distribution. The dipeptidases are responsible for most of the soluble-peptide-splitting activity in the cell. Over 10 years ago we began developing a simple and rapid spectrophotometric assay for dipeptidase activity (Josefsson & Lindberg 1965) and since then we have been very interested in these enzymes. Our studies have included investigations of their number, distribution, development, specificity and molecular properties, as well as their behaviour under physiological and pathological conditions. Our earlier results in this field were reviewed recently (Lindberg *et al.* 1975; Lindberg 1975).

The purification of glycylleucine dipeptidase (EC 3.4.13.2) from the pig (Norén *et al.* 1973) and the monkey intestine (Das & Radhakrishnan 1972) and of proline dipeptidase (EC 3.4.13.9) from pig intestine (Sjöström *et al.* 1973) made it possible to confirm that these enzymes are true dipeptidases, and are thus unable to hydrolyse any substrates other than dipeptides in their *laevo* form. Studies of their specificity revealed that glycylleucine dipeptidase, which has a wide specificity, is responsible for most of the intracellular dipeptidase activity, while proline dipeptidase has a rather narrow specificity, capable of splitting only dipeptides with the configuration X-Pro or X-Hyp. In our view, proline dipeptidase is the only known intestinal enzyme that can hydrolyse amino acyl-proline peptide bonds. Therefore this enzyme must be important in the final digestion process, since quantitatively important dietary proteins such as gliadins and collagens are rich in proline and hydroxyproline. Quantitative evaluation of the specificity data on these two dipeptidases in relation to the total soluble dipeptidase activity of the enterocyte also indicates that the cell must contain additional soluble-dipeptide-splitting enzymes with activity against glycylglycine and against dipeptides composed of aromatic and basic amino acids. Whether these enzymes represent other true dipeptidases or other peptidases is still unknown. Although the activities can be separated from the purified dipeptidases (Donlon & Fottrell 1973), a final answer has to await their further purification and characterization.

In spite of their different specificities, the two purified dipeptidases show great similarity in their chemical and physicochemical properties. Determination of their molecular weights by analytical ultracentrifugation as well as gel chromatography gave very similar figures for the two different enzymes in their native form, i.e. within a range of 104 000–115 000. Studies of their behaviour under denaturing conditions, using the same techniques, further revealed that each enzyme consists of two subunits with the same molecular weights, i.e. within a range of 52 000–63 000. The same close similarities between the two dipeptidase molecules were observed in studies of their Stokes radius, sedimentation coefficient and intrinsic viscosity, as shown in Table 2, which summarizes their physicochemical properties. A hydration of 20% was assumed for the dipeptidase molecules in solution, which is a common figure for globular proteins, and their physicochemical data were treated to get information about the shape of the molecules. This treatment showed that the two dipeptidases behave hydrodynamically as rigid prolate ellipsoids, with an axial ratio close to 5.

Both enzymes are inactivated by thiol reagents such as Hg^{2+} and *p*-hydroxymercuribenzoate, and they belong therefore to the SH enzymes (Norén 1974; Sjöström 1974). They are also inhibited by various SH-compounds and by EDTA, indicating that they are dependent on a metal ion for their proper function. Experiments on the proline dipeptidase with 1, 10-phenanthroline suggest that this metal ion may be zinc. The two enzymes follow Michaelis-Menten kinetics. Since their two subunits seem identical this finding suggests that each molecule has two similar active sites which are not cooperative in their function.

In chemical structure there is a slight difference in the amino acid composition of the two enzymes, with regard to their arginine, cysteine, glycine and lysine content (Norén & Sjöström 1974; Sjöström & Norén 1974). This is also reflected in their different pI values, which were found to be 5.1 for the glycylleucine dipeptidase and 4.4 for the proline dipeptidase. Determination

TABLE 2

Physicochemical properties of intestinal glycylleucine dipeptidase and proline dipeptidase

	Glycylleucine dipeptidase	*Proline dipeptidase*
Molecular weight		
Native	115 000	113 000
Denatured	52 000	53 000
$s^{\circ}_{20,w}$(S)	5.7	6.5
Stokes radius (nm)	4.2	4.2
$[\eta]$ (ml/g)	5.4	5.8
Axial ratio (20% hydration)	4.9	5.0

of free SH groups in the enzymes showed similar figures for the two molecules. In regard to the number of half-cystine residues of the molecules this means, however, that there are no or rather few disulphide bonds in the glycylleucine dipeptidase, while proline dipeptidase contains about 15 disulphide bonds per molecule. This difference signifies that the proline dipeptidase molecule has a much more rigid conformation than glycylleucine dipeptidase. That such a difference exists in the conformation of the two molecules is supported also by the much higher tendency towards aggregation observed for the proline dipeptidase molecules when stored in concentrated solutions, as well as by a higher enzymic stability in aqueous solutions. Furthermore, no N-terminal amino acid was detected in proline dipeptidase, while proline was easily detected as the N-terminal amino acid residues of glycylleucine dipeptidase.

With pure enzyme preparations, specific antisera against the two dipeptidases were raised in rabbits and the respective anti-dipeptidase IgG fractions of the sera were purified, using ammonium sulphate fractionation, ion-exchange chromatography and affinity chromatography. The immunological reactivity of the two anti-dipeptidase IgG fractions against the dipeptidases was tested by rocket immunoelectrophoresis and this showed that they were monospecific. Furthermore, no immunological cross-reaction was observed between the two enzymes (Fig. 1).

In the current discussion of the absorption and hydrolysis of the final products of dietary proteins, the physiological localization of the dipeptidases has been argued to be in the brush border (Ugolev 1972). The fact that dipeptidases are distributed mainly in the soluble fraction of the mucosa homogenate would then be explained by their rapid release during homogenization. Such a membrane hydrolysis of peptides, similar to the intestinal handling of the disaccharidases (Crane *et al.* 1970), has been impossible to exclude because previous experimental methods did not permit definite conclusions. An

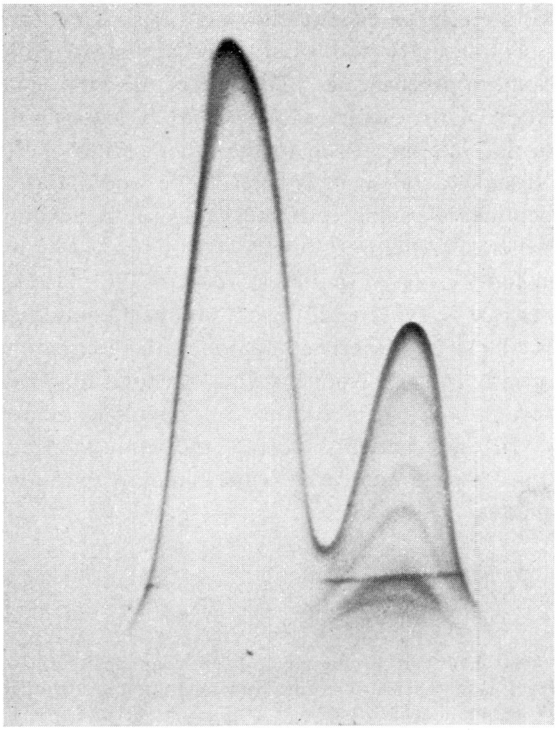

FIG. 4. Comparison of intestinal and brain dipeptidases by fused rocket immunoelectrophoresis. Purified intestinal glycylleucine dipeptidase (left well) and partially purified brain glycylleucine dipeptidase (right well) were run against a polyvalent antiserum raised from a partially purified intestinal glycylleucine dipeptidase preparation.

specific anti-dipeptidase IgG fractions against the intestinal dipeptidases to see whether they had immunological reactivity against corresponding dipeptidases of other tissues. Glycylleucine dipeptidase activity and proline dipeptidase activity from various tissues of the pig, including brain, liver and uterine tube, were therefore partially purified. The different dipeptidase preparations, together with the corresponding purified intestinal dipeptidases, were then subjected to fused rocket immunoelectrophoresis against polyvalent intestinal antiserum. Total identity between intestinal dipeptidases and the corresponding activity from the other tissues was observed in all cases. This is illustrated in Fig. 4, which shows an unbroken precipitation line between the two rockets. In addition, histological studies on liver tissue with the double-layer immunofluorescent staining technique, using the rabbit anti-dipeptidase IgG fractions raised with the enterocyte dipeptidases, showed specific fluorescence in the cytoplasm of the hepatocytes and epithelial cells of the bile ducts. Therefore it seems unlikely that the dipeptidases of the enterocytes are structurally different to the dipeptidases found in other tissues. The role of the enterocyte dipeptidases in the final digestion of the dietary proteins may be based only on their abundance in the intestine in comparison to the other tissues of the body (Josefsson *et al.* 1968). If such a statement is correct one would expect quantitative variations in dipeptidases in the intestinal mucosa, depending on the influx of exogenous proteins. Attempts were made to study this by varying the amount of protein food given to the intestine (Kim *et al.* 1973). However, a more definite answer can be expected if the amount of dipeptidases in normal functioning sections of the intestine is compared with the amount in sections of the intestine surgically isolated from the flow of exogenous proteins. Such studies are now in progress in our laboratory on patients who have been operated on for obesity. In these patients most of the jejunum–ileum has been withdrawn from food processing, and we can thus compare digestion-active and digestion-inactive mucosa.

References

ANDRIA, G., MARZI, A. & AURICCHIO, S. (1976) α-Glutamyl-β-naphthylamide hydrolase of rabbit small intestine. Localization in the brush border and separation from other brush border peptidases. *Biochim. Biophys. Acta 419*, 42-50

CORNELL, J. S. & MEISTER, A. (1976) Glutathione and γ-glutamyl cycle enzymes in crypt and villus tip cells of rat jejunal mucosa. *Proc. Natl. Acad. Sci. U.S.A. 73*, 420-422

CRANE, R. K., MALATHI, P., CASPARY, W. F. & RAMASWAMY, K. (1970) Evidence for a second glucose transport system in hamster small intestine for glucose released by brush border digestive enzymes. *Fed. Proc. 29*, 595-600

DAS, M. & RADHAKRISHNAN, A. N. (1972) Substrate specificity of a highly active dipeptidase from monkey small intestine. *Biochem. J. 128*, 463-465

DONLON, J. & FOTTRELL, P. F. (1973) Purification and characterization of one of the forms of peptide hydrolases from guinea-pig small intestinal mucosa. *Biochim. Biophys. Acta 327*, 425-436

JOSEFSSON, L. & LINDBERG, T. (1965) Intestinal dipeptidases I. Spectrophotometric determination and characterization of dipeptidase activity in pig intestinal mucosa. *Biochim. Biophys. Acta 105*, 149-161

JOSEFSSON, L. & SJÖSTRÖM, H. (1966) Intestinal dipeptidases IV. Studies on the release and subcellular distribution of intestinal dipeptidases of the mucosa cells of the pig. *Acta Physiol. Scand. 67*, 27-33

JOSEFSSON, L., NORÉN, O. & SJÖSTRÖM, H. (1968) Comparison of dipeptidase activity in different tissues of the pig. *Acta Physiol. Scand. 72*, 108-114

KATUNUMA, N. (1975) Regulation of intracellular enzyme levels by limited proteolysis. *Rev. Physiol. Biochem. Exp. Pharmacol. 72*, 83-104

KENNY, A. J., BOOTH, A. G., GEORGE, S. G., INGRAM, J., KERSHAW, D., WOOD, E. J. & YOUNG, A. R. (1976) Dipeptidyl peptidase IV. A kidney brush border serine peptidase. *Biochem. J. 155*, 169-182

KIM, Y. S., MCCARTHY, D. M., LANE, W. & FONG, W. (1973) Alterations in the levels of peptide hydrolases and other enzymes in brush border and soluble fractions of rat small intestinal mucosa during starvation and refeeding. *Biochim. Biophys. Acta 321*, 262-273

LINDBERG, T. (1975) Peptidases in intestinal disease, in *Peptide Transport in Protein Nutrition* (Matthews, D. M. & Payne, J. W., eds.), pp. 268-282, North-Holland, Amsterdam

LINDBERG, T., NORÉN, O. & SJÖSTRÖM, H. (1975) Peptidases of the intestinal mucosa, in *Peptide Transport in Protein Nutrition* (Matthews, D. M. & Payne, J. W., eds.), pp. 204-242, North-Holland, Amsterdam

MAROUX, S., LOUVARD, D. & BARATTI, J. (1973) The aminopeptidase from hog intestinal brush border. *Biochim. Biophys. Acta 321*, 282-295

MATTHEWS, D. M. (1972) Rates of peptide uptake by small intestine, in *Peptide Transport in Bacteria and Mammalian Gut* (*Ciba Found. Symp. 4*), pp. 71-88, Associated Scientific Publishers, Amsterdam

MATTHEWS, D. M. & PAYNE, I. W. (eds.) (1975) *Peptide Transport in Protein Nutrition*, North-Holland, Amsterdam

NORDSTRÖM, C., DAHLQUIST, A. & JOSEFSSON, L. (1968) Quantitative determination of enzymes in different parts of the villi and crypts of rat small intestine. Comparison of alkaline phosphatase, disaccharidases and dipeptidases. *J. Histochem. Cytochem. 15*, 713-721

NORÉN, O. (1974) Studies on a soluble dipeptidase from pig intestinal mucosa. Enzymatic properties. *Acta Chem. Scand. B 28*, 711-716

NORÉN, O. & SJÖSTRÖM, H. (1974) Studies on a soluble dipeptidase from pig intestinal mucosa. Structural properties. *Acta Chem. Scand. B 28*, 787-792

NORÉN, O., SJÖSTRÖM, H. & JOSEFSSON, L. (1973) Studies on soluble dipeptidase from pig intestinal mucosa I. Purification and specificity. *Biochim. Biophys. Acta 327*, 446-456

NORÉN, O., DABELSTEEN, E., SJÖSTRÖM, H. & JOSEFSSON, L. (1977) Histological localization of two dipeptidases in the pig small intestine and liver, using immunofluorescence. *Gastroenterology 72*, 87-92

PETERS, T. J. (1970) The subcellular localization of di- and tripeptide hydrolase activity in guinea pig small intestine. *Biochem. J. 120*, 195-207

PETERS, T. J. (1975) Subcellular localisation of intestinal peptide hydrolases, in *Peptide Transport in Protein Nutrition* (Matthews, D. M. & Payne, J. W., eds.), pp. 243-267, North-Holland, Amsterdam

SJÖSTRÖM, H. (1974) Enzymatic properties of pig intestinal proline dipeptidase. *Acta Chem. Scand. B 28*, 802-808

SJÖSTRÖM, H. & NORÉN, O. (1974) Structural properties of pig intestinal proline dipeptidase. *Biochim. Biophys. Acta 359*, 177-185

SJÖSTRÖM, H., NORÉN, O. & JOSEFSSON, L. (1973) Purification and specificity of pig intestinal prolidase. *Biochim. Biophys. Acta 327*, 457-470

UGOLEV, A. M. (1972) Membrane digestion and peptide transport, in *Peptide Transport in Bacteria and Mammalian Gut (Ciba Found. Symp. 4)*, pp. 123-137, Associated Scientific Publishers, Amsterdam

For discussion, see pp. *217-219.*

Endopeptidases in the brush border of the kidney proximal tubule

A. JOHN KENNY

Department of Biochemistry, University of Leeds, Leeds

Abstract Among the five peptidases known to be located in the microvillus membrane of the renal proximal tubule are two enzymes with endopeptidase activity. Neutral endopeptidase, a zinc-dependent enzyme, has a broad specificity comparable to that of thermolysin, and like the latter may be specifically inhibited by phosphoramidon. Dipeptidyl peptidase IV, a serine enzyme, is very sensitive to inhibition by diisopropyl phosphorofluoridate. It is also capable of endopeptidase activity, hydrolysing bonds involving the carboxyl group of proline.

My purpose is to describe two enzymes present in the kidney microvillus membrane, each possessing endopeptidase activity and each, though differing in the nature of the active site, susceptible to specific inhibitors. These two enzymes, neutral endopeptidase and microvillus membrane serine peptidase (dipeptidyl peptidase IV), are members of a group of five peptidases now known to be located in the brush border membrane (Table 1). All of them (with the exception of aminopeptidase A) have been purified and studied extensively in several laboratories (for review, see Kenny 1977) and are known to be intrinsic glycoproteins of the microvillus membrane. Indeed, aminopeptidase M is the principal glycoprotein on the microvillus surface. Three of the group, aminopeptidase M, neutral endopeptidase and serine peptidase, are among the major proteins of the membrane, each contributing about 4% of the protein of the microvillus, equivalent to 500-800 copies of the protein monomer on each microvillus (Booth & Kenny 1976). Since each of the group possesses a different substrate specificity they form a powerful battery of peptide hydrolases capable of degrading oligopeptides to free amino acids and a few dipeptides, of sequence X-Pro. In fact, the hydrolytic potential of this group of enzymes seems disproportionately high when compared with the amount of susceptible substrate (e.g. peptide hormones) likely to be present in the glomerular filtrate. This discrepancy merits further investigation of alternative and additional roles for these proteins.

TABLE 1

Kidney microvillus membrane peptidases

Enzyme	*EC No.*	*Active site*	*Assay substrate*	*Inhibitors*	*Sp. act. in kidney micro-villi (nmol min^{-1} (mg protein)$^{-1}$)*	*Subunit mol. wt.*	*Proportion of micro-villus protein (%)*	*Molecules per microvillus*
Neutral endopeptidase	3.4.24.-	Zn^{2+}	[^{125}I]iodo-insulin B chain	Chelating agents, Phosphoramidon	28.4	95 000	4.1	760
Microvillus membrane serine peptidase (dipeptidyl peptidase IV)	3.4.21.- (or 3.4.14.-)	Serine	Gly-Pro-2-NNap	DipF	1310	130 000	3.9	530
Aminopeptidase M	3.4.11.2	Zn^{2+}	Leu-2-NNap	Chelating agents	2350	160 000	4.1	450
Aminopeptidase A	3.4.11.7	Uncertain (probably metal ion)	α-Glu-2-NNap	Chelating agents	955	–	–	–
γ-Glutamyl transferase	2.3.2.2	Uncertain	γ-Glu-2-NNap + Gly-Gly	Iodoacetamide, Serine-borate	809	2 subunits* 46 000 22 000	–	–

The specific activities are for pig or rabbit and the data are from George & Kenny (1973); Booth & Kenny (1974, 1976, and unpublished) and Kenny *et al.* (1976) except for that marked*, from Tate & Meister (1976). 2-NNap, 2-naphthylamide.

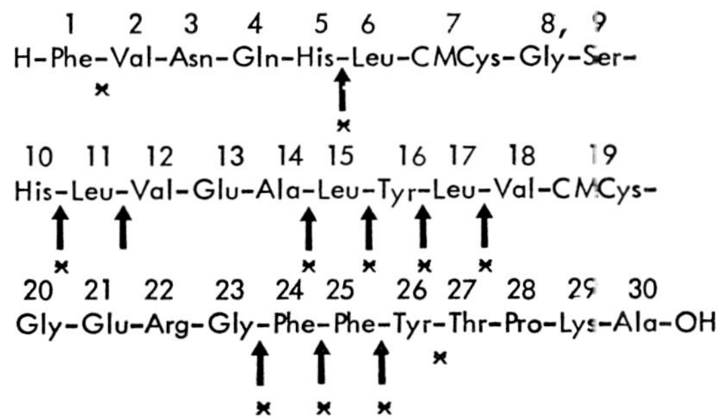

FIG. 1. Sequence of insulin B chain showing bonds hydrolysed by kidney neutral endopeptidase. Bonds attacked by the kidney enzyme are shown by arrows. For comparison those hydrolysed by protease I, *Aspergillus oryzae*, a microbial enzyme typical of the group EC 3.4.24.4, are marked with an asterisk.

NEUTRAL ENDOPEPTIDASE (EC 3.4.24.–)

The name served, in some early studies in this laboratory, to distinguish this activity from kidney lysosomal peptidases that hydrolysed the assay substrate, [^{125}I]iodo-insulin B chain, at pH 3.5 (Wong-Leung *et al.* 1968). The microvillus endopeptidase, which is maximally active at pH 6.0, has been investigated in some detail (Kerr & Kenny 1974*a*, *b*). In summary, it is a monomeric glycoprotein (10–20% carbohydrate) of 93 000 mol. wt. It contains one atom of Zn/mol, a finding that explains its sensitivity to inhibition by chelating agents and thiols. The bonds of insulin B chain found to be split by the purified enzyme are shown in Fig. 1.

In each case the bond hydrolysed is one involving the amino group of a hydrophobic amino acid residue. The same pattern of attack was observed with smaller peptides: Gly-Phe-Ala and Z-Gly-Phe-NH_2 were predictably hydrolysed, while Gly-Phe and Z-Gly-Phe were not attacked. The metal-dependence of this peptidase and this type of specificity are features shared with a group of microbial peptidases (see Matsubara & Feder 1971, for review) typified by thermolysin. A specific inhibitor, phosphoramidon, has been isolated from the culture medium of an actinomycetes: *Streptomyces tanashiensis* (Umezawa *et al.* 1972; Suda *et al.* 1973; Komiyama *et al.* 1975). Phosphoramidon, a hexose-phosphate derivative of a dipeptide, Leu-Trp (*N*-(α-L-rhamnopyranosyl-oxyhydroxyphosphinyl)-L-leucyl-L-tryptophan) has been found to be a potent competitive inhibitor of the kidney endopeptidase (Fig. 2, A. R. Young & A. J. Kenny, unpublished). The K_i value for phosphoramidon in the usual assay (containing 57 μM iodo-insulin B chain) was

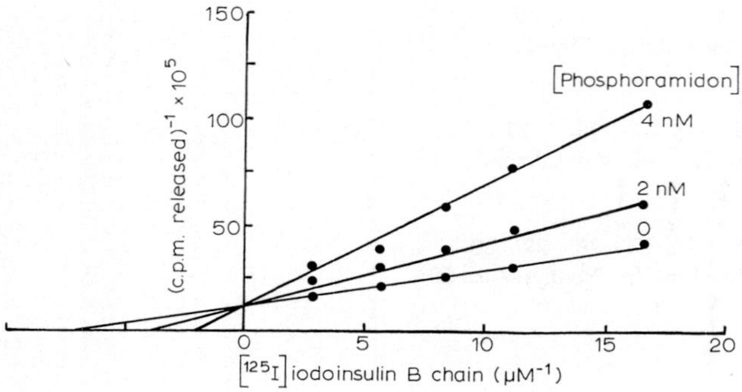

FIG. 2. Competitive inhibition of kidney neutral endopeptidase by phosphoramidon. Lineweaver-Burk plot. The substrate is [^{125}I]iodo-insulin B chain and the assay depends on the determination of trichloroacetic acid-soluble radioactivity. Phosphoramidon was present at concentrations of 2 nM and 4 nM.

2 nM. It is clear that the active site of the kidney enzyme closely resembles those of the microbial enzymes.

Although the broad nature of its bond specificity might suggest that the kidney neutral endopeptidase could degrade a wide range of proteins and peptides, this is not supported by experiment. It is noteworthy that while insulin B chain and glucagon are good substrates, insulin itself is not hydrolysed or inactivated, even on prolonged incubation with the purified enzyme. Nor were iodinated bovine serum albumin and iodinated casein degraded. On present evidence it is unlikely that the endopeptidase plays any role in the degradation of plasma proteins (as distinct from simple peptide hormones) in so far as they may cross the glomerular filter.

MICROVILLUS MEMBRANE SERINE PEPTIDASE (DIPEPTIDYL PEPTIDASE IV) (EC 3.4.14.– OR 3.4.21.–)

This enzyme is usually assayed with Gly-Pro-2-NNap (2-naphthylamide) as substrate (hence originally called glycylproline naphthylamidase: Hopsu-Havu & Glenner 1966). It is widely distributed among many tissues, but kidney cortex is the richest source. It is also present in the intestinal brush border, though at about 10% of the specific activity of that of the kidney. It is unique among kidney microvillus enzymes in being a serine enzyme and unique among serine peptidases in being a large protein intrinsic to the plasma membrane. Barth *et al.* (1974) first showed that it was inhibited by diisopropyl phosphorofluoridate (DipF). It is extremely sensitive to DipF, but much less so to other inhibitors of the class, e.g. diethyl *p*-nitrophenyl phosphate and phenyl-

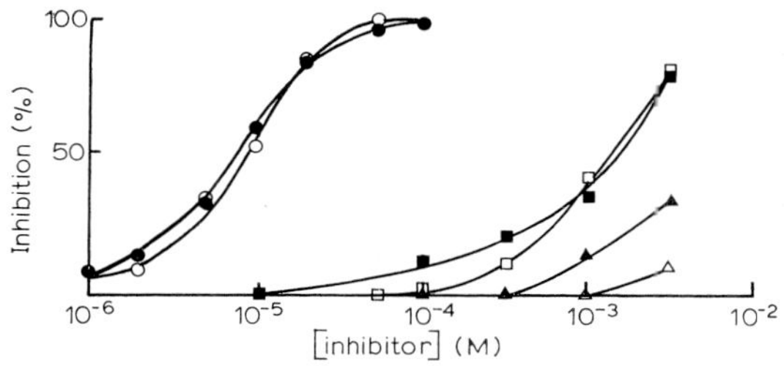

FIG. 3. Effect of inhibitors on microvillus membrane serine peptidase (dipeptidyl peptidase IV). ● ○, DipF; ■ □, diethyl *p*-nitrophenylphosphate; ▲ △, phenylmethylsulphonyl fluoride. Solid symbols (● ■ ▲) refer to purified enzyme; open symbols (○ □ △) refer to microvillus membrane.
The enzyme preparation was incubated for 1 h with inhibitors at the concentrations shown and assayed fluorimetrically with Gly-Pro-2-NNap as substrate.

methylsulphonyl fluoride (Fig. 3). The use of [^{32}P]DipF has been informative in several ways (Kenny *et al.* 1976). Autoradiography of microvillus proteins resolved by polyacrylamide gel electrophoresis in the presence of sodium dodecyl sulphate established that the peptidase was the only radioactive polypeptide in the membrane and that it had an apparent monomeric molecular weight of 130 000. Determination of the specific radioactivity of the labelled membrane and of the purified enzyme indicated that it constituted about 4% of the microvillus protein. Titration of the active site with the radioactive inhibitor demonstrated one labelled serine residue per monomer (130 000 daltons). When purified, the enzyme behaves as a dimer, both by gel filtration and in the analytical ultracentrifuge. Electron microscopy of negatively stained preparations has confirmed the dimeric structure. The specificity of the microvillus membrane serine peptidase is restricted to cleaving bonds involving the carboxyl group of either proline or alanine. In peptides with a free α-amino group the activity is typical of a dipeptidyl peptidase yielding dipeptides X-Pro and X-Ala. However the two types of substrate are hydrolysed at different rates: the ratio of the rates for Gly-Pro-2-NNap and Ala-Ala-2-NNap is about 30 : 1 for the pure enzyme and this ratio is constant during the final steps of purification. This observation and kinetic experiments support the view that the same enzyme catalyses both types of attack. However, it is becoming clear that the enzyme is not restricted to a step-by-step dipeptidyl peptidase attack, as is shown in Table 2. First, while short incubations of alanine oligopeptides gave the predicted products, longer incubation times yielded traces of products indicative of an anomalous attack. Secondly, the two N-blocked peptides gave clear evidence of an endopeptidase attack, though in these experiments

TABLE 2

Hydrolysis of oligopeptides by microvillus membrane serine peptidase

Peptide	*Products identified*	*Remarks*
Gly-Pro-Ala	Gly-Pro + Ala	
Gly-Ala-Ala	Gly-Ala + Ala	
Val-Ala-Ala-Phe	Val-Ala + Ala-Phe	
$(Ala)_3$	$(Ala)_2$ + Ala	
$(Ala)_4$	$(Ala)_2$	Ala + $(Ala)_3$ also observed after 2.5 h. incubation
$(Ala)_5$	$(Ala)_2$ + $(Ala)_3$	Ala + $(Ala)_4$ also observed after 2.5 h incubation
Z-Gly-Pro-Leu-Gly-Pro	Z-Gly-Pro + Leu-Gly-Pro	Substrate/enzyme = 10/1(wt/wt) 5 h incubation
Z-Gly-Pro-Gly-Gly-Pro-Ala	Z-Gly-Pro + Gly-(Gly)-Pro + Ala	Substrate/enzyme = 3/1 (wt/wt) 4 h incubation

Except where noted, substrate/enzyme = 25/1 (wt/wt) and incubation time 15 or 30 min.
Control incubations containing DipF yielded no products.

larger amounts of enzyme and longer incubation times were used. It is too early to assess the importance of the endopeptidase activity of this enzyme: it may be a minor function. However it is relevant to note that *in vivo* concentrations of enzymes may, in the microenvironment in which they act, be many times higher than those commonly used *in vitro*. In the case of the microvillus surface, the volume of the microenvironment may be no more than a few nanometres thick and hence the major peptidases may function at millimolar concentrations.

ACKNOWLEDGEMENT

This work was supported by grants from the Medical Research Council.

References

BARTH, A., SCHULZ, H. & NEUBERT, K. (1974) Studies on the purification and characterization of dipeptidyl aminopeptidase IV. *Acta Biol. Med. Ger. 32*, 157-174

BOOTH, A. G. & KENNY, A. J. (1974) A rapid method for the preparation of microvilli from rabbit kidney. *Biochem. J. 142*, 575-581

BOOTH, A. G. & KENNY, A. J. (1976) Proteins of the kidney microvillus membrane. Identification of subunits after sodium dodecyl sulphate/polyacrylamide gel electrophoresis. *Biochem. J. 159*, 395-407

GEORGE, S. G. & KENNY, A. J. (1973) Studies on the enzymology of purified preparations of brush border from rabbit kidney. *Biochem. J. 134*, 43-57

HOPSU-HAVU, V. K. & GLENNER, G. G. (1966) A new dipeptide naphthylamidase hydrolysing glycyl-prolyl-β-naphthylamide. *Histochemie 7*, 197-201

KENNY, A. J. (1977) Proteinases associated with cell membranes, in *Proteinases of Mammalian Cells and Tissues* (Barrett, A. J., ed.), pp. 393-443, ASP Biological and Medical Press, North-Holland, Amsterdam

KENNY, A. J., BOOTH, A. G., GEORGE, S. G., INGRAM, J., KERSHAW, D., WOOD, D. J. & YOUNG, A. R. (1976) Dipeptidyl peptidase IV, a kidney brush border serine peptidase. *Biochem. J. 157*, 169-182

KERR, M. A. & KENNY, A. J. (1974*a*) The purification and specificity of a neutral endopeptidase from rabbit kidney brush border. *Biochem. J. 137*, 477-488

KERR, M. A. & KENNY, A. J. (1974*b*) The molecular weight and properties of a neutral metallo-endopeptidase from rabbit kidney brush border. *Biochem. J. 137*, 489-495

KOMIYAMA, T., SUDA, H., AOYAGI, T., TAKEUCHI, T., UMEZAWA, H., FUJIMOTO, K. & UMEZAWA, S. (1975) Studies on the inhibitory effect of phosphoramidon and its analogs on thermolysin. *Arch. Biochem. Biophys. 171*, 727-731

MATSUBARA, H. & FEDER, J. (1971) Other bacterial, mold and yeast proteases, in *The Enzymes*, vol. 3 (Boyer, P. D., ed.), pp. 765-786, Academic Press, New York

SUDA, H., AOYAGI, T., TAKEUCHI, T. & UMEZAWA, H. (1973) A thermolysin inhibitor produced by actinomycetes: phosphoramidon. *J. Antibiot. (Tokyo) 26*, 621-623

TATE, S. S. & MEISTER, A. (1976) Subunit structure and isozymic forms of γ-glutamyl transpeptidase. *Proc. Natl. Acad. Sci. U.S.A. 73*, 2599-2603

UMEZAWA, S., TATSUTA, K., IZAWA, O. & TSUCHIYA, T. (1972) A new microbial metabolite phosphoramidon (isolation and structure). *Tetrahedron Lett. 1*, 97-100

WONG-LEUNG, Y. L., GEORGE, S. G., APARICIO, S. G. R. & KENNY, A. J. (1968) The intracellular localization of some particulate peptidases in the kidney of the rat and rabbit. *Biochem. J. 110*, 5P-6P

Discussion of the two preceding papers

Semenza: Is phosphoramidon available commercially, or is there any simpler synthetic derivative which would be equally effective?

Kenny: It is not yet on the market. But *N*-phosphoryl-L-leucyl-L-tryptophan is reported to be 10 times more potent than the natural product containing rhamnose in inhibiting thermolysin (Komiyama *et al.* 1975), and we have found this to be so with the kidney enzyme.

Semenza: It now seems that most membrane-bound hydrolases have subunit molecular weights ranging from 100 000 to 130 000. I only know of a single exception to this rule (trehalase; Sasajima *et al.* 1975). Does this have any profound meaning or is it just chance that it is like that?

Kenny: I think the phlorizin-binding protein is smaller.

Semenza: I was referring just to hydrolases.

Kenny: Perhaps all transport proteins need to be of that size and also, perhaps, dimeric. On the other hand, soluble hydrolases may be relatively small. I am struck by the contrast in size between most serine proteinases and the only known membrane-bound enzyme of this class, dipeptidyl peptidase IV, in which the monomer is four times as big as trypsin. This suggests to me that such a protein has an intra- or transmembrane function.

Rhodes: I wonder whether Drs Norén, Kim or Maroux have had an opportunity to test the dipeptidyl substrate with their enzyme preparations? Dr Kenny has shown that it is clearly there in the intestine, associated with the membrane.

Kim: We haven't tested it.

Norén: We tested the hydrolysis of Gly-Pro-β-naphthylamide with solubilized brush borders. The enzyme is there and I think it furnishes the oligopeptide transport system with dipeptides. Perhaps this is its physiological function.

Dr Kenny, you said (p. 212) that PMSF did not inhibit the enzyme but this may be a matter of the experimental conditions. We found complete inactivation of the intestinal dipeptidyl peptidase when we incubated the preparation for 24 h. It is a matter of the reaction velocity.

Kenny: Yes, but there is a big difference in the rate of inhibition by these reagents: we found that diethyl *p*-nitrophenyl phosphate is 100 times and phenylmethylsulphonyl fluoride 1000 times less potent than diisopropyl phosphorofluoridate. I am not saying that it won't inhibit if you incubate it long enough.

Norén: Do you consider the dipeptidyl peptidase as an endopeptidase or as a dipeptidyl aminopeptidase? You discussed this matter recently (Kenny *et al.* 1976).

Kenny: I am satisfied that it possesses endopeptidase activity, but I do not know how significant this aspect may be. In our experiments, we have used rather high ratios of enzyme to substrate to demonstrate endopeptidase activity, and we have not quantified this in comparison with the rates for exopeptidase activity. In each experiment the endopeptidase activity was totally inhibited by diisopropyl phosphorofluoridate, so if one attributed this activity to a contaminant it must be another serine peptidase, and I don't know of any others with the required specificity.

Radhakrishnan: You mentioned that the inhibitor probably binds to the zinc via the phosphate group.

Kenny: That is an assumption rather than established fact.

Radhakrishnan: Do any other phosphate compounds inhibit?

Kenny: Orthophosphate inhibits the enzyme, as it does a number of other metalloenzymes. Putting the dipeptide (Leu-Trp) on as well makes it a much better inhibitor, presumably because the amino acid side chains direct it to the Zn^{2+} at the active site.

Kim: Does your dipeptidyl enzyme work on trialanine? Some dipeptidyl enzymes cannot act on trialanine as a substrate.

Kenny: Yes, it can.

Kim: There seems to be some difference between tetraalanine and trialanine as substrates. For instance dipeptidyl enzyme isolated from liver can act on tetraalanine but not on trialanine.

Kenny: It was Professor Barth in Halle (Barth *et al.* 1974) who first showed it to be a serine peptidase. And he also showed that it attacked a series of alanine peptides, $(Ala)_3$, $(Ala)_4$ and $(Ala)_5$. For the most part, the products were those predicted for a dipeptidyl peptidase but some minor departures from prediction may be found which suggest that the substrate may slip one residue to give a small yield of, for example, $(Ala)_3$ and Ala from $(Ala)_4$.

Kim: Professor Josefsson, have you purified your dipeptidyl hydrolase?

Josefsson: Dr Norén spoke about aspartic aminopeptidase in his paper (pp. 177–191). We are trying to purify all the membrane-bound peptidases including the disaccharidases.

Radhakrishnan: With your proline dipeptidase have you tried any sarcosine peptides?

Josefsson: Yes. It has a very low activity against glycylsarcosine.

Radhakrishnan: With the monkey intestinal enzyme there is a detectable activity. This can be expected since glycylsarcosine has a very similar structure to Gly-Pro (Krishna Baksi & Radhakrishnan 1974).

Josefsson: There is a very small amount of activity against glycylsarcosine in the gut.

References

BARTH, A., SCHULTZ, H. & NEUBERT, K. (1974) Untersuchungen zur Reinigung und Charakterisierung der Dipeptidylaminopeptidase. (Eng. Abstr.) *Acta Biol. Med. Ger. 32*, 157-174

KENNY, A. J., BOOTH, A. G., GEORGE, S. G., INGRAM, J., KERSHAW, D., WOOD, E. J. & YOUNG, A. R. (1976) Dipeptidyl peptidase IV. A brush border serine peptidase. *Biochem. J. 157*, 169-182

KOMIYAMA, T., AOYAGI, T., TAKEUCHI, T. & UMEZAWA, H. (1975) Inhibitory effects of phosphoramidon on neutral metallopeptidases and its application to affinity chromatography. *Biochem. Biophys. Res. Commun. 65*, 352-357

KRISHNA BAKSI & RADHAKRISHNAN, A. N. (1974) Purification and properties of prolidase (imidodipeptidase) from monkey small intestine. *Indian J. Biochem. Biophys. 11*, 7-11

SASAJIMA, K., KAWACHI, T., SATO, S. & SUGIMURA, T. (1975) Purification and properties of α; α-trehalase from the mucosa of rat small intestine. *Biochim. Biophys. Acta 403*, 139-146

Membrane and intracellular hydrolysis of peptides: differentiation, role and interrelations with transport*

A. M. UGOLEV, N. M. TIMOFEEVA, L. F. SMIRNOVA, P. DE LAEY, A. A. GRUZDKOV, N. N. IEZUITOVA, N. M. MITYUSHOVA, G. M. ROSHCHINA, E. G. GURMAN, V. M. GUSEV, V. A. TSVETKOVA and G. G. SHCHERBAKOV

I. P. Pavlov Institute of Physiology, Academy of Sciences of the USSR, Leningrad

Abstract The possible relative importance of the membrane and intracellular peptidases of the enterocytes in splitting dietary peptides to amino acids has been analysed. On the anoxic criterion, membrane hydrolysis was found to be predominant. Model experiments revealed cooperative interactions between the membrane enzyme and the coupled transport system. This cooperativity allows the main characteristics of oligomer transport to be described in terms of membrane digestion. Comparison of the behaviour of membrane and intracellular peptidases under different conditions has shown that the former are largely involved in digestion and the latter in intracellular metabolism. It is suggested that the efficiency of the membrane system is high enough to account for the hydrolysis of protein, especially taking into account the stimulation of enzyme processes in the brush border that occurs after ingestion of protein and carbohydrate. A sequential model based on the concept of three interacting enzyme layers (glycocalyx, lipoprotein membrane and cytosol) is presented.

At an earlier Ciba Foundation symposium (Ugolev 1972*a*) membrane digestion was considered the most probable mechanism of peptide absorption in the small intestine in mammals. The scheme proposed was briefly as follows: oligomers enter the glycocalyx space with the water flow and there undergo hydrolysis by adsorbed pancreatic enzymes. After they reach the lipoprotein membrane surface, dipeptides and tripeptides are split by enteric surface enzymes integrated with transport systems—the so-called 'enzyme transport ensembles'. The features of the latter are such that monomers formed during hydrolysis gain advantages in transport over free monomers.

In recent years the membrane concept has received further development and new support. At the same time much progress has been made in the development of the concept of active transport of peptides. According to this concept, peptides are transported through membrane by special systems

* Professor Ugolev was unable to attend the meeting and the paper was presented by Dr P. De Laey.

TABLE 1

Roles of membrane (M) and cytosol (C) enzymes in peptide hydrolysis (recent data from different authors)

No.	*Authors*	*Predominant system (membrane or cytosol hydrolysis)*
1	Addison *et al.* 1972	C
2	Adibi 1975	C
3	Asatoor *et al.* 1970	C
4	Auricchio *et al.* 1971	M*
5	Becciolini *et al.* 1974	M*
6	Caspary 1972	M
7	Cheeseman & Parsons 1974	C, M
8	Cheng *et al.* 1971	C, M,
9	Das & Radhakrishnan 1974	C
10	Fern *et al.* 1969	M
11	Fogel & Adibi 1974	C
12	Fujita *et al.* 1972	M
13	Gray & Cooper 1971	C, M
14	Heizer *et al.* 1972	C, M
15	Hellier *et al.* 1971	C, M
16	Kim *et al.* 1973	C, M
17	Lane *et al.* 1975	C, M
18	Lindberg *et al.* 1975	C, M
19	Maroux & Louvard 1976	M*
20	Matthews 1972, 1975	C>M
21	Nicholson *et al.* 1973	C, M
22	Noack *et al.* 1975	C, M
23	Nordström *et al.* 1968	C, M
24	Norén *et al.* 1971	C, M
25	Peters 1972, 1975	C, M
26	Rhodes & Folscroft 1972	M
27	Rubino *et al.* 1971	C, M
28	Sigrist-Nelson 1975	C
29	Silk 1974	C, M
30	Smirnova *et al.* 1976	M
31	Smyth 1972	C, M
32	Takesue & Nishi 1976	M*
33	Ugolev 1972*a*, *b*	M
34	Wojnarowska & Gray 1975	M*

* Cytosol hydrolysis was not considered in these papers.

and then hydrolysed in the cytosol. The concept of peptide transport is now widely accepted (Table 1). Why is this so? Table 2 summarizes phenomena confirming the concept of peptide transport and dispelling that of membrane digestion. Some of the arguments seem striking. Let us begin by discussing them, then consider all the data presented, and conclude by assessing the relative roles of intracellular and membrane digestion in peptide hydrolysis.

TABLE 2

Evidence for peptide transport and intracellular hydrolysis of peptides

No.	Criteria	References (from Table 1)
1	The demonstration of intact peptides in enterocytes, in the serosal fluid of everted sacs and in the blood	10, 28, 31
2	Active accumulation of poorly hydrolysed peptides in the enterocyte	1, 20
3	Increased rates of peptide absorption compared to equivalent free amino acids	2, 20, 27, 28, 29, 31
4	Absence of competition for the transport mechanism between peptides and amino acids and distinct competition between peptides	7, 9, 20, 27, 28, 31
5	Absence or reduction of competition for the transport mechanism between amino acid constituents of a peptide	20
6	Normal peptide uptake in the free amino acid transport defects of Hartnup disease and cystinuria	3, 15, 20
7	The inhibition of active transport and hydrolysis of d peptides by intact but not homogenized mucosa: (*a*) in the presence of dinitrophenol; (*b*) under anoxia	8, 31
8	The inhibition of free Met uptake by L-amino acid oxidase (in concentrations of 0.1–0.2 mg/ml) but absence of this effect on the transport of Met from L-Met-L-Met	8, 20
9	Decrease in dipeptide uptake (after acidification of the medium) correlates with reduction in transport rather than hydrolysis	2, 11
10	Increase in the rate of dipeptide hydrolysis by mucosal homogenates but not intact mucosa in the presence of Co^{2+}	31
11	Different sites for amino acid and peptide uptake along the small intestine	20
12	Low content of dipeptidase activities in the brush border of enterocytes compared to cytosol	18, 25, 31

BASIC EVIDENCE FOR PEPTIDE TRANSPORT FOLLOWED BY INTRACELLULAR HYDROLYSIS

The evidence is presented in Table 2. At least some of this evidence unambiguously demonstrates that peptide transport through the membrane, followed by hydrolysis, is a basic mechanism in the final stages of protein assimilation. In fact, accumulation of poorly hydrolysed peptides in the cell against an electrochemical gradient is unquestionable proof of active transport (Addison *et al.* 1972; Matthews 1972, 1975; Matthews & Adibi 1976). This process, however, may be accomplished by a mechanism of minor importance which plays only a small part in the process of peptide assimilation.

The anoxic criterion (Cheng *et al.* 1971; Smirnova *et al.* 1976) is universal and can be used for evaluating the role of membrane and cytosol hydrolases

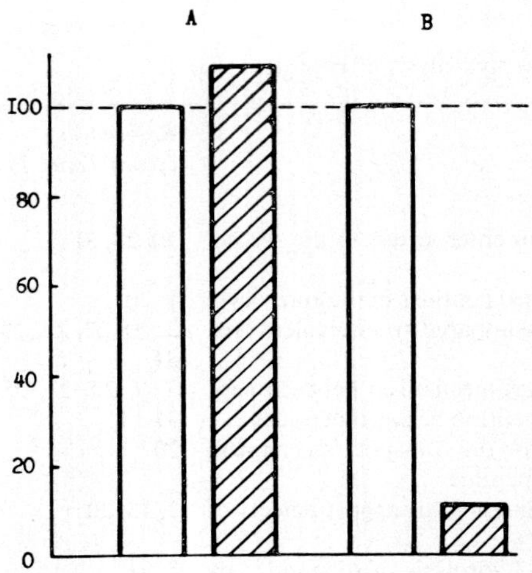

FIG. 1. The effect of anoxia on membrane (A) and intracellular (B) hydrolysis of peptides (mathematical experiment). Accumulation of hydrolysis products in the medium under oxygenation (open columns), taken as 100%, and under anoxia (shaded columns).

in splitting any peptide. The inhibition of active transport by anoxia will result in a concomitant decrease in hydrolysis if hydrolysis follows transport. In contrast, anoxia will not influence the rate of membrane hydrolysis preceding transfer through a membrane.

Cheng *et al.* (1971) demonstrated that the inhibition of active transport by anoxia produced an equivalent inhibition of hydrolysis of L-Met-L-Met, suggesting that the process takes place in the cytosol. The results of a special experimental test (Smirnova *et al.* 1976) and a mathematical analysis in our laboratory (Fig. 1) demonstrated that this criterion is unequivocal and fruitful for differentiating between membrane and intracellular hydrolysis.

We investigated the effect of anoxia on the transport and hydrolysis of seven dipeptides by intact everted intestinal rings and their homogenates in rats. The choice of dipeptides (Gly-L-Leu, Gly-L-Val, Gly-L-Ala, L-Leu-Gly, Gly-L-Phe, Gly-Gly, L-Ala-L-Ala) permits the different dipeptidase activities of the enterocytes to be characterized. No inhibitory effect was revealed in these cases. However, under the same conditions we were able to observe complete inhibition of the active transport of free and peptide glycine (Fig. 2). This means that in the above cases (and probably others with similar mechanisms) hydrolysis precedes transport through the membrane. It is important to emphasize that the anoxic criterion, in contrast to many others, gives good reason for unambiguous conclusions.

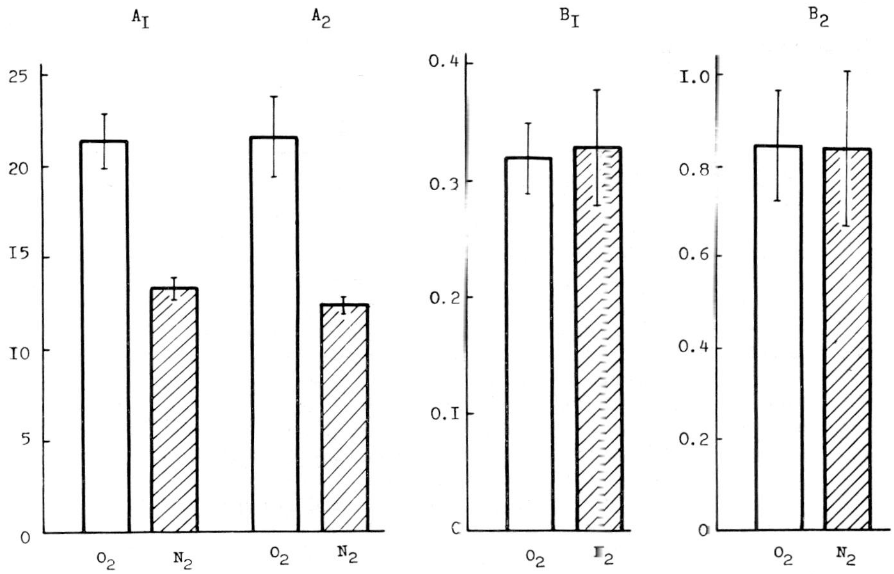

FIG. 2. The effect of anoxia on glycine uptake from free glycine (A_1) and L-leucylglycine (A_2) solutions and on the dipeptide hydrolysis by intact (B_1) and homogenized (B_2) intestine of rats. A: concentration of glycine in intracellular fluid (in mM); B: rate of dipeptide hydrolysis in μmol glycine/(cm min). Open columns: under O_2; shaded columns: under N_2. Vertical bars: ± S.E.M.

ENZYME TRANSPORT COOPERATIVITIES AND SOME CHARACTERISTICS OF PEPTIDE TRANSPORT

It is noteworthy that the same criteria are used by supporters of both the membrane and the intracellular concepts (Tables 2 and 3). It is believed that only peptide transport can account for phenomena Nos. 3 and 5 (Table 2). However, there exists an alternative membrane concept, that is the hypothesis of the 'enzyme transport ensemble' which has made advances in recent years (Ugolev 1964, 1968, 1972*a*, *b*; Crane 1968: Semenza 1971; Ugolev & De Laey 1973; Ramaswamy *et al.* 1974; Maroux & Louvard 1976). For instance, phenomena Nos. 3 and 5 (Table 2) can be explained by either concept. These phenomena occur in disaccharide absorption, in which membrane hydrolysis of dimers followed by the transport of monomers is the only physiological mechanism (Crane 1968; Semenza 1971; Ugolev 1972*b*; Ramaswamy *et al.* 1974; Takesue & Nishi 1976).

Thus, phenomena of facilitated transport of dimers, such as a decrease in competition between constituent monomers or a change in kinetic constants,

TABLE 3

Evidence for membrane hydrolysis of peptides*

No.	*Criteria*	*References (from Table 1)*
1	Increased rates of peptide absorption compared to equivalent free amino acids	33
2	Absence or reduction of competition for the transport mechanism between amino acid constituents of a peptide	33
3	Anoxia-produced inhibition of the active transport of peptides and free amino acids but absence of anoxia-produced inhibition of dipeptide hydrolysis by intact and homogenized intestinal mucosa	30
4	Inhibition of Met uptake from L-Met-L-Met in mucosa by L-amino acid oxidase (in a concentration of 0.5 mg/ml)	8, 20
5	Presence of peptide hydrolases splitting both dipeptides (up to 20%) and tripeptides (up to 80%) in the isolated brush border fraction	4, 12, 14, 22, 26, 34
6	Immunological identification of aminopeptidases and disaccharidases on the surface of the brush border vesicles	32
7	Solubilization of peptide hydrolases by papain and Triton X-100 from the surface of the brush border vesicles	19
8	Different sites for amino acid and peptide uptake along the small intestine	17
9	Normal peptide uptake in free amino acid transport defects during Hartnup disease and cystinuria	33

* The evidence discussed at the preceding symposium (Ugolev 1972*a*) is not included in this table.

may be characterized both in terms of intracellular digestion and of membrane digestion closely coupled with transport of monomers (Fig. 3).

With the use of the anoxic criterion, it became obvious that the features of L-Met-L-Met uptake could be accounted for by cytosol hydrolysis, and those of Gly-L-Leu or L-Leu-Cly by membrane hydrolysis.

However, characterization of phenomena Nos. 4 and 6 (Table 2) in terms of membrane digestion encounters serious difficulties. These are the absence of competition between amino acids and peptides, and the ability to take up affected amino acids from peptides in the amino acid transport defects of Hartnup disease and cystinuria. In fact, why do the transport systems of ensembles not transfer free amino acids? These difficulties arise largely because investigations of the kinetics of membrane enzymes have failed to reveal cooperative interactions between enzymes and carriers. However, very recently, in studying hydrolysis of sucrose, maltose and starch in the rat with the accumulating preparation of mucosa (Ugolev *et al.* 1970*b*) in conditions in which active transport was preserved, we found the sigmoid kinetics of hydrolysis and

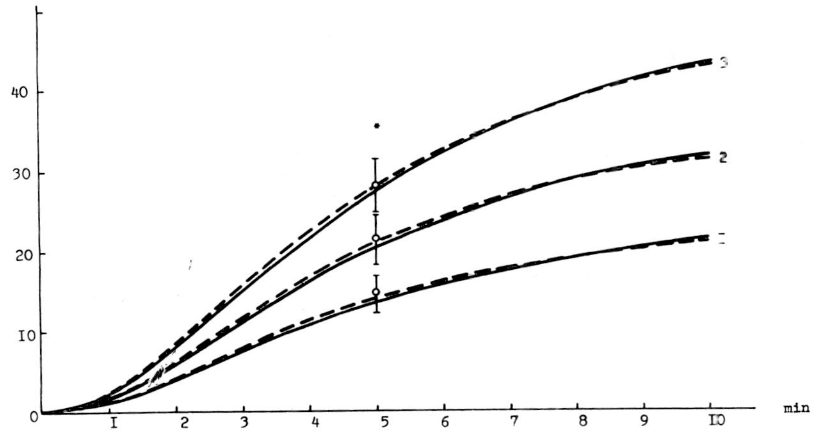

FIG. 3. Similarities in kinetics of monomer accumulation during intracellular (solid lines) and membrane (dashed lines) hydrolysis of dimers (mathematical model and experimental data). Abscissa: time of incubation (in min): Ordinate: accumulation of monomers in intracellular fluid (μmol/ml). 1, 2 and 3 – concentrations of dimers in the medium (5, 20 and 40 μmol/ml, respectively). *Data from Cheng *et al.* (1971) are denoted by circles.

transport which is characteristic of cooperative systems (Ugolev *et al.* 1976) (Fig. 4). It should be pointed out that solubilized enzymes and enzymes of isolated brush border give hyperbolic kinetics.

We found that the inhibition of active transport by anoxia or specific blockage of the glucose carrier by phlorizin led to complete disappearance of sigmoidality. These important phenomena suggest the existence of cooperative interactions in the enzyme-carrier system of the brush border.

The scheme in Fig. 5 shows that when substrate binds with the active site of an enzyme a direct induced fit takes place (Koshland 1970). Moreover, the conformation of the contact site of a carrier simultaneously changes. The latter phenomenon can be called 'allosteric induced fit' in a carrier by the enzyme. This model is able to explain not only phenomena of facilitated transport of monomers from dimers and polymers but also absence of affinity of the enzyme-integrated carrier for monomers. In addition, this model may help to account for the findings in Hartnup disease and cystinuria (Asatoor *et al.* 1970; Hellier *et al.* 1971; Matthews 1972, 1975).

BEHAVIOUR OF MEMBRANE AND CYTOSOL PEPTIDASE SYSTEMS UNDER DIFFERENT CONDITIONS

This approach (similar to the 'black box' method) helps us to understand into which of two aspects of regulation the cytosol and membrane peptidase

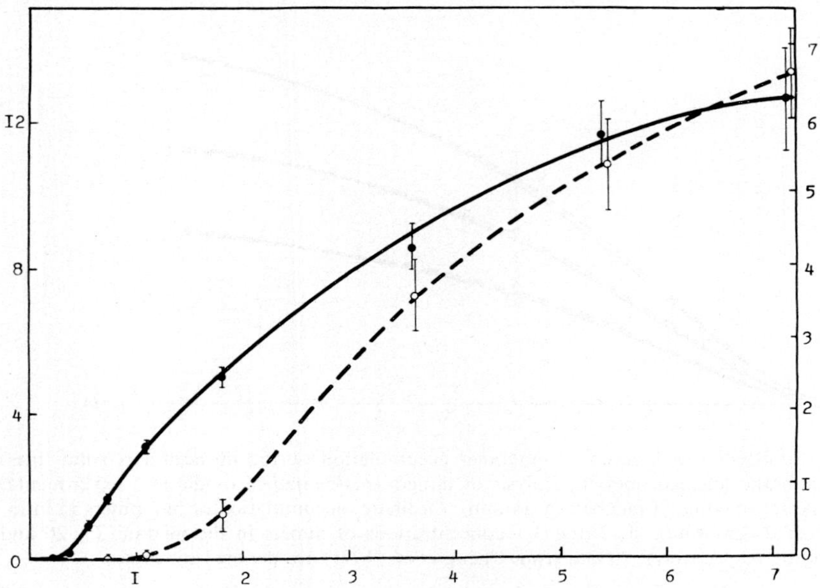

FIG. 4. The effect of different substrate concentrations on the rate of starch (solid line) and sucrose (dashed line) hydrolysis by rat accumulating mucosa preparations under conditions in which active transport is preserved. Abscissa: substrate content of medium (in g/ml). Ordinate: to the left – sucrose hydrolysis (in mmol glucose cm^{-1} min^{-1} 10^{-5}); to the right – starch hydrolysis (in mmol glucose cm^{-1} min^{-1} 10^{-5}). Vertical bars: ± S.E.M.

systems are incorporated: (1) digestion proper, or (2) intracellular catabolism of proteins as shown in the classical study by Smith (1951). Variations in their behaviour make it possible to evaluate the physiological importance of different groups of peptidases.

(1) Significant inhibition of membrane carbohydrase and membrane peptidase activities in mammals has been produced by emotional stress (immobilization). In contrast, the dipeptidase activity of the cytosol remained almost unaffected (Fig. 6) (Ugolev *et al.* 1969; Ugolev 1972*b*).

(2) Membrane invertase activity in rats was increased shortly after feeding and decreased during fasting. A reduction in the brush border peptidase activity was shown by Kim *et al.* (1973) in fasting, whereas the level of cytosol dipeptidases remained unchanged during five days of fasting. The similarity in the behaviour of cytosol peptidases in fasting and under stress may reflect their involvement in intracellular catabolism.

(3) The comparison of cytosol and membrane systems of the enterocytes during their differentiation and migration from crypt to villus tip opens up

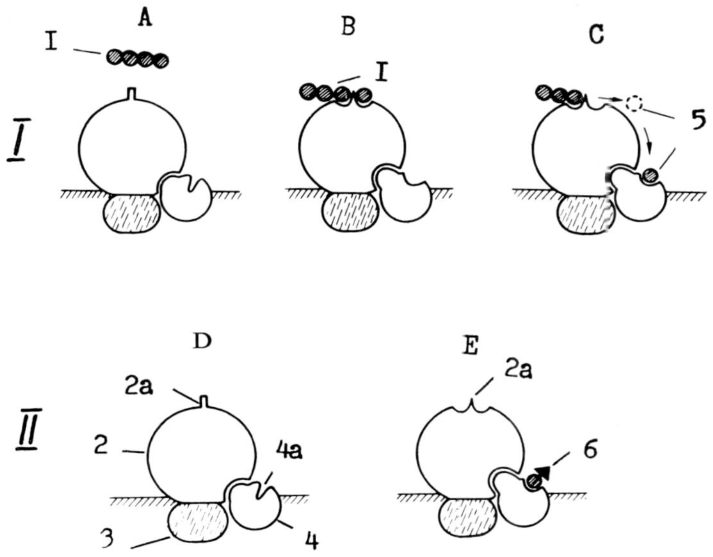

FIG. 5. Simplified scheme of allosteric interaction between the enzymic and transport parts of the ensemble. *I*: allosteric effect of the enzyme on entry into the transport system. *II*: allosteric effect of the transport system on enzyme activity. A: low affinity of the carrier contact site for hydrolysis product; B and C: high affinity of the carrier contact site for hydrolysis product; D: catalytic enzyme centre is less active; E: catalytic enzyme centre is more active. 1 – substrate (polymer), 2 – enzyme, hydrophilic part, 3 – enzyme hydrophobic part, 4 – carrier, 5 – product (monomer), 6 – phlorizin, 2a – active centre of enzyme, 4a – monomer-binding site of the carrier.

great possibilities. Anoxic and other criteria, as well as direct estimation, have shown that enzyme activity, in particular dipeptidase and disaccharidase activities, in intact mucosa is produced by membrane enzymes. We developed a technique which allows intact and homogenized enterocytes in different parts of the villus to be compared (Ugolev *et al.* 1970*a*; Ugolev 1972*b*), while other authors (Nordström *et al.* 1968; Fortin-Magana *et al.* 1970) investigated only homogenized slices. Fig. 7 indicates that in rats, except in the crypt, invertase activity is accounted for by surface enzymes and is maximal in functioning parts of the villus. Peptidase activity reflects the existence of two relatively autonomous systems (cytosol and membrane). The cytosol system predominates in the crypt-basal parts. In the middle-apical parts the activities are equal and in the villus region membrane peptidases predominate. A distinct absorption gradient has been established along the villus length. In particular, amino acids and monosaccharides in hamsters are accumulated mainly by the cells close to the villus tip and very little at its base (Kinter & Wilson 1965). Max-

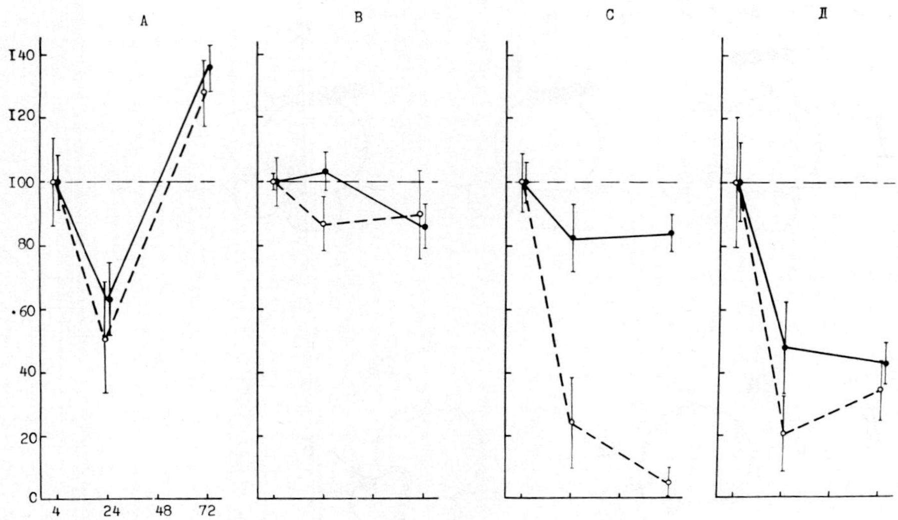

FIG. 6. Effect of stress (immobilization) on enzyme activities of intact (dashed line) and homogenized (solid line) intestinal mucosa in rats. A: sucrose hydrolysis; B: maltose hydrolysis; C: glycyl-L-leucine hydrolysis; D: tetraglycine hydrolysis. Abscissa: time after immobilization (in hours). Ordinate: enzyme activity as a percentage of the control, taken as 100. Vertical bars: ± S.E.M.

imum γ-glutamyltranspeptidase activity in rats has been shown to occur near the villus tip (Cornell & Meister 1976).

It is interesting that we found maximum dipeptidase activity in this region. On the other hand, cytosol peptidases are predominantly found in the zones of intensive metabolism of intrinsic proteins.

The conclusion can be drawn that the behaviour of cytosol peptidases may be accounted for by their predominant participation in the intracellular metabolism of proteins, whereas the behaviour of membrane peptidases reflects their digestive functions.

THE EFFECT OF SURFACE MODIFIERS ON PEPTIDASE ACTIVITY

We have recently found that food components which are substrates for certain surface enzymes may serve as powerful modifiers for other enzymes (Ugolev & De Laey 1973; Ugolev *et al.* 1975). In particular, tributyrin is capable of producing significant inhibition of dipeptidase and alkaline phosphatase acitivity (40–70 %) as early as after the first five minutes of incubation. The effect is not produced by mechanical blockage of the membrane surface

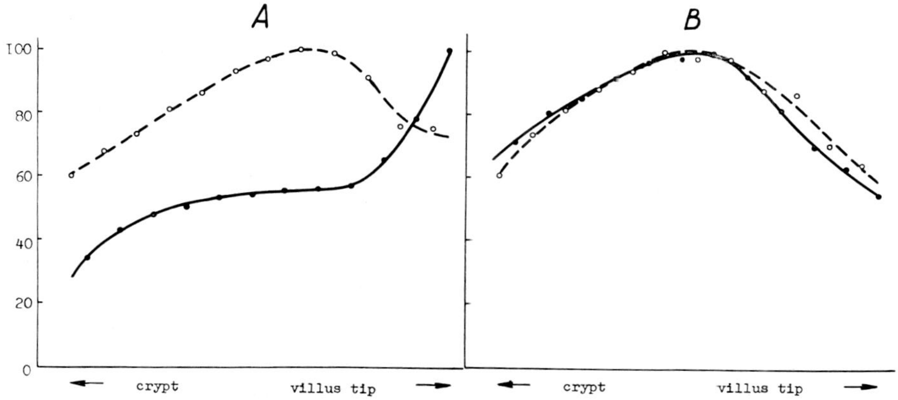

FIG. 7. Distribution of glycyl-L-leucine dipeptidase (A) and invertase (B) activities in intact (dashed line) and homogenized (solid line) transverse sections (thickness 25 μm) of the crypt–villus system of rat small intestine. Abscissa: sections in the direction from crypts to villus tips. Ordinate: % enzyme activity, with maximal activity in each experiment taken as 100%.

since invertase activity is unaffected. We have recently found that rat intestinal mucosa contains unidentified substances which we call antipermeins. Affecting the luminal side they are able to depress both glucose and amino acid transport as well as dipeptidase activity.

DISTRIBUTION OF ENZYMES BETWEEN THE BRUSH BORDER AND CYTOSOL

A number of authors (Rhodes *et al.* 1967; Peters 1972, 1975, and others) have reported that the brush border contains only 5–10%, and the cytosol 90–95%, of dipeptidases. This distribution suggests that practically all dipeptides penetrate through the cytoplasm unsplit.

With the advent of new data the concept of membrane digestion has been re-established. Thus, using a refined technique of isolation of the brush border, Fujita *et al.* (1972) showed that dipeptidase activity in this area amounts to 15%. The estimation Fujita *et al.* made on the basis of this evidence demonstrates that membrane dipeptidase activity is high enough to meet the protein requirements of the organism. However, data are now available which suggest that brush border activity is much greater than was previously believed. Peters (1975) showed that the brush border contains 60% of tripeptidases and practically all the tetra-, penta- and hexapeptidase activities. Thus, considering all the data, the conclusion may be drawn that about 40–60% of peptide bonds are hydrolysed at the brush border surface.

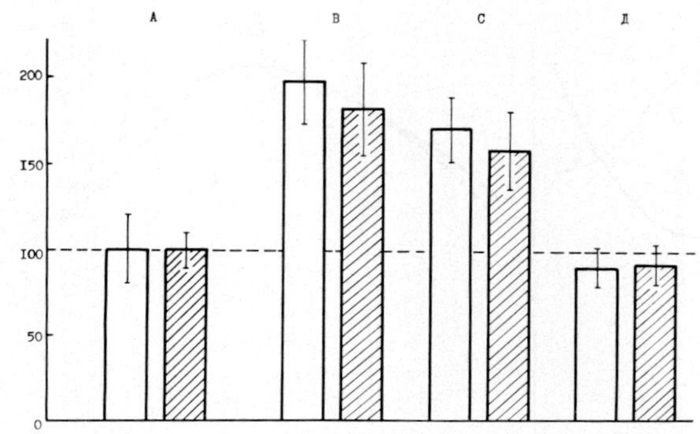

FIG. 8. The effect of feeding on the invertase activity of intact (shaded columns) and homogenized (open columns) intestinal mucosa in Wistar rats. A: fasted rats (control); B: 2 hours after feeding with meat; C: 2 hours after feeding with 40% glucose solution; D: 2 hours after feeding with butter. Enzyme activity in % of the control value taken as 100. Vertical bars: ± S.E.M.

Moreover, some enzymes with distinct dipeptidase activity are localized in the brush border. For instance, isolated arylamidase is capable of reproducing almost the whole dipeptidase spectrum of the brush border (Noack *et al.* 1975). Finally, it has been established (Komissarchik & Ugolev 1970; Ugolev 1972*a*,*b*) that isolation of the brush border results in disappearance of glycocalyx together with adsorbed pancreatic proteases.

It should be noted that investigation of enzyme activity after feeding, but not in the fasting state, is important for the understanding of the natural digestive process. Fig. 8 shows that in rats the marker invertase activity of microvillus membrane was twice as high after ingestion of meat and sugar as before. Along with this, a surprising effect of feeding was the increase in number and stability of brush borders released per unit weight of homogenate (Fig. 9).

As a whole, the relationship of cytosol and membrane peptidase activities appears to be shifted towards the membrane system side. In addition to the findings reported above, digestive potentials of the enterocyte are increased 1.5–2-fold after feeding. In summary, the membrane peptidase system proved to be capable not only of satisfying the optimal requirements of the organism for protein but also of providing an essential functional reserve.

'SEQUENTIAL' MODEL OF PEPTIDE UPTAKE

Thus, (1) the universal anoxic criterion demonstrates that membrane hydrolysis rather than cytosol hydrolysis of peptides is predominant.

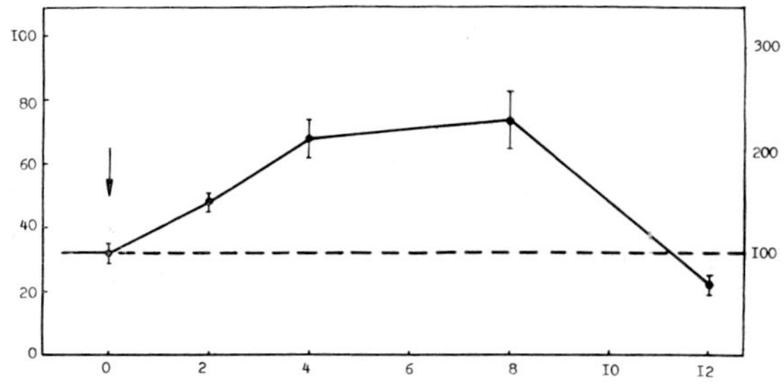

FIG. 9. Effect of feeding on the number of brush borders obtained from mucosa of Wistar rats. Abscissa: time after feeding (in hours). Ordinate: to the left – number of brush borders (in millions/g wet weight); to the right – number of brush borders (as percentage of the control value of 100). Solid line: fed rats; dashed line: fasted rats (control).

(2) New data on enzyme-transport cooperativities resolve earlier difficulties and allow the characteristics of peptide absorption to be described in terms of the membrane digestion concept.

(3) New findings on the amount of peptidases in the brush border surface and mathematical analysis show that peptidase activities are adequate for the nutritional requirement for protein.

(4) Stimulation of the brush border processes by feeding caused a 1.5–2-fold increase in the digestive power of the digestive-absorptive surface.

(5) Study of the behaviour of cytosol and membrane peptidases suggests that the former are associated with intracellular catabolism of proteins and the latter with digestive processes.

The arguments cited do not give a complete estimate of the role played by membrane digestion. This aim can be partly achieved by using the mathematical 'sequential' model schematically represented in Fig. 10. This model takes into account the 'functional topography' of the enterocyte (we use a very good term introduced by Professor D. H. Smyth), since the efficiency of an enzyme depends largely on its localization. Thus, localization of membrane enzymes at entrances into the transport channels of a membrane increases the probability of their contact with a substrate. From this point of view, the reliability of the lipoprotein enzyme barrier appears to be higher than that of the glycocalyx and much higher than that of the cytosol. However, each of the three enzyme layers (barriers) supplement each other as a result of their different specificities. Let us assume that the efficiency of hydrolysis is the same and that the rate of passage of substances through all the compartments of the enterocyte is similar. In these conditions, membrane hydrolysis is of greater importance than the relative amount of enzymes in this zone. In most cases a certain amount of peptides, as shown in

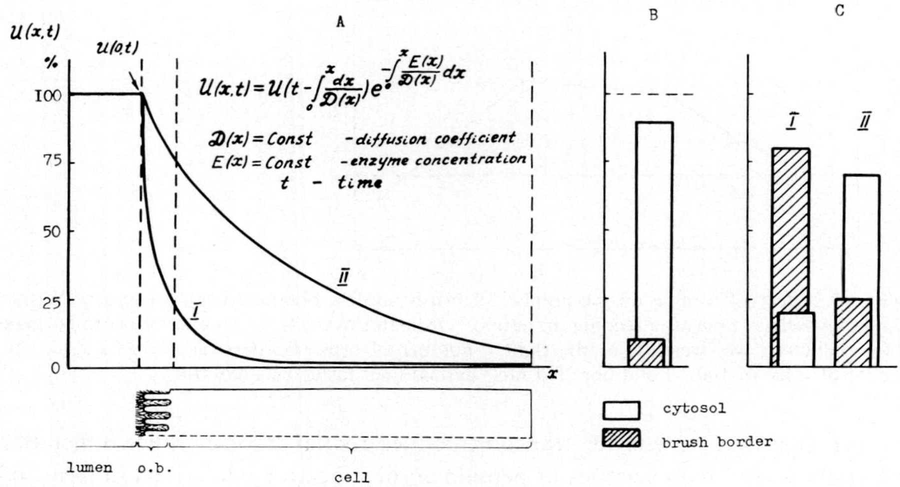

FIG. 10. Sequential model of peptide uptake. A: decrease in concentration of peptide during movement through the cell (as percentage of initial concentration). B: enzyme activity in the brush border and cytosol (as percentage of total activity). C: the level of dipeptide hydrolysis in the brush border and cytosol (as percentage of total hydrolysis). I: high enzyme activity (or slow movement); II: low enzyme activity (or rapid movement).

the scheme (Fig. 10), will penetrate through the first and second barriers. Penetration of peptides into the cell increases with both rate of flow and reduction in membrane hydrolysis. Thus, cytosol peptidases are not only involved in catabolism of proteins but may also function as the third barrier in the way of dietary peptides. The exponential kinetics of the process in the 'sequential' model suggests that the membrane-enzyme system has a much greater role than the intracellular one. At the same time cytosol enzymes perform an important and probably vital function as the third peptide barrier and reserve digestive zone.

OVERALL VIEW OF PEPTIDE UPTAKE

It would be unreasonable to deny that peptide transport and cytosol hydrolases have an important and possibly vital function during the final stages of protein uptake. Our conclusion is that membrane and cytosol systems interact and supplement each other. Indeed, active transport of some dipeptides, especially those which are poorly hydrolysed, is well documented. After overloads small peptides tend to enter the blood. The question arises of why peptide transport remains and disaccharide transport disappears in the process

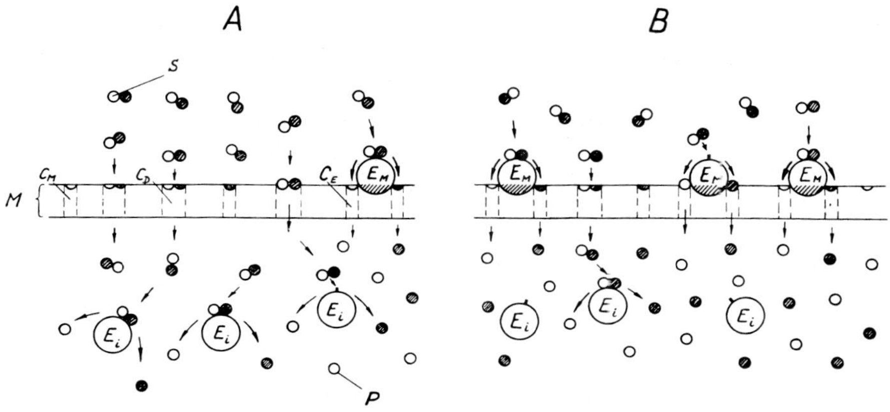

FIG. 11. The relationship between peptide transport and membrane hydrolysis during assimilation of dietary nitrogen. A: with intracellular hydrolysis predominant; B: with membrane hydrolysis predominant; S: dipeptides; P: monomers; M: membrane; E: intracellular enzymes; E_M: membrane enzymes; C_D: transport system for dipeptides; C_M: transport system for free amino acids; C_E: enzyme-bound transport system.

of evolution from prokaryocytes to eukaryocytes. Why did cytosol hydrolysis of disaccharides disappear and that of peptides persist? Organisms which maintain a mechanism of peptide transport, even on a limited scale, have advantages in the process of natural selection. We can visualize at least three instances: (1) absorption of conjugated bile salts; (2) absorption of folic acid derivatives; (3) absorption of poorly hydrolysed peptides containing essential amino acids.

It is obvious that the loss of small quantities of essential amino acids from slowly hydrolysed peptides will lower the nutritive value of proteins. Hence, it is understandable that transport systems for oligomers containing amino acids have been maintained and developed in evolution. As follows from the foregoing, the bulk of dietary nitrogen is transferred by the enzyme transport ensemble of the membrane. The above speculations and data are represented in a simplified form in the scheme shown in Fig. 11.

Several types of fundamental mechanisms which originated at the beginning of life have retained their role (though in modified form) in higher animals up to the present. These are: (1) amino acid transport systems; (2) the peptide transport system; (3) enzymes splitting oligomers into monomers; (4) enzyme transport systems responsible for hydrolysis and transport.

A relationship between various mechanisms of peptide uptake is shown in Fig. 11, schemes A and B. In one of them the transport of oligomers is shown to be predominant, while in the other transport of monomers is predominant, since the enzyme transport ensembles prevail. We believe that a change from scheme A (prevalence of peptide transport and cytosol digestion) to scheme **B** reflects evolution of the mechanism of protein nutrition from prokaryocytes to higher animals with their intensive metabolism and perfected homeostasis. One cannot, however, rule out the possibility that the change in the relative roles of membrane and intracellular hydrolysis occurs to a greater or lesser extent in the process of evolution or even individual adaptation as is the case in bacteria.

References

ADDISON, J. M., BURSTON, D. & MATTHEWS, D. M. (1972) Evidence for active transport of the dipeptide glycylsarcosine by hamster jejunum *in vitro*. *Clin. Sci. (Oxf.) 43*, 907-911

ADIBI, S. A. (1975) Dipeptide absorption and hydrolysis in human small intestine, in *Peptide Transport in Protein Nutrition* (Matthews, D. M. & Payne, J. W., eds.), pp. 147-166, North-Holland, Amsterdam

ASATOOR, A. M., CHENG, B., EDWARDS, K. D. G., LAND, A. F., MATTHEWS, D. M., MILNE, M. D., NAVAB, F. & RICHARDS, A. J. (1970) Intestinal absorption of two dipeptides in Hartnup disease. *Gut 11*, 380-387

AURICCHIO, S., PIERRO, M. & ORSATTI, M. (1971) Assay of peptidase activities of intestinal brush border membrane with L-amino acid oxidase. *Anal. Biochem. 39*, 15-23

BECCIOLINI, A., CARIAGGI, P., ARGANINI, L., CASTAGNOLI, P. & DEGIULI, G. (1974) Effect of 200, 650 and 1200 R on the intestinal disaccharidases and dipeptidases. *Acta Radiol. 13*, 141-152

CASPARY, W. F. (1972) Intestinal transport mechanism for amino acids derived from dipeptides. *Eur. J. Clin. Invest. 2*, 278

CHEESEMAN, C. I. & PARSONS, D. S. (1974) Intestinal absorption of peptides: peptide uptake by small intestine of *Rana pipiens*. *Biochim. Biophys. Acta 373*, 523-526

CHENG, B., NAVAB, F., LIS, M. T., MILLER, T. N. & MATTHEWS, D. M. (1971) Mechanisms of dipeptide uptake by rat small intestine *in vitro*. *Clin. Sci. (Oxf.) 40*, 247-259

CORNELL, J. S. & MEISTER, A. (1976) Glutathione and γ-glutamyl cycle enzymes in crypt and villus tip cells of rat jejunal mucosa. *Proc. Natl. Acad. Sci. U.S.A. 73*, 420-422

CRANE, R. K. (1968) Absorption of sugars, in *Handb. Physiol. Sect.* 6; *Alimentary Canal*, vol. 3: *Intestinal Absorption* (Code, C. F. & Heidel, W., eds.), pp. 1323-1351, American Physiological Society, Washington, D.C.

DAS, M. & RADHAKRISHNAN, A. N. (1974) Studies on the uptake of glycyl-L-leucine by strips of monkey-small intestine. *Indian J. Biochem. Biophys. 11*, 12-16

FERN, E. B., HIDER, R. C. & LONDON, D. R. (1969) The sites of hydrolysis of dipeptides containing leucine and glycine by rat jejunum *in vitro*. *Biochem. J. 114*, 855-861

FOGEL, M. R. & ADIBI, S. A. (1974) Assessment of the role of brush border peptide hydrolases in luminal disappearance of dipeptides in man. *J. Lab. Clin. Med. 84*, 327-333

FORTIN-MAGANA, R., HURWITZ, R., HERBST, J. J. & KRETCHMER, N. (1970) Intestinal enzymes: indicators of proliferation and differentiation in the jejunum. *Science (Wash. D.C.) 167*, 1627-1628

FUJITA, M., PARSONS, D. S. & WOJNAROWSKA, F. (1972) Oligopeptidases of brush border membranes of rat small intestinal mucosal cells. *J. Physiol. (Lond.) 227*, 377-394

GRAY, G. M. & COOPER, H. L. (1971) Protein digestion and absorption. *Gastroenterology 61*, 535-544

HEIZER, W. D., KERLEY, R. L. & ISSELBACHER, K. J. (1972) Intestinal peptide hydrolases differences between brush border and cytoplasmic enzymes. *Biochim. Biophys. Acta 264*, 450-461

HELLIER, M. D., PERRETT, D., HOLDSWORTH, C. D. & THIRUMALAI, C. (1971) Absorption of dipeptides in normal and cystinuric subjects. *Gut 12*, 496-497

KIM, Y. S., MCCARTHY, D. M., LANE, W. & FONG, W. (1973) Alterations in the levels of peptide hydrolases and other enzymes in brush borders and soluble fractions of rat small intestinal mucosa during starvation and refeeding. *Biochim. Biophys. Acta 321*, 262-273

KINTER, W. B. & WILSON, T. H. (1965) Autoradiographic study of sugar and amino acid absorption by everted sacs of hamster intestine. *J. Cell Biol. 25*, 19-39

KOMMISSARCHIK, Ya. J. & UGOLEV, A. M. (1970) Ultrastructure and possible functional importance of glycocalyx of the microvilli of intestinal cells. *Dokl. (Proc.) Acad. Sci. U.S.S.R. 194*, 731-733

KOSHLAND, D. E. (1970) Structure and control. in *The Enzymes* (Boyer, P. D., ed.), vol. 1, pp. 342-397, Academic Press, New York

LANE, A. E., SILK, D. B. & CLARK, M. L. (1975) Absorption of two proline containing peptides by rat small intestine *in vitro*. *J. Physiol. (Lond.) 248*, 143-149

LINDBERG, T., NORÉN, O. & SJÖSTRÖM, H. (1975) Peptidases of the intestinal mucosa, in *Peptide Transport in Protein Nutrition* (Matthews, D. M. & Payne, J. W., eds.), pp. 204-242, North-Holland, Amsterdam

MAROUX, S. & LOUVARD, D. (1976) On the hydrophobic part of aminopeptidase and maltase which bind the enzyme to the intestinal brush border membrane. *Biochim. Biophys. Acta 419*, 189-195

MATTHEWS, D. M. (1972) Rates of peptide uptake by small intestine, in *Peptide Transport in Bacteria and Mammalian Gut (Ciba Found. Symp. 4)*, pp. 71-92, Associated Scientific Publishers, Amsterdam

MATTHEWS, D. M. (1975) Absorption of peptides by mammalian intestine, in *Peptide Transport in Protein Nutrition* (Matthews, D. M. & Payne, J. W., eds.), pp. 61-146, North-Holland, Amsterdam

MATTHEWS, D. M. & ADIBI, S. A. (1976) Peptide absorption *Gastroenterology 71*, 151-161

NICHOLSON, J. A., MCCARTHY, D. & KIM, Y. S. (1973) Differential response of intestinal brush border and cytosol peptide hydrolase activities to variation of dietary protein content. *Gastroenterology 64*, A-95/778

NOACK, R., FRIEDRICH, M., PROLL, J. & UHLIG, J. (1975) Verdauung und Resorption von Proteinen. *Nahrung 19*, 891-901

NORDSTRÖM, C., DAHLQVIST, A. & JOSEFSSON, L. (1968) Quantitative determination of enzymes in different parts of the villi and crypts of rat small intestine. Comparison of alkaline phosphatase, disaccharidases and dipeptidases. *J. Histochem. Cytochem. 15*, 713-721

NORÉN, O., SJÖSTRÖM, H. & JOSEFSSON, L. (1971) Preparation of highly purified glycyl-L-leucine dipeptidase from pig intestinal mucosa. *Acta Chem. Scand. 25*, 1913-1915

PETERS, T. J. (1972) Subcellular fractionation of the enterocyte with special reference to peptide hydrolases, in *Peptide Transport in Bacteria and Mammalian Gut (Ciba Found. Symp. 4)*, pp. 107-115, Associated Scientific Publishers, Amsterdam

PETERS, T. J. (1975) The subcellular localisation of intestinal peptide hydrolases, in *Peptide Transport in Protein Nutrition* (Matthews, D. M. & Payne, J. W., eds.), pp. 243-267, North-Holland, Amsterdam

RAMASWAMY, K., MALATHI, P., CASPARY, W. F. & CRANE, R. K. (1974) Studies on the transport of glucose from disaccharides by hamster small intestine *in vitro*. II. Characteristics of the disaccharidase-related transport system. *Biochim. Biophys. Acta 345*, 39-48

RHODES, J. B. & FOLSCROFT, J. (1972) Hydrolysis of small gliadin peptides by hog intestinal microvillus membranes. *Gastroenterology 62*, 865 (Abstr.)

RHODES, J. B., EICHHOLZ, A. & CRANE, R. K. (1967) Studies on the organization of the brush border in intestinal epithelial cells. IV. Aminopeptidase activity in microvillus membranes of hamster intestinal brush borders. *Biochim. Biophys. Acta 135*, 959-965

RUBINO, A., FIELD, M. & SHWACHMAN, H. (1971) Intestinal transport of amino acids residues of dipeptides. 1. Influx of the glycine residue of glycyl-L-proline across mucosal border. *J. Biol. Chem. 246*, 3542-3548

SEMENZA, G. (1971) Digestion and absorption of sugars in the intestine, in *XXV International Congress of Physiological Sciences*, vol. 8, pp. 175-176, Munich

SIGRIST-NELSON, K. (1975) Dipeptide transport in isolated intestinal brush border membrane. *Biochim. Biophys. Acta 394*, 220-226

SILK, D. B. A. (1974) Peptide absorption in man. *Gut 15*, 494-501

SMIRNOVA, L. F., TIMOFEEVA, N. M. & UGOLEV, A. M. (1976) On the subcellular localization of dipeptide hydrolysis in small intestine. *Dokl. (Proc.) Acad. Sci. U.S.S.R. 228*, 992-994

SMITH, E. L. (1951) The specificity of certain peptidases, in *Advances in Enzymology* (Nord, F. F., ed.), vol. 12, pp. 191-257, Interscience, New York

SMYTH, D. H. (1972) Peptide transport by mammalian gut, in *Peptide Transport in Bacteria and Mammalian Gut (Ciba Found. Symp. 4)*, pp. 59-66, Associated Scientific Publishers, Amsterdam

TAKESUE, Y. & NISHI, Y. (1976) Topographical relationship between sucrase and leucine β-naphthylamidase on the microvilli membrane of rabbit intestinal mucosal cells. *J. Biochem. (Tokyo) 79*, 479-488

UGOLEV, A. M. (1964) Interactions of digestive and absorptive processes in intestinal cell, in *Physiology and Pathology of Absorption in the Gastrointestinal Tract*, pp. 7-20, Odessa

UGOLEV, A. M. (1968) *Physiology and Pathology of Membrane Digestion*, Plenum Press, New York

UGOLEV, A. M. (1972*a*) Membrane digestion and peptide transport, in *Peptide Transport in Bacteria and Mammalian Gut (Ciba Found. Symp. 4)*, pp. 123-143, Associated Scientific Publishers, Amsterdam

UGOLEV, A. M. (1972*b*) [*Membrane Digestion. Polysubstrate Processes, Organization and Regulation*] Izd. 'Nauka', Leningrad

UGOLEV, A. M. & DE LAEY, P. (1973) A concept of enzymic hydrolysis on cell membranes. *Biochim. Biophys. Acta 300*, 105-128

UGOLEV, A. M., TIMOFEEVA, N. M., IEZUITOVA, N. N., DOMBROVSKAYA, M. P. & RAHIMOV, K. R. (1969) Influence of some stress agents on the digestive functions on intestinal epithelium. *Dokl. (Proc.) Acad. Sci. U.S.S.R. 188*, 489-492

UGOLEV, A. M., IEZUITOVA, N. N., KUSHAK, R. I. & SHCHERBACOV, G. G. (1970*a*) Distribution of invertase and dipeptidase activity in the crypt-villus system. *Nahrung 14*, 437-452

UGOLEV, A. M., ZIGURE, D. R. & NURKS, E. E. (1970*b*) Accumulating preparation of mucosa – a new method of investigation on initial stages of the transfer of substances through intestine cell. *Sechenov Physiol. J. U.S.S.R. 56*, 1638-1641

UGOLEV, A. M., GRUZDKOV, A. A., DE LAEY, P., EGOROVA, V. V., IEZUITOVA, N. N., KOLTUSHKINA, G. G., TIMOFEEVA, N. M., TULYAGANOVA, E. CH., TSVETKOVA, V. A., CHERNYAKHOVSKAYA, M. YU. & SHCHERBAKOV, G. G. (1975) Substrate interactions on the intestinal mucosa: a concept for the regulation of intestinal digestion. *Br. J. Nutr. 34*, 205-220

UGOLEV, A. M., NURKS, E. E. & GURMAN, E. G. (1976) Some features of kinetics of accumulation of the free glucose and glucose formed in the hydrolysis of disaccharides in accumulating preparations of rat small intestinal mucosa. *Dokl. (Proc.) Acad. Sci. U.S.S.R. 231*, 1018-1020

WOJNAROWSKA, F. & GRAY, G. M. (1975) Intestinal surface peptide hydrolases: identification and characterization of three enzymes from rat brush border. *Biochim. Biophys. Acta 403*, 147-160

Discussion

Smyth: I would accept that we can have membrane digestion provided that it means that the products of digestion do not escape into the luminal fluid. We fundamentally agree but we use different words.

[*Ugolev:** I take this opportunity to express my admiration for Professor Smyth, whose investigations and ideas had a great influence on many lines of physiology. I think that we are approaching a common terminology. In my view spatial criteria seem to be preferable to start with. In this connection, let me remind you that our classification of different types of digestion (Ugolev 1965) is now becoming accepted, with slight modifications.

'At present no universally satisfactory classification of digestion is available. The one used in this review is based on relevant literature dealing with various types of digestion.

'There are two general types of digestion. 1) Extracellular digestion is caused by enzymes secreted by cells and acting outside the cells and is best described as extracellular or distant digestion; it may be: *a*) cavital, when foods are split in special digestive cavities, or *b*) extracavital, when hydrolytic enzymes are secreted into the medium surrounding the organism or are released directly into a food substrate. 2) Cellular digestion is characterized by destruction of food by enzymes associated with cell structures; it may be subdivided into: *a*) intracellular and *b*) membrane digestion.

'Intracellular digestion may occur in special intracellular cavities, designated then as vacuole or intracellular extraplasmatic digestion; in cytoplasm, as cytoplasmic digestion; in special subcellular structures, as granular digestion; and on the internal surface of cell membranes, as submembrane digestion. The latter, found in yeasts, probably also occurs in higher animals.

'As membrane digestion occurs on the external surface of cell membranes it resembles extracellular digestion, but in many respects it is similar to intracellular digestion since it is due to the action of enzymes associated with the cell structures...' (Ugolev 1965, pp. 555-556).

Considering the progress made in the study of membrane digestion, a subdivision into glycocalyx digestion, surface membrane digestion, intracellular digestion and submembrane digestion may seem more appropriate.]

Semenza: Were the data for the Hill plot of amylase obtained after incubation for 1 h?

De Laey: Yes.

---

* Professor Ugolev's comments in this discussion were added after the meeting.

Norén: You mentioned that the brush border enzymes are organ-specific. I think the contrary is true. The experimental evidence so far for the brush border enzymes and the kidney enzymes shows that they are rather similar in their inhibition patterns, and they are immunologically identical, as far as we know. Therefore so far there is no evidence for any organ-specific peptidase.

De Laey: You are referring to intestinal and kidney peptidases, and I agree. From the pathological point of view we know that they belong to the same family. If one is affected, the other will be affected too. When I speak about organ-specificity I mean differences between, e.g. liver and muscle, rather than between intestine and kidney.

[*Ugolev:* Many of the kidney and intestinal brush border enzymes such as peptidases are very similar. Their immunological identity appears to be related to the structure of the hydrophilic part of the enzyme molecule. It is of interest that regulatory properties have recently been shown (Ugolev *et al.* 1975) to be non-identical. Thus, tributyrin, a powerful inhibitor of intestinal glycylleucine dipeptidase, has been demonstrated to have no effect on the kidney enzyme. We have recently found that the hydrophobic subunit is responsible for the regulatory properties (Gozitte *et al.* 1976).]

Silk: I was very interested by the concept that membrane hydrolysis of peptides may precede transport. However, one problem of extrapolating *in vitro* enzyme data to the *in vivo* situation is that *in vitro* assays are usually done at optimal pH, which is not necessarily the pH to which the respective enzymes are exposed *in vivo*. This may especially be the case for brush border peptidases, for which *in vitro* assays have an optimal pH in the 7.0–8.0 range (Kim & Brophy 1976), whereas *in vivo* they are exposed to a more acid pH (Lucas *et al.* 1975).

De Laey: The microclimate will dictate the velocity at which hydrolysis might happen. I wouldn't have thought that pH was so important physiologically. The digestive capacity, as Professor Matthews has shown, is about 10–20 or even 100 times higher than the capacity to absorb amino acids. Yet none of us has constant diarrhoea. Even at suboptimal conditions for the enzyme I don't think it would affect the digestive capacity.

[*Ugolev:* With reference to the functions of the membrane-bound enzyme, our ideas on microclimate and indexes such as pH have undergone significant changes as a result of (1) ultramicroheterogeneity of the surface, (2) interactions between the enzyme and the locus of support, (3) specific properties of the interphases. It may be suggested that the enzyme–membrane complex has formed during evolution as an optimal system. Really, the kinetic characteristics deteriorate after solubilization of a number of the brush border enzymes and improve after reconstruction within the composition of liposomes. Thus, there exist conditions at the microvillus surface which provide a high efficiency of naturally acting enzymes. The data have been published and the physicochemical aspects analysed (Ugolev & Koltushkina 1975).]

Matthews: I know Professor Ugolev has accepted that peptide transport occurs in the intestine. The only controversy is about whether peptide transport is trivial and membrane digestion, as he calls it, is important. The root of the disagreement is related to the fact that two processes are involved in peptide absorption: hydrolysis in the brush border region followed by uptake of amino acids and peptide transport followed by intracellular hydrolysis. Evidence suggesting a dual mechanism appears in the early papers of Newey & Smyth, and since about 1970 I have maintained that there are two processes (Cheng *et al.* 1971). What I think is happening is that Professor Ugolev does ingenious experiments in which he always manages to emphasize the brush border hydrolysis component (see Matthews 1975). I agree with his interpretation in so far as some hydrolysis of many peptides clearly occurs on the luminal side of the brush border membrane, but what I cannot see is the evidence for coupled transport. I think this brush border hydrolysis releases amino acids in free solution. I don't see why he thinks that there is coupled transport, i.e. that amino acids released from peptides are taken up by special sites not available to free amino acids.

Why is the transport of an unhydrolysable peptide such as carnosine or Gly-Sar inhibited by an extremely wide variety of other peptides which are readily hydrolysed? Why do peptides such as Gly-Gly, Gly-Leu, Gly-Pro and Pro-Gly all behave as if they shared a common transport system in spite of being hydrolysed by different hydrolases?

Professor Ugolev has a scheme (Fig. 5) in which a peptide becomes attached to a membrane-bound hydrolase and this modifies an adjacent amino acid uptake site so that it will readily accept the amino acid liberated. If the site is modified in this way by attachment of peptides to the hydrolase, why doesn't it also readily accept other amino acids which are around it in free solution? In other words, it seems to me that this scheme would enable free amino acids to compete for transport with peptide-bound amino acids, whereas in fact amino acid transport and peptide transport appear to be independent processes.

[*Ugolev:* Professor Matthews has put several difficult questions. Certainly, even though the membrane localization of hydrolysis is demonstrable, there are no direct methods yet for surveying the further path of hydrolysis products from the final enzyme into the water phase or onto the enzyme-bound carrier. In the first case the products released will be absorbed as an amino acid mixture and exhibit competition under certain conditions. However, it has been experimentally shown, in particular by our group, that during membrane hydrolysis of some peptides competition between the amino acids released may be avoided, and facilitated transport occurs (Smirnova & Ugolev 1972). The latter depends on the enzyme-transport interactions, such as a direct transfer of amino acids from the final enzyme to the entrance into the transport system. The data on enzyme-carrier cooperativities first obtained in our laboratory are given in our paper.

If a carrier coupled with the enzyme is periodically activated by this enzyme at the step when hydrolysis products are formed, then a predominant absorption of monomers released by the enzyme, rather than those from the liquid phase, will be achieved. Besides, there might be other insufficiently known conditions which facilitate the enzyme-transport cooperativities, such as approximation of the bonding sites of the enzyme and carrier at the conformational transitions probably utilizing the energy of hydrolysis, and so on. It is important that we succeeded in demonstrating the kinetic properties of cooperativity in the enzyme–transport complex (see paper).

It is impossible to answer the questions about inhibition of unhydrolysable peptides and about the behaviour of certain peptides in terms of the membrane concept. However, many peptides inhibit competitively the membrane hydrolysis of other peptides, which suggests an important role for membrane digestion.]

Josefsson: I think the anoxia experiments fit very well with the other evidence. The bad thing is that the substrates used are hydrolysed by membrane-bound peroxidases. All those substrates are split by the enzyme(s) you told us about, Dr De Laey. If you expected an effect with the anoxia experiments you should have used glycylproline, then you would know it was not interacting with the membrane aminopeptidase.

De Laey: That was done by Professor Matthews.

Adibi: Questions about obesity always interest nutritionists, who ask whether the mucosal cells of obese people are 'super' in terms of transport or hydrolysis. I have looked for this in human obesity and I am sorry to report that we have not found it. Did you actually find a high specific activity on the membrane for transport or hydrolysis in the intestine of obese animals?

De Laey: Yes, but we observed it only during a very short while when the body was gaining weight rapidly. Afterwards, obesity is a static condition. We see obese people and we see lean people, but we never see people becoming obese. They have already stabilized and changed the gut to conform to their nutritional patterns. We can't study this in human beings, only in animals.

References

CHENG, B., NAVAB, F., LIS, M. T., MILLER, T. N. & MATTHEWS, D. M. (1971) Mechanisms of dipeptide uptake by rat small intestine *in vitro*. *Clin. Sci. (Oxf.) 40*, 247-257

GOZITIE, I. K., KOLTUSHKINA, G. G., EGOROVA, V. V. & UGOLEV, A. M. (1976) Comparative kinetic characterization of some brush border enzymes solubilized from the membranes by means of detergents and proteases, in *Digestive Enzymes*. Communications of the 1st bilateral symposium, USSR-Czechoslovakia, Uzhgorod

KIM, Y. S. & BROPHY, E. J. (1976) Rat intestinal brush border membrane peptidases 1. Solubilization, purification and physiochemical properties of two different forms of the enzyme. *J. Biol. Chem. 251*, 3199-3205

LUCAS, M. L., SCHNEIDER, W., HAVERICH, F. J. & BLAIR, J. A. (1975) Direct measurement by pH-microelectrode of the pH microclimate in rat proximal jejunum. *Proc. R. Soc. Lond. B Biol. 192*, 39-48

MATTHEWS, D. M. (1975) Intestinal absorption of peptides. *Physiol. Rev. 55*, 537-608

SMIRNOVA, L. F. & UGOLEV, A. M. (1972) Relationship between transport of some glycine-containing peptides and equivalent mixtures of constituent amino acids in the rat small intestine. *Dokl. Akad. Nauk. U.S.S.R. 206*, 763-765

UGOLEV, A. M. (1965) Membrane (contact) digestion. *Physiol. Rev. 45*, 555-595

UGOLEV, A. M. & KOLTUSHKINA, G. G. (1975) Membrane digestion and enzyme microvillus apparatus, in *Structure and Functions of Biological Membranes*, pp. 276-306, Izd. 'Nauka', Academy of Sciences of USSR, Moscow

UGOLEV, A. M., GRUZDKOV, A. A., DE LAEY, P., EGOROVA, V. V., IEZUITOVA, N. N., KOLTUSHKINA, G. G., TIMOFEEVA, N. M., TULYAGANOVA, E. CH., TSVETKOVA, V. A., CHERNYAKHOVSKAYA, M. YU. & SHCHERBAKOV, G. G. (1975) Substrate interactions on the intestinal mucosa: a concept for regulation of intestinal digestion. *Br. J. Nutr. 34*, 205-220

Intestinal hydrolysis of disaccharides and peptides: comparison of hydrolases and perfusion studies

J. B. RHODES, C. ARVANITAKIS and J. FOLSCROFT

Departments of Medicine and Physiology, University of Kansas Medical Center, College of Health Sciences and Hospital, Kansas City, Kansas

Abstract This study compares the hydrolysis of gliadin peptides by intestinal tissue fractions and compares the hydrolysis of disaccharides and two tripeptides with substrate disappearance and accumulation of products during intestinal perfusion.

First, hog intestine was fractionated to isolate microvillus membranes and a sucrase-free cytoplasmic fraction. Partially purified wheat gliadin was hydrolysed with pepsin and peptide fractions containing two to eight amino acids were combined to provide a peptide mixture substrate. Hog microvillus membranes hydrolysed the peptide mixture at appreciable rates—46% of the rate of sucrase. The peptide mixture was extensively (93%) hydrolysed. The cytoplasmic fraction hydrolysed the mixture at a slower rate, but more extensively. Although membranes rapidly and extensively hydrolyse the peptide mixture, these studies do not indicate the site of nutritional peptide hydrolysis.

Secondly, *in vitro* hamster tissue hydrolysis of sucrose, maltose, leucylglycylglycine and prolylglycylglycine was compared with substrate disappearance and accumulation of product during *in vivo* intestinal perfusion. Sucrose, maltose and Leu-Gly-Gly are rapidly hydrolysed by hamster microvillus membranes and product appeared in the lumen during perfusion. In sharp contrast, Pro-Gly-Gly was poorly hydrolysed by membranes and little product was recovered during perfusion, but it disappeared at a rate comparable to maltose.

It is concluded that some peptides, i.e. Leu-Gly-Gly, like sucrose and maltose, are hydrolysed by microvillus membranes with subsequent product transport. Other peptides, i.e. Pro-Gly-Gly, are transported intact across the membrane and predominantly hydrolysed by cytoplasmic enzymes.

The cellular site for nutritional peptide hydrolysis remains controversial. Three theories have been proposed. One theory proposes that peptides are transported into intestinal epithelial cells and hydrolysed by cellular or cytoplasmic peptidases (Newey & Smyth 1960; Rubino *et al.* 1971; Adibi 1975; Matthews 1975). Others believe peptide hydrolysis occurs in the brush border area, with subsequent product transport (Ugolev 1965; Rhodes 1968; Gray & Cooper 1971; Rhodes & Folscroft 1972; Kim *et al.* 1974). Another possibility is that large oligopeptides of three or more amino acids are hydrolysed at the membrane to free amino acids and small peptides (molecular weight less than 250) which are transported into the cell where final hydrolysis occurs (Peters *et al.* 1972; Peters 1973; Adibi *et al.* 1975).

Our approach to the problem has been twofold. First, a gliadin peptide mixture containing two to eight amino acids is prepared. The peptide mixture is incubated with isolated hog microvillus membranes and a cytoplasmic fraction so that we can compare the hydrolytic potential of both fractions. Secondly, the hamster tissue hydrolysis of disaccharides and tripeptides is compared to substrate disappearance and product accumulation during intestinal perfusion studies.

METHODS

Hog tissue preparation

Young weaned hogs of mixed breeds weighing over 25 kg are dewormed and fed commercial hog chow for 10–14 days in the animal quarters. A positive weight gain is documented and no parasites are detected by microscopic examination of the faeces. Jejunal segments are removed at death or at operation. Hogs are killed by a stunning hammer blow to the head and the aorta is transected with a long knife passed behind the sternum. Operated hogs are tranquillized with phencyclidine 2 mg/kg and anaesthetized with intravenous pentobarbital (pentobarbitone) 15–20 mg/kg through an ear vein. Normal saline is infused and additional pentobarbitone injected if required. Respiration is assisted with a motor-driven pump through an endotracheal tube. The abdomen is prepared and opened aseptically. Jejunal segments of 40–60 cm are removed, the bowel is rejoined by end-to-end anastomosis and the abdomen is closed. Recovery is rapid and uneventful.

The removed jejunal segments are rinsed with normal saline and tissue fractions prepared by the method of Eichholz & Crane (1965) with minor modifications. Brush borders are prepared in 5 mM-EDTA, pH 7.4, and nuclei are removed by saline agglutination (Forstner *et al.* 1968). Microvillus membranes are disrupted with freshly prepared 1.0 M-Tris, pH 7.4, and separated by centrifugation on a glycerol gradient (Eichholz & Crane 1965).

A sample of the first supernatant from the brush border preparation is centrifuged at 26 400 *g* for 30 min to obtain a sucrase-poor supernatant to represent the cytoplasmic fraction. Tissue fractions are dialysed overnight (18 h) in 5 mM-EDTA, pH 7.4 at 4 °C.

Gliadin peptide preparation

Partially purified gliadin is prepared from commercial crude gliadin. Protein fractionation is monitored with cellulose acetate electrophoresis and aluminum lactate–lactic acid buffer, 17 mM, pH 3 (Woychick *et al.* 1961) at room temperature in a Beckman Microzone cell. The bands are fixed and stained with 10% sulphosalicylic acid containing Coomassie Blue dye, 0.25%. Crude gliadin (2 g) is slowly dissolved in 100 ml 0.03 M-acetic acid, pH 3.0, at room temperature and centrifuged at 17 000 *g* for 30 min at 4 °C. The gel precipitate migrates poorly as glutenin. Sodium chloride is added to the supernatant to 0.2 M and centrifuged at 133 600 *g* for 2 h. The supernatant contains rapidly migrating albumins and globulins while the precipitate contains proteins migrating to the α, β and γ positions. This final precipitate contains 40–50% of the original protein.

The precipitate enriched in α, β and γ gliadins is dissolved in 50 ml 0.03 M-acetic acid, placed in a Visking 18/32 dialysis sack and 5 mg pepsin added. The dialysis sack is suspended in 400 ml 0.03 M-acetic acid at 37 °C and the surrounding solution constantly stirred for 4–6 h. The dialysate is reduced to dryness by a rotating vacuum condenser and taken up in water.

Samples (5 ml) are passed through a 2.5 × 30 cm Sephadex G-25 column equilibrated and eluted with 0.03 M-acetic acid. Fractions of 10 ml are collected and monitored by ninhydrin (Matheson & Tattrie 1964). Peptide fractions are hydrolysed by being boiled in 2.5 M-NaOH for 2 h, cooled and brought to pH 5.0 with 2.5 M-acetic acid and again analysed with ninhydrin. Leucine is the standard. Fractions which contain two to eight amino acids are combined, again concentrated by vacuum and taken up in water. This peptide mixture is used as the peptide substrate. A sample is acid hydrolysed and the amino acids are determined by quantitative column chromatography.

Hog enzyme assays

Protein, sucrase (β-fructofuranosidase, EC 3.2.1.26) and leucyl-naphthylamidase are measured by standard methods (Eichholz & Crane 1965; Rhodes *et al.* 1967). Activity is expressed as nmol substrate hydrolysed (per minute per mg protein). Percentage distribution of activity in the various tissue fractions is expressed as a percentage of the similarly treated whole homogenate. Tissue hydrolysis of the gliadin peptide mixture is measured by the increase in ninhydrin colour as nmol leucine equivalents (per minute per mg protein). Hydrolytic rates are determined between 10–50% increase in colour above the initial substrate blanks where the reaction rate is nearly linear.

Microvillus membranes are used to establish maximal conditions for hydrolysis. Divalent cations (final concentration 1 mmol/l) in the chloride form are added before the buffer. Phosphate buffer, 50 mM, is used. Heat inactivation studies are carried out at 60 °C for varying times. The results are analysed by Student's t-test.

Hamster tissue fractions and intestinal perfusion

The methods used for hamster tissue fractionation and intestinal perfusion have been published (Arvanitakis *et al.* 1976). In brief, hamster microvillus membranes are prepared, but the fractions are dialysed overnight against 10 mM phosphate buffer, pH 7.4. The cytoplasmic fraction is prepared by high speed centrifugation to precipitate particles and sucrase. Zinc chloride (final concentration 1 mmol/l) is added to enzyme just before the buffer and substrate.

Lightly anaesthetized adult hamsters are perfused from the ligament of Treitz to the terminal ileum. The solution contains 20 mM substrate, sodium chloride to isotonicity (290 $\pm$ 10 mosM) and 0.5% polyethyleneglycol 4000 (PEG) as the non-absorbable water-soluble marker. After a 30-min equilibration period, three 15-min collections are obtained at a flow rate of 0.4 ml/min. Disaccharides and products are measured by chemical and enzymic methods. Peptides and products are determined by quantitative amino acid chromatography.

RESULTS

Hog tissue fractions

Table 1 summarizes the percentage distribution of protein, sucrase and leucyl-naphthylamidase in operated hogs to establish the method in this species. Microvillus membranes contain 1.6% of the protein with 22% of total sucrase and 15% total leucyl-naphthylamidase. A rheostat setting of 40 with a Waring Blendor time of 45 s gives reproducible results with highest yields. These conditions are somewhat milder than for hamster preparations and imply that hog tissue is more easily disrupted. This distribution is similar to previous reports with hamsters (Miller & Crane 1961; Holt & Miller 1962; Eichholz & Crane 1965). The distributions are more variable in 18 individual hog preparations of mixed breeds, but the general patterns are comparable in individual hogs.

TABLE 1

Percentage distribution of protein, sucrase and leucyl-naphthylamidase (LNA'ase) in operated hog tissue fractions*

Fraction	*Protein*	*Sucrase*	*LNA'ase*
Whole homogenate	95 ± 6	110 ± 9	104 ± 9
1st supernatant	83 ± 5	44 ± 8	54 ± 7
2nd supernatant	6 ± 0.4	10 ± 3	11 ± 2
3rd supernatant	2 ± 0.2	8 ± 3	5 ± 0.6
Brush borders	5 ± 0.6	47 ± 7	34 ± 5
Microvillus membranes	1.6 ± 0.6	22 ± 5	15 ± 3

* Mean ± S.E.M.; $n = 6$

Gliadin peptides

Fig. 1 shows the elution pattern of pepsin-hydrolysed gliadin from Sephadex G-25 columns as determined with ninhydrin. The average number of amino acids in the peptide fractions determined after alkaline hydrolysis is indicated. Fractions averaging two to eight amino acids (120–200 ml) are combined and constitute the gliadin peptide mixture. The average amino acid content of the peptide mixture is 3.4 ± 0.3 amino acids. Two-dimensional chromatography of the mixture reveals at least 27 different peptides. The mixture is stable in solution at 4 °C for four to five weeks. Blank values decrease at five to six weeks, and fresh substrate is required.

Table 2 compares the acid hydrolysed amino acid content of the gliadin peptide mixture with previous reports of purified α- (Bernardin *et al.* 1967) and γ-gliadin (Huebner *et al.* 1967). The results are comparable for arginine, threonine, serine, glycine, alanine and methionine. Other amino acids are present in lesser amounts, especially glutamine and proline, which constitute 50% of the total amino acids in gliadin. The amino acid composition of the gliadin peptide mixture is not representative of whole gliadin. However, pepsin hydrolyses many peptide bonds and is likely to yield peptides with many different N-terminal amino acids.

Hog microvillus membranes are incubated with the gliadin peptide mixture to establish maximal rates of hydrolysis and partially characterize the activity. Dialysis for 18 h against 5 mM-EDTA, pH 7.4, inhibits the activity to only 10%. The activity is maximally restored by cobalt, closely followed by zinc which restores 93 ± 5% of the activity. Manganese restores half of the activity while magnesium and calcium have little effect. While EDTA inhibition has not been noted by all investigators in other species, Maroux *et al.* (1973) report that two

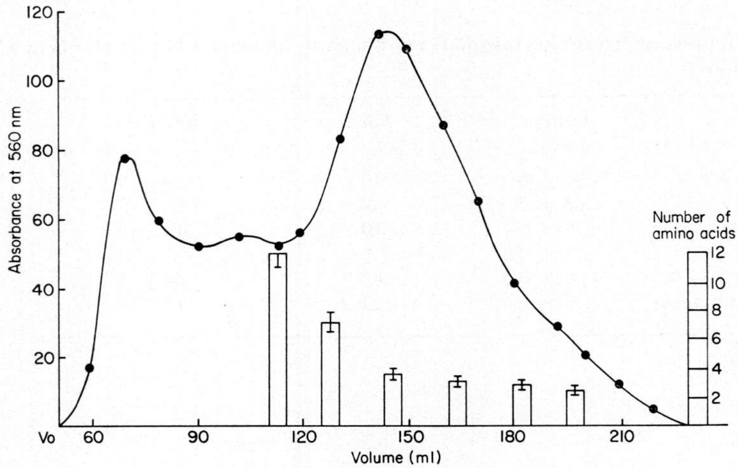

FIG. 1. Elution pattern of peptic digest of gliadin from a 2.5 × 30 cm Sephadex G-25 column equilibrated and eluted with 0.03 M-acetic acid. Fractions of 10 ml are collected. The line is the absorbance of the ninhydrin colour. The bars are the number of amino acids after alkaline hydrolysis. Mean ± S.E.M.; $n = 6$.

TABLE 2

Amino acid composition of gliadin peptide mixture compared to purified α- and γ-gliadin (mmol/100g)

	α-Gliadin	*γ-Gliadin*	*Peptic peptides*
Lys	5	5	9
His	19	15	10
Arg	18	15	14
Asp	25	20	15
Thr	13	18	18
Ser	47	38	39
Glu	330	317	107
Pro	120	148	28
Gly	23	25	28
Ala	23	25	26
1/2 Cys	18	12	1
Val	42	43	28
Met	7	12	6
Ile	39	37	27
Leu	71	62	43
Tyr	29	16	10
Phe	32	38	14
Try	11	4	NT

NT-not tested

molecules of zinc are associated with each molecule of hog brush border aminopeptidase.

The cobalt-activated enzyme has a broad pH maximum from 6.7 to 7.5 in phosphate buffer. Maroux & Louvard (1975) report a comparable pH curve. A pH of 7.4 is selected for the standard assay. The membrane activity is stable (93 $\pm$ 5%) after 60 °C for 12–18 min. In contrast, the cytoplasmic activity is heat-labile and only 21 $\pm$ 4% is present after 6–12 min. These results are comparable to a previous report in rat tissue (Heizer *et al.* 1972). These differences in heat stability are not present in all species, especially not the rabbit and guinea pig (Peters 1972).

The rate of hydrolysis of gliadin peptide mixtures by selected tissue fractions is compared to sucrase and leucyl-napththylamidase in Table 3. Sucrase is low in the cytoplasmic fraction, which contains only 1% of the total sucrase. The cytoplasmic fraction is essentially free of microvillus membrane particles. Sucrase activity is increased 15-fold in the microvillus membranes compared to the whole homogenate. Cytoplasmic leucyl-naphthylamidase is considerably more active than sucrase. There are cytoplasmic enzymes which hydrolyse leucyl-naphthylamide and the earlier literature has been reviewed (Rhodes 1968). Like sucrase, leucyl-naphthylamidase is increased 14-fold in microvillus membranes. This distribution is comparable to previous reports (Holt & Miller 1962; Rhodes *et al.* 1967).

The tissue distribution of gliadin peptide hydrolysis differs from that of sucrase and leucyl-naphthylamidase. The rate of gliadin peptide hydrolysis by the cytoplasmic fraction is comparable to that by the whole homogenate. The cytoplasm contains enzymes known to hydrolyse dipeptides and tripeptides (Peters 1970). Microvillus membranes actively hydrolyse the gliadin peptides, and the activity is increased 22-fold over the whole homogenate. The activity is brisk: 46% the rate of sucrase. Gliadin peptide hydrolysis by microvillus membranes increases as the substrate concentration increases (Fig. 2). At low substrate concentration with a 240 min incubation, 96 $\pm$ 14% of the gliadin peptides are hydrolysed to free amino acids as estimated by alkaline hydrolysis. The percentage of hydrolysis decreases as the substrate concentration increases, due to product inhibition. This high activity and extensive hydrolysis broadens the aminopeptidase activity of microvillus membranes (Rhodes *et al.* 1967; Rhodes 1968; Maroux *et al.* 1973; Maroux & Louvard 1975; Kim *et al.* 1976; Shoaf *et al.* 1976). These results support the theory that microvillus membranes participate in nutritional digestion of food peptides.

However, as noted in Table 4, the cytoplasmic fraction is also capable of extensively hydrolysing the gliadin peptide mixture to 109 $\pm$ 17%, that is, more completely than microvillus membranes, but at a slower rate (Table 3).

TABLE 3

Sucrase, leucyl-naphthylamidase and gliadin peptide mixture hydrolysis by hog tissue fractions*, nmol/(min mg)

Fraction	*Sucrase* *(n = 18)*	*LNA'ase* *(n = 18)*	*Gliadin peptides* *(n = 6)*
Whole homogenate (WH)	100 ± 5	140 ± 70	32 ± 15
Cytoplasm	1 ± 1	16 ± 5	35 ± 12
Brush borders	840 ± 600	1000 ± 680	NT
Microvillus membranes (MM)	1530 ± 730	1950 ± 730	710 ± 210
Increase (fold) MM/WH	15	14	22

* Mean ± S.E.M.
NT – not tested

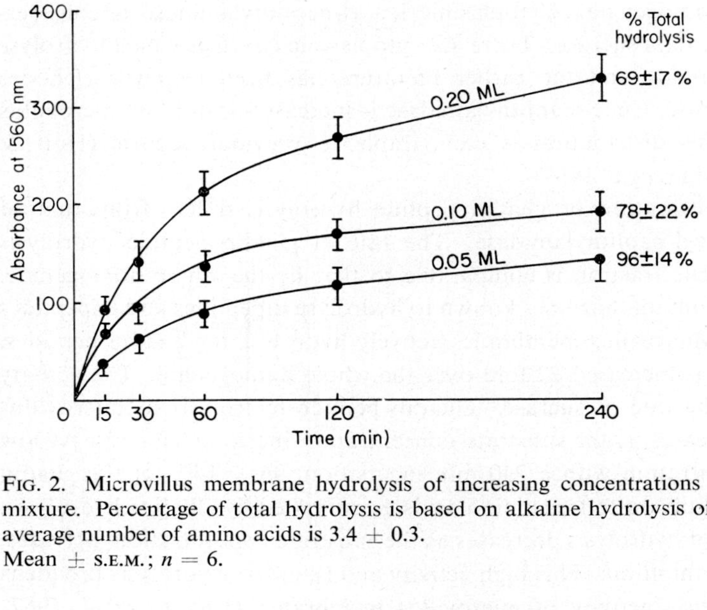

FIG. 2. Microvillus membrane hydrolysis of increasing concentrations of gliadin peptide mixture. Percentage of total hydrolysis is based on alkaline hydrolysis of the mixture. The average number of amino acids is 3.4 ± 0.3.
Mean ± S.E.M.; $n = 6$.

The overestimate of the percentage hydrolysis of the gliadin peptide mixture is not due to enzyme (protein) autohydrolysis. It is probably a combination of alkaline degradation of amino acids and the presence of N-terminal amino acids which give low colour with ninhydrin, for example proline. Cytoplasmic peptidases are capable of hydrolysing the gliadin peptide mixture more extensively than microvillus membranes. These results support the transport

TABLE 4

Percentage hydrolysis of increasing amounts of gliadin peptide mixture by microvillus membranes and cytoplasmic fraction*

Substrate (ml)	*Microvillus membranes (%)*	*Cytoplasmic fraction (%)*
0.05	96 ± 14	109 ± 17
0.10	78 ± 22	85 ± 17
0.20	69 ± 17	76 ± 12

* Mean ± S.E.M.; $n = 5$

TABLE 5

Hydrolysis of substrates by hamster tissue fractions*, nmol/(min mg)

Fraction	*Sucrose*	*Maltose*	*Leu-Gly-Gly*	*Pro-Gly-Gly*
Whole homogenate (WH)	70 ± 7	240 ± 22	743 ± 163	135 ± 27
Cytoplasmic fraction	3 ± 0.3	23 ± 1	1032 ± 242	453 ± 88
Microvillus membranes (MM)	2285 ± 262	7590 ± 1202	4650 ± 957	7 ± 3
Increase (fold) MM/WH	33	33	6	0.05

* Mean ± S.E.M.; $n = 6$

and cytoplasmic hydrolysis theory. These studies do not differentiate the cellular locus of gliadin peptide hydrolysis.

Hamster tissue hydrolysis and intestinal perfusion

We recently reported the hydrolysis of sucrose, maltose, leucylglycylglycine and prolylglycylglycine in hamster tissue and the disappearance of substrate and accumulation of product during intestinal perfusion (Arvanitakis *et al.* 1976). Table 5 lists the results of the enzymic activity. As in the hog preparation, sucrase is associated with microvillus membranes with minimal activity in the cytoplasmic fraction. Membrane specific activity is 33-fold greater than in the whole homogenate. Maltase (α-glucosidase, EC 3.2.1.20) has a similar distribution, but with more activity in the cytoplasmic fraction. Like sucrase, maltase activity is 33-fold greater in the membrane than in the whole homogenate. The distribution of Leu-Gly-Gly hydrolysis is different. This activity is increased in the cytoplasmic fraction and increased sixfold in microvillus membranes. Sucrase, maltase and Leu-Gly-Gly hydrolysis are considerably increased in microvillus membranes when compared to hydrolysis in the whole homogenate. In sharp contrast, Pro-Gly-Gly hydrolysis is negligible in the microvillus membranes and increased threefold in the cytoplasmic fraction. Thus, maximal rates for sucrase, maltase and Leu-Gly-Gly hydrolysis reside with microvillus membranes while Pro-Gly-Gly hydrolysis is associated with the cytoplasmic fraction and is just detectable in the membranes. These results are statistically significant ($P < 0.001$).

Perfusion studies

The results of intestinal perfusion studies in six hamsters with each of the four substrates are summarized in Fig. 3. The results are the measured substrate and products in the effluent from each hamster intestine. For sucrose, 360 μmol are infused and 300 μmol (83 %) are recovered. Twenty-four μmol fructose and 1 μmol glucose are recovered. The recovered sucrose and fructose account for 90 % of the infused substrate; 10 % disappeared. The maltase results are similar, but more product is recovered. These results are generally comparable with previous studies in the rat (Dahlqvist & Thompson, 1963*a*, *b*) and man (Gray & Ingelfinger 1965; Gray & Santiago 1966; McMichael *et al.* 1967). These studies have generally been interpreted as membrane hydrolysis

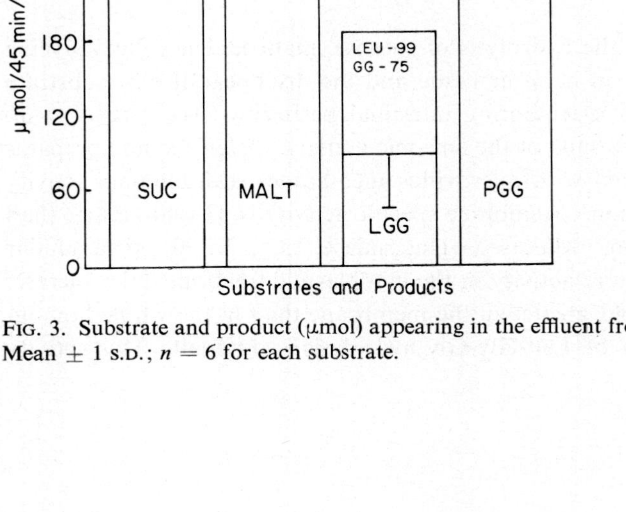

FIG. 3. Substrate and product (μmol) appearing in the effluent from each intestine in 45 min. Mean ± 1 S.D.; $n = 6$ for each substrate.

of the substrate, which is the rate-limiting step, followed by product transport (Modigliani *et al.* 1973).

The results of Leu-Gly-Gly perfusion are similar to the results obtained with sucrose and maltose, in that considerable product was obtained (Fig. 3). These results are similar to those for glycine oligopeptides in guinea pigs (Peters 1973), alanylglycylglycine (Silk *et al.* 1974) and leucylleucylleucine (Adibi *et al.* 1975) in man. That is, the oligopeptides and tripeptides disappeared with considerable accumulation of product in the lumen. The results favour digestion and transport mechanisms similar to the disaccharides: hydrolysis with subsequent product transport. However, Leu-Gly-Gly differs from sucrose and maltose, in that proportionately more of the substrate was accounted for by product. The *in vitro* rate of hydrolysis of Leu-Gly-Gly is not well correlated with the perfusion results. Further, the recovered substrate and product account for only 50% of the infused substrate, rather than the 90% observed with the disaccharides. At least two explanations are possible. The *in vitro* rate of Leu-Gly-Gly hydrolysis underestimates the membrane activity present in the perfusion studies, or a considerable amount of Leu-Gly-Gly may cross the membrane intact.

For Pro-Gly-Gly, with essentially no membrane hydrolysis, about 10% of this substrate disappeared during perfusion. This disappearance is comparable to the disaccharides. In sharp contrast, essentially no product was obtained in the effluent. We interpret these results with Pro-Gly-Gly (low membrane hydrolysis, essentially no product in the effluent) as showing that this tripeptide is transported intact across the membrane. The transport process may be passive, or perhaps active as is the case with glycylproline (Rubino *et al.* 1971). Hydrolysis of Pro-Gly-Gly probably occurs by cytoplasmic enzymes.

We conclude:

(A) The digestive disaccharidases are localized to microvillus membranes.

(B) Microvillus membrane aminopeptidase rapidly and extensively hydrolyses a mixture of gliadin peptides containing two to eight amino acids.

(C) Mechanisms of nutritional peptide digestion include:

- (1) Membrane hydrolysis with peptide transport (e.g. sucrose, maltose and Leu-Gly-Gly)
  - (a) *Rapid* membrane hydrolysis
  - (b) *Product* accumulation in the lumen
  - (c) *Tend* to leave the lumen rapidly
- (2) Peptide transport and cytoplasmic hydrolysis (e.g. Pro-Gly-Gly)
  - (a) *Negligible* membrane hydrolysis
  - (b) *Minimal* product accumulation in the lumen
  - (c) *May* leave the lumen at rates comparable to disaccharides.

Our results support the theory of Peters (1973) that peptide digestion occurs predominantly at the membrane with subsequent product transport. Certain peptides, not hydrolysed by the membrane, or small peptides, may cross the membrane intact.

ACKNOWLEDGEMENTS

A large portion of these studies were supported by the National Institutes of Health (AM-12329). The support of the J. B. Reynolds Foundation is gratefully acknowledged.

References

ADIBI, S. A. (1975) Dipeptide absorption and hydrolysis in human small intestine, in *Peptide Transport in Protein Nutrition* (Matthews, D. M. & Payne, J. W., eds.), pp. 147-166, North-Holland, Amsterdam

ADIBI, S. A., MORSE, E. L., MASILAMANI, S. S. & AMIN, P. M. (1975) Evidence for two different modes of tripeptide disappearance in human intestine. *J. Clin. Invest. 56*, 1355-1363

ARVANITAKIS, C., RUHLEN, J., FOLSCROFT, J. & RHODES, J. B. (1976) Digestion of tripeptides and disaccharides: relationship with brush border hydrolases. *Am. J. Physiol. 231*, 87-92

BERNADIN, J. E., KASARDA, D. D. & MECHAM, D. K. (1967) Preparation and characterization of alpha gliadin. *J. Biol. Chem. 242*, 445-450

DAHLQVIST, A. & THOMPSON, D. L. (1963*a*) The digestion and absorption of sucrose by the intact rat. *J. Physiol. (Lond.) 167*, 193-209

DAHLQVIST, A. & THOMPSON, D. L. (1963*b*) The digestion and absorption of maltose and trehalose by the intact rat. *Acta Physiol. Scand. 59*, 111-125

EICHHOLZ, A. & CRANE, R. K. (1965) Studies of the organization of the brush border in intestinal epithelial cells. I. Tris disruption of isolated hamster brush borders and density gradient separation of fractions. *J. Cell Biol. 26*, 687-691

FORSTNER, G. G., SABESIN, S. M. & ISSELBACHER, K. J. (1968) Rat intestinal microvillus membranes. Purification and biochemical characterization. *Biochem. J. 106*, 381-390

GRAY, G. M. & COOPER, H. L. (1971) Protein digestion and absorption. *Gastroenterology 61*, 535-544

GRAY, G. M. & INGELFINGER, F. J. (1965) Intestinal absorption of sucrose in man: the site of hydrolysis and absorption. *J. Clin. Invest. 44*, 390-398

GRAY, G. & SANTIAGO, A. M. (1966) Disaccharide absorption in normal and diseased human intestine. *Gastroenterology 51*, 489-498

HEIZER, W. D., KERLEY, R. L. & ISSELBACHER, K. J. (1972) Intestinal peptide hydrolases. Differences between brush border and cytoplasmic enzymes *Biochim. Biophys. Acta 264*, 450-461

HOLT, J. H. & MILLER, D. (1962) The localization of phosphomonesterase and aminopeptidase in brush borders from intestinal epithelial cells. *Biochim. Biophys. Acta 58*, 239-243

HUEBNER, F. R., ROTHFUS, J. A. & WALL, J. S. (1967) Isolation and chemical comparison of different gamma-gliadins from hard red winter wheat flour. *Cereal Chem. 44*, 221-229

KIM, Y. S., NICHOLSON, J. A. & CURTIS, K. J. (1974) Intestinal peptide hydrolases: peptide and amino acid absorption. *Med. Clin. North Am. 58*, 1397-1412

KIM, Y. S., BROPHY, E. J. & NICHOLSON, J. A. (1976) Rat intestinal brush border membrane peptidases. II. Enzymatic properties, immunochemistry, and interactions with lectins of two different forms of the enzyme. *J. Biol. Chem. 251*, 3206-3212

MAROUX, S. & LOUVARD, D. (1975) The aminopeptidase from hog intestinal brush border. *Digestion 13*, 180

MAROUX, S., LOUVARD, D. & BARUTTI, J. (1973) The aminopeptidase from hog intestinal brush borders. *Biochim. Biophys. Acta 321*, 282-295

MATHESON, A. T. & TATTRIE, B. L. (1964) A modified YEMM and COCKING ninhydrin reagent for peptidase assay. *Can. J. Biochem. 42*, 95-103

MATTHEWS, D. M. (1975) Absorption of peptides by mammalian intestine, in *Peptide Transport in Protein Nutrition* (Matthews, D. M. & Payne, J. W., eds.), pp. 61-146, North-Holland, Amsterdam

MCMICHAEL, H. B., WEBB, J. & DAWSON, A. M. (1967) The absorption of maltose and lactose in man. *Clin. Sci. (Oxf.) 33*, 135-145

MILLER, D. & CRANE, R. K. (1961) The digestive function of the epithelium of the small intestine. II. Localization of disaccharide hydrolyses in the isolated brush border portion of intestinal epithelial cells. *Biochim. Biophys. Acta 52*, 293-298

MODIGLIANI, R., RAMBAUD, J. C. & BERNIER, J. J. (1973) The method of intraluminal perfusion of the human small intestine. II. Absorption studies in health. *Digestion 9*, 264-290

NEWEY, H. & SMYTH, D. H. (1960) Intracellular hydrolysis of dipeptides during intestinal absorption. *J. Physiol. (Lond.) 152*, 367-380

PETERS, T. J. (1970) The subcellular localization of di- and tripeptide hydrolase activity in guinea pig small intestine. *Biochem. J. 120*, 195-203

PETERS, T. J. (1972) Subcellular fractionation of the enterocyte with special reference to peptide hydrolases, in *Peptide Transport in Bacteria and Mammalian Gut (Ciba Found. Symp. 4)*, pp. 107-115, Associated Scientific Publishers, Amsterdam

PETERS, T. J. (1973) The hydrolysis of glycine oligopeptides by guinea pig intestinal mucosa and by isolated brush borders. *Clin. Sci. Mol. Biol. 45*, 803-816

PETERS, T. J., DONLON, J. & FOTTRELL, P. F. (1972) The subcellular localization and specificity of intestinal peptide hydrolyases, in *Transport Across the Intestine* (Burland, W. L. & Samuel, P. D., eds.), pp. 153-167, Churchill Livingstone, Edinburgh

RHODES, J. B. (1968) Peptidases of the small intestine, in *Handb. Physiol. Sect. 6: Alimentary Canal*, vol. 5, pp. 2589-2604, American Physiological Society, Washington, D.C.

RHODES, J. B. & FOLSCROFT, J. (1972) Hydrolysis of small gliadin peptides by hog intestinal microvillus membranes. *Gastroenterology 62*, 865

RHODES, J. B., EICHHOLZ, A. & CRANE, R. K. (1967) Studies on the organization of the brush border in intestinal epithelial cells. IV. Aminopeptidase activity in microvillus membranes of hamster intestinal brush borders. *Biochim. Biophys. Acta 135*, 959-965

RUBINO, A., FIELD, M. & SHWACHMAN, H. (1971) Intestinal transport of amino acid residues of dipeptides. I. Influx of the glycine residue of glycyl-L-proline across mucosal border. *J. Biol. Chem. 246*, 3542-3548

SHOAF, C. R., BERKO, R. M. & HEIZER, W. D. (1976) Isolation and characterization of four peptide hydrolases from brush border of rat intestinal mucosa. *Biochim. Biophys. Acta 445*, 694-719

SILK, D. B. A., PERRETT, D., WEBB, J. P. W. and CLARK, M. L. (1974) Absorption of two tripeptides by the human small intestine: a study using a perfusion technique. *Clin. Sci. Mol. Med. 46*, 393-402

UGOLEV, A. M. (1965) Membrane (contact) digestion. *Physiol. Rev. 45*, 555-595

WOYCHIK, J. H., BOUNDY, J. A. & DIMLER, R. J. (1961) Starch gel electrophoresis of wheat gluten proteins with concentrated urea. *Arch. Biochem. Biophys. 94*, 477-482

Discussion

Semenza: You said that the pigs were dewormed. Are there any small intestines available in large amounts free of coccidiosis, worms and other parasites?

Rhodes: The major problem with young feeder hogs is *Ascaris suis*. These can be easily eradicated with a deworming medicine. We dewormed the animals, fed them for 14 days and documented a positive weight gain. No pathological organisms were found on microscopic examination of the faeces. Parasites are more easily controlled in small mammals.

Semenza: This is the problem that we all have. Will we have to buy calves, hogs or horses and keep them in the laboratory until we are sure they are healthy before we can use the small intestine?

Baker: The only way to get over this problem would be to use gnotobiotic animals, and what relevance do they have to the problem? Even if there are no worms there are bacteria, so this problem is always present.

Radhakrishnan: In gliadin peptide hydrolysis you said that hydrolysis was 96% with the low substrate concentration and 69% with the high substrate concentration. You ascribed this to product inhibition. Is this really so?

Rhodes: We have little evidence for that, and we didn't check it extensively.

Radhakrishnan: How did you measure the size of the peptides with two to eight residues? You used Sephadex, and I think the aromatic amino acid-containing peptides will be retarded on Sephadex, irrespective of the size.

Rhodes: We anticipated that and didn't find it. It is possible that peptides containing aromatic amino acids were hidden in some of the later fractions. We did not explore this possibility. While recovery from the column was about 90–95%, some peptide material was retained on the column. Tyr and Phe were relatively low in the peptide mixture as determined by acid hydrolysis and amino acid chromatography (Table 2, p. 250), and are perhaps retained on the column.

Radhakrishnan: How do you know the size of the peptides?

Rhodes: We did not calibrate the column with peptides of known size and composition. We did not further examine each fraction to isolate or determine the number of peptides in each fraction. We were after a representative sample of nutritional food peptides.

We estimated the average number of amino acids in the peptide fractions by comparing the ninhydrin colour before and after alkaline hydrolysis, using leucine as the standard. At least two problems are present with these methods. Alkaline hydrolysis destroys some amino acids, which will tend to underestimate the number of amino acids in the fractions. Certain N-terminal groups, such as proline, do not react well with the ninhydrin method used, which would tend to underestimate the initial colour before hydrolysis. I think the number of amino acids per peptide is within the range of what we consider physiological.

Silk: In your *in vivo* perfusion model you are using the accumulation of breakdown products to support the idea that surface hydrolysis of the tripeptide occurs in rather the same way that appearance of hydrolytic products during maltose and sucrose preparation is taken to indicate that surface hydrolysis of those sugars precedes the transport step. In your experiments I understand that the entire length of intestine was perfused. A problem that Dr Adibi and I have always had with our *in vivo* perfusion experiments is that higher concentrations of hydrolytic products accumulate during ileal peptide perfusion than during the jejunal studies. Have you performed any jejunal and ileal perfusions with this peptide?

Rhodes: We have not separately perfused the upper and lower portions of the small intestine, so we have no information to this point. Our primary interest was in the whole small bowel. The upper part of the small bowel does most of the work normally because there are more villi or mass, and the distal part plays a minor digestive role normally. If the upper jejunum is resected, the ileum will hypertrophy to compensate for the loss. Amino acid transport is best shown distally in the ileum. If you are a teleologist that makes sense, because the ileum is the main sump for pulling out everything that remains in the lumen.

The lack of correlation between the membrane enzymic activity and the accumulation of products during perfusion, especially with Leu-Gly-Gly, is disturbing. While we have some ideas about this, a good explanation is not available yet.

The efflux of products after intracellular hydrolysis bothers us a bit. We were surprised with Pro-Gly-Gly where little product was recovered. Our enzymic results strongly favour intracellular hydrolysis of this tripeptide. In ring and gut sac studies there is quite a bit of efflux and one can demonstrate the counter transport that Professor Semenza talked about. In our studies, with the blood supply intact, I wonder whether efflux is as much of a problem as we have been led to believe. As we digest our lunch, there is a considerable influx of products. Perhaps efflux plays a smaller part in normal digestion than we have previously thought.

Silk: I and my co-workers originally considered that surface hydrolysis played a more important role during ileal than jejunal peptide transport as we detected higher concentrations of free amino acids in gut contents aspirated during ileal than jejunal perfusions (Silk *et al.* 1974). I am no longer sure however that this conclusion is valid, as our recent studies suggest that there could be significant efflux of free amino acids out of the ileal mucosal cell after hydrolysis of peptides by cytoplasmic peptidases (Silk *et al.* 1976).

Baker: There aren't more villi in the proximal gut of the rat than in the distal gut. If anything, there are fewer.

Parsons: The surface area of the intestinal mucosa per unit length is greater in the upper than in the lower intestine.

Nicholson: Dr Rhodes, you compared the activity of the gliadin peptidases in the microvillus membrane with the activity in the cytosol. You purified the microvillus membranes so presumably there was a considerable loss of microvillus peptidase activity. Did you make a correction for that loss?

Rhodes: No.

Nicholson: So the actual rate of hydrolysis by the microvillus membrane compared with the cytosol may be as much as five times more than your estimate?

Rhodes: Yes.

Nicholson: Do the percentage activities represent a comparison between the activities in the microvillus membrane preparation and those in the cytosol?

Rhodes: No; if you calculate the total activity of peptide hydrolysis, the total cytoplasmic activity would exceed the membrane activity. As we have heard, the cytoplasm contains many dipeptidases and tripeptidases.

Nicholson: Even for gliadin peptides?

Rhodes: Yes.

Baker: But your figures there were 109% $\pm$ 17 and 96% $\pm$ 14. Those were standard errors and I don't think they are significant.

Rhodes: Correct. Both the microvillus membranes and the cytoplasm extensively hydrolyse the gliadin peptide mixture. They were not statistically compared.

In the gliadin peptide mixture and hog tissue studies, we are comparing the ability to hydrolyse. The cytoplasmic fraction readily hydrolyses the peptide mixture, which is not surprising when one considers the considerable number and variety of cytoplasmic peptidases. The major point we wish to make is that isolated microvillus membranes rapidly and extensively hydrolyse a mixture of at least 27 gliadin peptides. In order to enter the cell or cytoplasmic area, a peptide must cross this active hydrolytic membrane. Our studies do not differentiate the cellular site for nutritional peptide hydrolysis, in my opinion, but emphasize the hydrolytic potential of the membrane.

Matthews: Interpreted in the simplest way, your results indicate that Pro-Gly-Gly is a tripeptide which enters the absorptive cells by peptide transport, while Leu-Gly-Gly is a good example of one which undergoes hydrolysis in the microvillus region. If one goes along with Ugolev's concept of membrane digestion coupled to amino acid transport, then why, with a peptide which is hydrolysed at the membrane, do you get such tremendous back-diffusion of amino acids into the lumen? It must be a funny sort of transport mechanism that releases so much backwards. This is the sort of thing which makes me believe that hydrolysis at the brush border releases amino acids in free solution.

Rhodes: I am not sure of the correct answer to that, but we have thought about it. Most perfusion studies, such as ours, are performed with relatively rapid flow rates. The hamster intestine was distended, and the outflow volume almost equalled the inflow. There was little change in polyethylene glycol concentration, with less than 10% absorption of the fluid. We guess that the distension reduces the contact of the substrate with the membrane and perhaps thins out the unstirred water layer and may reduce the glycocalyx during the 30-min equilibration period. Substrate concentration is another important luminal factor. The contact of the substrate with the membrane hydrolase is reduced by the high flow rate. As the membrane hydrolases extend morphologically from the membrane and are luminal to the transport site, the availability of the product for transport is reduced. The disaccharide results are compatible with this explanation; 80% was recovered intact, 10% recovered as product and 10% disappeared. This is not the case in normal digestion and transport, where essentially all of the substrate is digested and absorbed. This is compatible with your concept of the hydrolytic products being released in free solution, under these experimental conditions. Under other more physiological conditions with lower flow rates, hydrolysis may favour transport, as others have shown. The apparent back-diffusion may be membrane hydrolysis and the high flow rate sweeps the products away before they are captured by the transport mechanisms. Pro-Gly-Gly disappears into the cell intact and is hydrolysed in the cell, with essentially no efflux.

As previously mentioned, back flux of product, in normal digestion with low flow rates and an intact blood supply, may be the exception rather than the rule.

Baker: Did you tie off or occlude the lumen at all?

Rhodes: We tied it off proximally, to avoid pancreatic fluid. We washed it through for 30 min.

Parsons: The difference is that if the blood supply is intact, there is no accumulation within the epithelial region, whereas with classical *in vitro* preparations there is accumulation. This must be a very important factor to be borne in mind when comparing results from *in vivo* studies with those obtained *in vitro*.

Baker: Did you try any inhibition studies with this?

Rhodes: No; that is why I was interested in learning how we might inhibit that enzyme activity.

Baker: Did you try to look at surface hydrolysis resulting in free amino acids, which might inhibit peptide transport?

Rhodes: No, except as occurred in the total system. Free leucine, for example, was present in the effluent.

Adibi: You found that disappearance of Leu-Gly-Gly was greater than that of maltose, while there was much more hydrolytic potential in the membrane against maltose than against Leu-Gly-Gly. But if you take into account the hydrolytic potential of whole homogenate, the activity is greater against Leu-Gly-Gly than against maltose. Does this mean Leu-Gly-Gly disappearance is a function of both membrane and intracellular enzyme?

Rhodes: Yes, the test tube hydrolysis of Leu-Gly-Gly was less than the disaccharides. We were surprised to find that there was so much hydrolysis of Leu-Gly-Gly during perfusion. There are differences between the test tube data and the perfusion results. This is similar to your experience as well as that of Drs Silk and Nicholson. There are at least two possibilities that we considered. First, the *in vitro* method that we used may underestimate the hydrolytic potential of the membranes for Leu-Gly-Gly. Secondly, half of the Leu-Gly-Gly may leave the lumen intact. There are more similarities between Leu-Gly-Gly and sucrose and maltose–and we think we know how those work–than there are with Pro-Gly-Gly. The differences between the two tripeptides were statistically significant at the 0.001 level.

De Laey: In your preparation, Dr Rhodes, there might be quite a substantial release of membrane enzymes within the medium. One is dealing with a very complex system where there is luminal digestion and membrane digestion. I suppose that the presence of very high levels of Gly-Gly in this medium might be a consequence of that solubilization of membrane enzymes.

Rhodes: We have checked the effluent for enzyme activity, as have Dr Adibi, Dr Silk and several others. The luminal washout of activity is very low, and would only account for less than 5% of the hydrolysis observed. The amount of enzyme efflux during the experiments is very low.

Silk: The concentrations of hydrolytic products recovered at the end of any perfusion experiment is determined not only by rates of hydrolysis but also by rates of recapture. Thus it may not be valid to compare rates of glucose appearance during maltose perfusion with rates of appearance of free leucine during Leu-Gly-Gly perfusion because, once released, glucose and leucine may be reabsorbed at different rates.

Gardner: The comment about vascular perfusion reducing the intracellular accumulation and thus affecting the back flux may be extremely significant.

In my system without vascular perfusion I am certain that there is a substantial back flux of free amino acids, proline being the most dramatic example. I estimate that on average about 40% of the peptides that are absorbed go back into the lumen as free amino acids (Gardner 1975). An extension of this is that *in vitro*, without vascular perfusion, if one gets unusually high tissue concentrations one may be prejudicing one's views on intracellular hydrolysis of peptides. Therefore, if one were to reduce the intracellular peptide concentrations by a vascular perfusion method (e.g. that of Fisher & Gardner 1974), one might find that intracellular hydrolysis was substantially less important than the passage of intact peptides. Would you agree with that, Dr Parsons?

Parsons: Yes, I would agree. I don't think all the accumulation is intracellular. Some must also be extracellular.

Matthews: A very good way of minimizing superficial hydrolysis is to reduce pH, as Dr Adibi has shown (Fogel & Adibi 1974). When we studied uptake of Gly-Leu by rings of everted hamster jejunum *in vitro* at pH 7, we found a grossly uneven uptake. Far more Leu was taken up than Gly, which we attributed to a strong element of superficial hydrolysis with preferential uptake of free Leu. When we reduced the pH to 5, we found that uptake of Gly and Leu from Gly-Leu was equal and hydrolysis, as measured by appearance of free Gly and Leu in the medium, became very slight (see p. 90, Fig. 6). This approach might be valuable, Dr Rhodes.

Rhodes: You have raised a point that should be emphasized. With sucrose and maltose, up to 90% of the substrate could be accounted for in the effluent. Only half of the Leu-Gly-Gly is accounted for, while half of it is not. Our studies can't differentiate the possible occurrence of two mechanisms acting at the same time. Perhaps Leu-Gly-Gly hydrolysis accounts for half of the observed disappearance, while intact transport and intracellular hydrolysis account for the other half. We cannot exclude the possibility of some intact transport on the basis of these studies.

References

FISHER, R. B. & GARDNER, M. L. G. (1974) Dependence of intestinal glucose absorption on sodium, studied with a new arterial infusion technique. *J. Physiol. (Lond.) 241*, 235-260

FOGEL, M. R. & ADIBI, S. A. (1974) Assessment of the role of brush border peptide hydrolases in luminal disappearance of dipeptides in man. *J. Lab. Clin. Med. 84*, 327-333

GARDNER, M. L. G. (1975) Absorption of amino acids and peptides from a complex mixture in the isolated small intestine of the rat. *J. Physiol. (Lond.) 253*, 233-256

SILK, D. B. A., WEBB, J. P. W., LANE, A. E., CLARK, M. L. & DAWSON, A. M. (1974) Functional differentiation of human jejunum and ileum: a comparison of the handling of glucose, peptides and amino acids. *Gut 15*, 444-449

SILK, D. B. A., NICHOLSON, J. A. & KIM, Y. S. (1976) Relationships between mucosal hydrolysis and transport of two phenylalanine dipeptides. *Gut 17*, 870-877

Clearance of dipeptides from plasma: role of kidney and intestine

SIAMAK A. ADIBI

Gastrointestinal and Nutrition Unit of Montefiore Hospital, University of Pittsburgh School of Medicine, Pittsburgh

Abstract The metabolic fate of glycylglycine and glycyl-L-leucine administered intravenously to rats, and the processes involved in the clearance of these dipeptides from plasma, have been studied. The dipeptides disappeared rapidly from plasma without being lost in the urine or accumulating in body tissues, but concentrations of glycine or leucine, or both, increased markedly in plasma and body tissues. All tissues except blood had considerable hydrolase activity against Gly-Gly and Gly-Leu, with kidney and intestine being far more active than liver and muscle in this regard. When uptake of glycylsarcosine, which is resistant to hydrolysis, was investigated after intravenous injection, its disappearance from plasma was slower than that of Gly-Gly, but Gly-Sar accumulated in all tissues examined (liver, muscle, renal cortex and intestinal mucosa). Its concentration in the renal cortex far exceeded (over 14-fold) that in other tissues, and a considerable portion of the injected Gly-Sar dose appeared in the urine.

Bilateral nephrectomy did not have a remarkable effect on the plasma half-life of Gly-Leu or Gly-Gly but markedly prolonged that of Gly-Sar. Nephrectomy also allowed greater tissue accumulation of Gly-Sar and recovery of Gly-Gly and Gly-Leu from tissues. The plasma half-life of Gly-Gly was also not remarkably affected by total enterectomy. These studies suggest that (*a*) clearance of dipeptides from plasma is chiefly a function of tissue uptake and hydrolysis, and (*b*) among dipeptides, those resistant to hydrolysis depend to a great extent on kidneys for their clearance from plasma.

In recent years, there has been much progress in our knowledge of the transport and hydrolysis of dipeptides by intestinal mucosa (Adibi 1975, 1976; Matthews & Adibi 1976). In contrast, there have been few studies of dipeptide transport and hydrolysis by other tissues. Although some dipeptides have been identified in systemic or portal plasma in a variety of experimental and physiological situations (Adibi 1971; Adibi *et al.* 1975; Boullin *et al.* 1973; Perry *et al.* 1967; Prockop *et al.* 1962), it is as yet unclear whether the circulating dipeptides are lost in the urine or taken up and hydrolysed by tissues to enrich their pools of amino acids. Several years ago we became interested in this question and began a series of experiments on the metabolic fate of dipeptides administered intravenously to rats (Adibi & Krzysik 1977; Adibi *et al.* 1977).

DYNAMICS OF DIPEPTIDE CLEARANCE FROM PLASMA

In the initial experiments we investigated the dynamics of the clearance of dipeptides from rat plasma by measuring plasma concentrations of glycylglycine and glycyl-L-leucine before and at intervals after the intravenous injection of these dipeptides. These two dipeptides were selected because: (*a*) both glycylglycine and glycylleucine were being used extensively as model dipeptides for our studies of intestinal transport in human intestine (Adibi 1971; Adibi & Soleimanpour 1974), and (*b*) according to previous investigations in the intestine (Fogel & Adibi 1974), glycylglycine represents those dipeptides with a rather slow rate of hydrolysis, while glycylleucine represents dipeptides with a relatively rapid rate of hydrolysis. The amount of glycylglycine or glycyl-L-leucine injected was kept constant at 0.5 μmol/g body weight. Although glycylglycine and glycylleucine were not detectable in plasma in the basal state, both of these dipeptides attained high concentrations in plasma immediately after the injection (Fig. 1). Plasma concentration of glycylleucine decreased sharply by 10 min after the injection, while glycylglycine decreased more gradually. By 40 min, neither glycylglycine nor glycylleucine were detectable in plasma.

MECHANISMS OF DIPEPTIDE CLEARANCE FROM PLASMA

The following mechanisms were considered during our attempt to account for the disappearance of the above dipeptides from plasma: (*a*) excretion into urine, (*b*) hydrolysis within the blood-stream, (*c*) accumulation in tissues, and (*d*) hydrolysis by tissues.

To investigate the first possibility, we collected urine for 40 min after intravenous injections of dipeptides into rats (the time interval during which dipeptides disappeared from plasma, as shown in Fig. 1). Only trace amounts of glycylglycine and glycylleucine were recovered. Hydrolysis within the blood-stream was examined by *in vitro* assay of peptide hydrolase activities of whole blood against glycylglycine and glycylleucine. Blood lacked any hydrolase activity against glycylglycine, and had only trace activity against glycylleucine (Krzysik *et al.* 1975). To investigate tissue accumulation, we analysed liver, renal cortex, gastrocnemius muscle and intestinal mucosa for glycylglycine and glycylleucine content after injection of these dipeptides. Neither dipeptide was detected in these tissues. When we measured total cytoplasmic peptidase

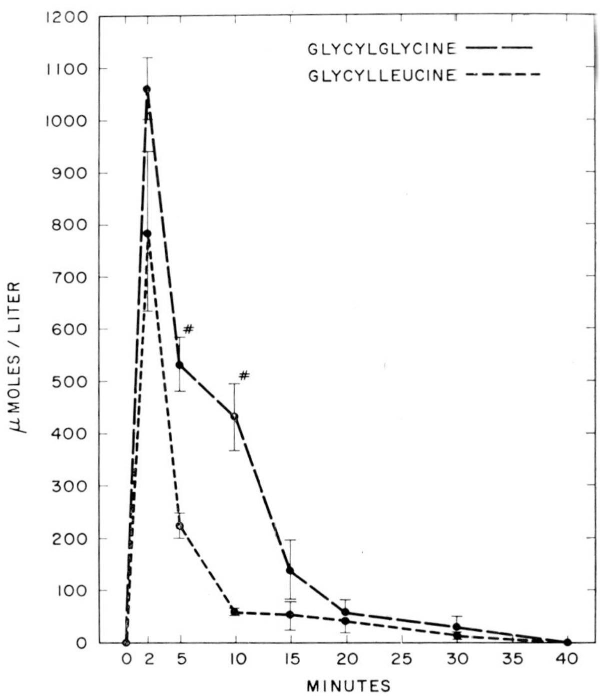

FIG. 1. Plasma concentrations (mean ± S.E.M., 5 rats) of glycylglycine and glycylleucine at intervals after intravenous injection of each dipeptide (0.5 μmol/g body weight). None of these dipeptides were detected in plasma of saline-injected rats. # Indicates that at that interval the concentration of glycylleucine was significantly different ($P < 0.01$) from that of glycylglycine. (From Adibi *et al.* 1977.)

activity of liver, gastrocnemius muscle, renal cortex and intestinal mucosa all these tissues contained considerable dipeptide hydrolase activities against glycylglycine and glycylleucine (Krzysik *et al.* 1975). These activities were more than sufficient for hydrolysis of the amount of dipeptides injected. From these observations, we concluded that excretion into the urine and hydrolysis within the blood-stream are not quantitatively important, but that tissue uptake followed by rapid intracellular hydrolysis is the principal mechanism for rapid disappearance of glycylglycine and glycylleucine from plasma.

To investigate further the possibility that there is, indeed, tissue uptake of dipeptides, we studied the plasma clearance and tissue accumulation of glycylsarcosine, which is known to be resistant to hydrolysis (Smith 1951) but is actively transported by the intestinal mucosa (Addison *et al.* 1972). Again, rats were injected with 0.5 μmol/g body weight of glycylsarcosine and plasma

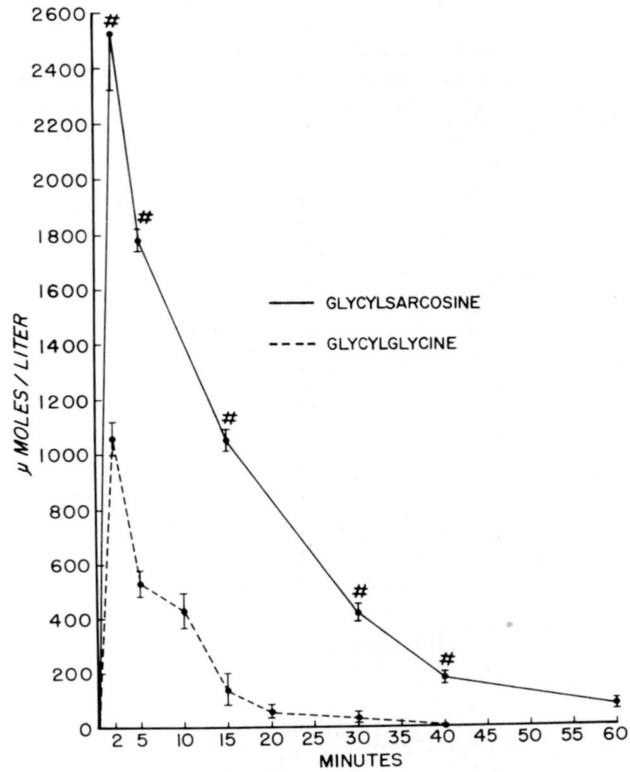

FIG. 2. Plasma concentrations (mean ± S.E.M., 6 rats) of glycylglycine and glycylsarcosine at intervals after intravenous injection of each dipeptide (0.5 μmol/g body weight). # Indicates that at that interval the concentration of glycylsarcosine was significantly ($P < 0.01$) greater than that of glycylglycine. (From Adibi *et al.* 1977.)

concentrations of this dipeptide before and at intervals after the injection were determined (Fig. 2). Although the plasma disappearance of glycylsarcosine was slower than that of its natural analogue (glycylglycine), by 60 min it was almost entirely cleared from plasma. Unlike glycylglycine and glycylleucine, which could not be detected in tissue, we found that glycylsarcosine accumulated in all tissues, including renal cortex, intestinal mucosa, skeletal muscle and liver (Adibi *et al.* 1977). Concentrations of glycylsarcosine which accumulated in the renal cortex were not only greater than those found in other tissues, but were several times greater than the plasma concentration. These observations supported the concept that there is tissue uptake of glycylglycine and glycylleucine; but rapid hydrolysis by intracellular enzymes is probably responsible for our failure to detect them in the intracellular fluid. Similarly,

Nützenadel & Scriver (1976) have independently concluded from a recent study that there is uptake of dipeptides by the tissues. By *in vitro* studies, they have shown that slices of kidney cortex, intact hemi-diaphragm and scraped mucosal epithelium of jejunum all transport carnosine (β-alanyl-L-histidine) in unhydrolysed form to the cell interior. Peptide transport by brain slices (Abraham *et al.* 1964; Yamaguchi *et al.* 1970) and by cultured fibroblasts (Grahl-Nielsen *et al.* 1974) has also been reported from other laboratories.

ROLE OF KIDNEY AND INTESTINE IN THE CLEARANCE OF DIPEPTIDES FROM PLASMA

The above observations raised the possibility that kidney may be the organ which is chiefly responsible for the clearance of dipeptides from plasma. Among tissues examined, the rates of dipeptide uptake from plasma (Adibi *et al.* 1977) and of dipeptide hydrolysis (Krzysik *et al.* 1975) were greater by the renal cortex than by the liver, skeletal muscle or intestinal mucosa. In addition, kidneys have relatively high rates of blood perfusion and have the unique function of continuous filtration and reabsorption of plasma constituents. To investigate the importance of kidney in plasma dipeptide clearance, we measured the plasma half-lives of glycylglycine, glycylleucine and glycylsarcosine in bilaterally nephrectomized rats and compared them with those of control rats. The half-lives of the dipeptides in plasma were determined from a straight line drawn by the least squares method on a semi-logarithmic graph of the mean plasma concentration (log scale) plotted against the elapsed time after the injection (linear scale). The half-lives of glycylglycine and glycylleucine were not changed remarkably by nephrectomy, but that of glycylsarcosine was markedly increased (over fourfold) by nephrectomy (Fig. 3).

The prolongation of the plasma half-life for glycylsarcosine after nephrectomy can be attributed to the excretion of a considerable portion of this dipeptide in the urine, but there is scarcely any urinary excretion of glycylglycine or glycylleucine (Adibi *et al.* 1977). Glycylsarcosine is not a naturally occurring dipeptide, and the impairment of its clearance from plasma by nephrectomy may not be physiologically relevant. However, there are naturally occurring dipeptides which, like glycylsarcosine, are not efficiently hydrolysed by tissue peptide hydrolases. For example, hydroxyprolyl-, β-aspartyl-, and γ-glutamyl-dipeptides have been recovered in the urine of man and rat after either a protein meal or increased breakdown of endogenous protein (Milne 1971). Therefore, under physiological conditions, kidneys play an important role in the clearance of dipeptides which are relatively resistant to hydrolysis by other tissues.

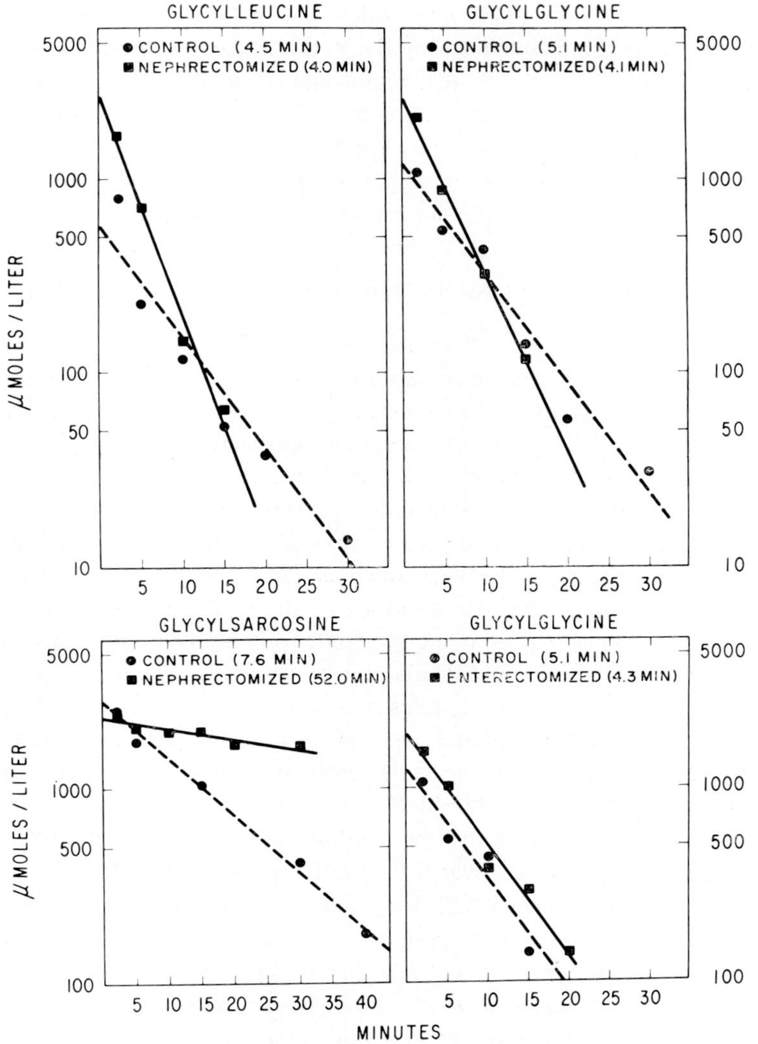

FIG. 3. Plasma concentrations (mean values) of glycylleucine, glycylglycine and glycylsarcosine at various intervals after their injections into sham-operated (control, 6 rats), nephrectomized (6, 6, and 4 rats, respectively), and enterectomized (4 rats) animals. Concentrations were plotted on the semi-logarithmic scale against time on the linear scale. Calculated half-lives of the dipeptides in plasma are shown in parentheses beside the descriptions of the surgical treatments. (From Adibi & Krzysik 1977.)

The failure of nephrectomy to alter the plasma half-lives of glycylglycine and glycylleucine can be explained by the considerable hydrolase activity against these dipeptides in the extrarenal tissues (Krzysik *et al.* 1975); and further, in the absence of kidneys, other tissues assume a larger role in the uptake of circulating dipeptides. The concentrations of glycylsarcosine in the muscle, liver and intestinal mucosa of nephrectomized rats were higher than those in the control rats (Adibi & Krzysik 1977). In addition, glycylglycine and glycylleucine, which were not detected in tissues of control rats, were recovered in one or more of these tissues in nephrectomized rats (Adibi & Krzysik 1977).

The specific activities of dipeptidases in the small intestine far exceed those of other tissues except the kidney (Krzysik *et al.* 1975). To investigate the importance of the small intestine in plasma dipeptide clearance, we determined the plasma half-life of glycylglycine in control and totally enterectomized rats. As shown in Fig. 3, enterectomy was without a noticeable effect on the plasma half-life of glycylglycine.

These studies indicate that because of the widespread distribution and large tissue potential of dipeptidase activities, the body is well equipped for handling circulatory dipeptides. This large potential protects the organism from the loss of a valuable source of amino acids should the function of any organ be compromised (Fig. 3).

ENRICHMENT OF AMINO ACID POOLS BY DIPEPTIDES

It is now well established that protein nutrition can be adequately maintained by amino acids administered intravenously. For dipeptides to serve as a substrate for protein nutrition, they must enrich the tissue pools of amino acids with their residues. Therefore, from a nutritional point of view, it was important to determine whether intravenously administered dipeptides could enrich the amino acid pools in tissues and whether this would be as efficient as injecting the corresponding free amino acid mixtures. To investigate this possibility, rats were injected with either saline, glycylglycine, glycine, glycylleucine, equimolar mixtures of glycine and leucine, or glycylsarcosine. The concentrations of glycine, leucine or sarcosine, or all three amino acids, were determined before and at intervals after the injections in the plasma, liver, gastrocnemius muscle, intestinal mucosa and renal cortex of these rats. The amount of dipeptide injected was 0.5 μmol/g body weight, and the amount of amino acid injected was identical to the concentration in dipeptide form. Both glycylleucine and its corresponding amino acid mixture markedly increased the concentrations of glycine and leucine in plasma (Figs. 4 & 5).

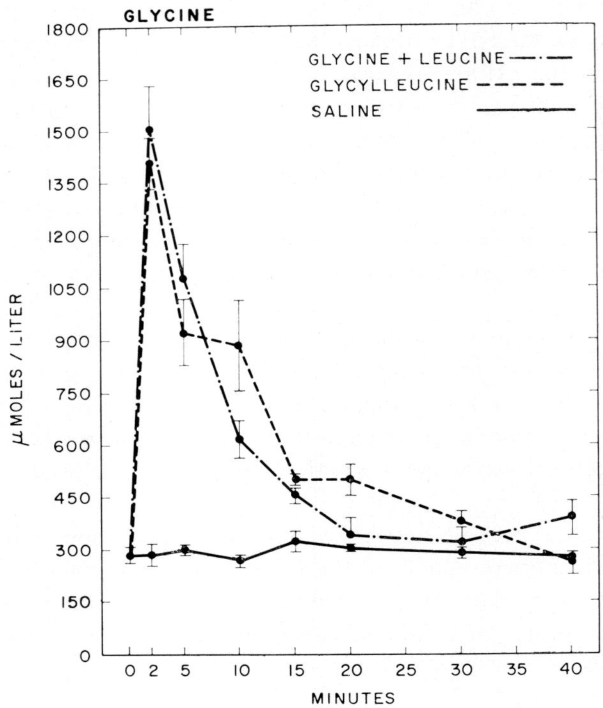

FIG. 4. Plasma concentrations (mean ± S.E.M., 6 rats) of free glycine before (0 min) and at intervals after intravenous injection of either glycylleucine (0.5 μmol/g body weight), equimolar mixture (0.5 μmol each/g body weight) of glycine plus leucine, or saline solution. The differences in glycine concentrations after injection of glycylleucine and the corresponding free amino acid mixture were not statistically significant. (From Adibi *et al.* 1977.)

However, there were no significant differences in the levels of leucine or glycine at any interval after the injection of dipeptides or amino acids. Saline injection did not cause any remarkable alteration in plasma concentrations of leucine or glycine (Figs. 4 & 5).

The plasma glycine concentrations before and after injection of either glycylsarcosine, glycylglycine or an equivalent amount of glycine are summarized in Fig. 6. The initial increase in glycine concentrations was significantly greater after glycine than after glycylglycine. After 2 min the raised plasma glycine levels fell gradually in glycylglycine-injected rats, while there was a sharp fall in glycine-injected rats. However, after the first 10 min there were no longer any significant differences between the levels of free glycine in the two groups of rats. By 40 min the level of glycine approached the basal level

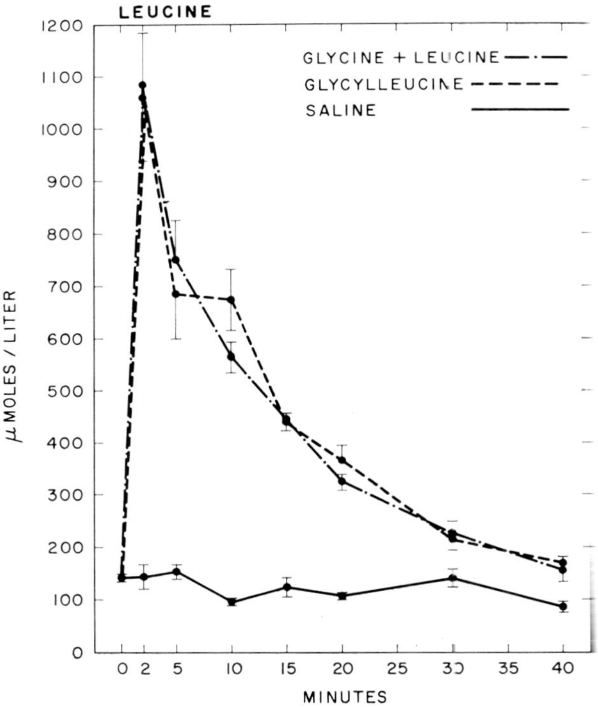

FIG. 5. Plasma concentrations (mean ± S.E.M., 6 rats) of free leucine before (0 min) and at intervals after intravenous injection of either glycylleucine (0.5 μmol/g body weight), equimolar mixture (0.5 μmol each/g body weight) of glycine plus leucine, or saline solution. The differences in leucine concentrations after injection of glycylleucine and the corresponding amino acid mixture were not statistically significant. (From Adibi *et al.* 1977.)

in both groups. At none of the intervals examined was sarcosine detected in plasma; however, there was a slight increase in plasma concentration of glycine after glycylsarcosine injection (Fig. 6).

The concentrations of free glycine and free leucine in liver, skeletal muscle and kidney before and at intervals after the injection of either saline, glycylleucine or the corresponding free amino acid mixture are shown in Fig. 7. At 5 min, the time when there was the sharpest decrease in the plasma concentration of glycylleucine (Fig. 1), there was a significant increase in the concentrations of both free glycine and free leucine in all three tissues. These expansions in pool sizes were similar whether the rats were injected with glycylleucine or the corresponding free amino acid mixture. At 30 min after the injection, when the glycylleucine had almost disappeared from the plasma

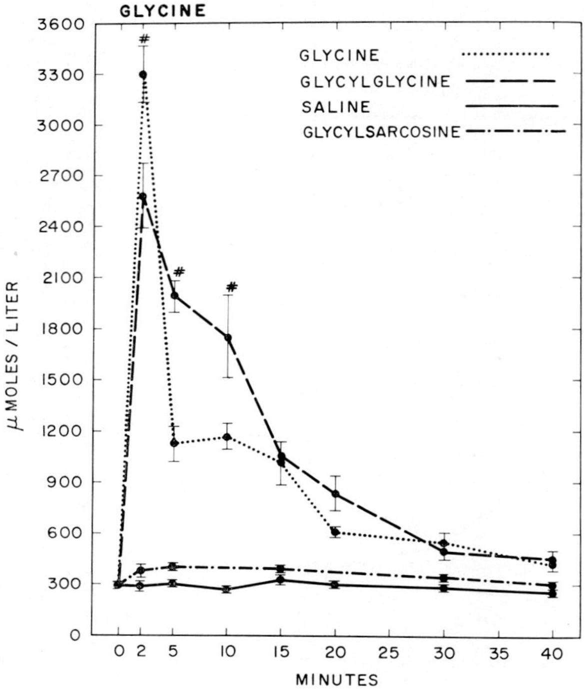

FIG. 6. Plasma concentrations (mean ± S.E.M., 6 rats) of free glycine before (0 min) and at intervals after intravenous injection of either glycylglycine (0.5 μmol/g body weight), glycine (1.0 μmol/g body weight), glycylsarcosine (0.5 μmol/g body weight), or saline solution. # Indicates that at that interval there was a significant difference ($P < 0.01$) between the concentrations of free glycine in the plasma of glycylglycine- and glycine-injected rats. (From Adibi *et al.* 1977.)

(Fig. 1) and the plasma concentrations of both free glycine and free leucine approached basal levels (Figs. 4 & 5), the tissue concentrations of free glycine and free leucine had fallen below the preceding high concentrations. However, except for glycine in the kidney, the concentrations of both these amino acids at 30 min were still higher in the tissues of the rats injected with amino acids and dipeptides than in those injected with saline. The tissue concentrations of glycine and leucine were not significantly changed after saline injection (Fig. 7). Glycylglycine and glycine injection similarly enriched free glycine pools in the liver, muscle and kidney.

Analysis of tissues for glycine and sarcosine content after intravenous injection of glycylsarcosine produced striking results (Fig. 8). Sarcosine was not detected in the liver or in the muscle, but was found in considerable con-

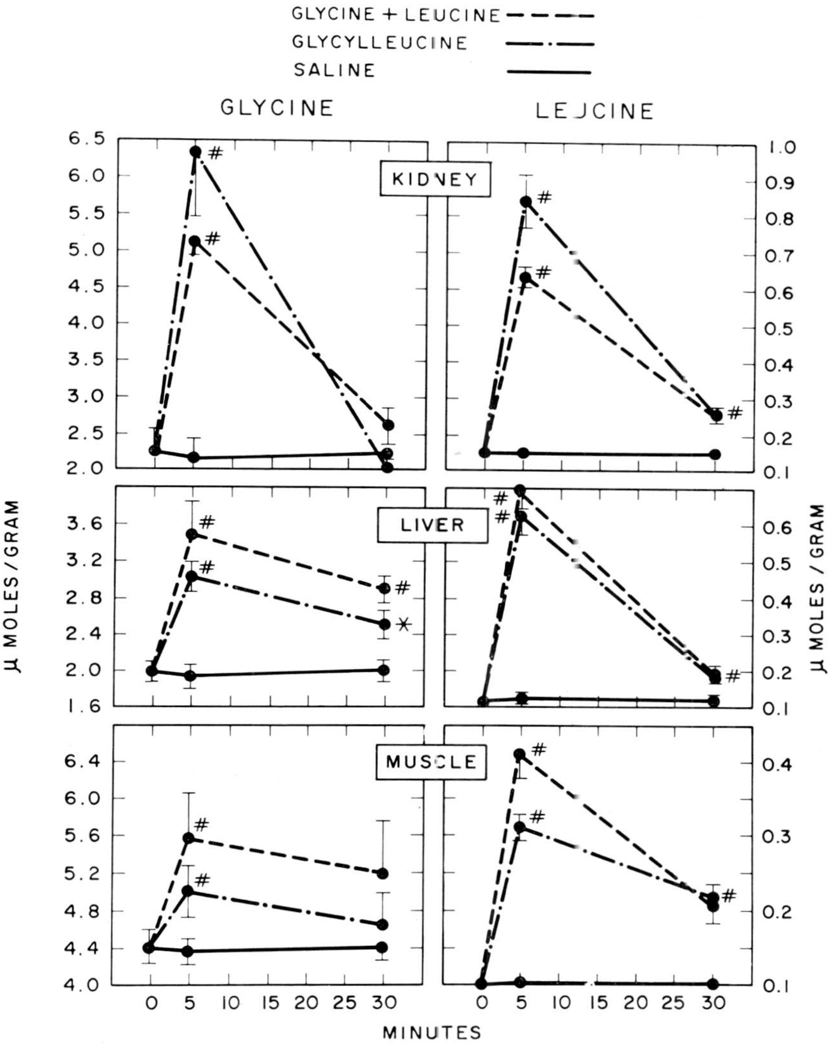

FIG. 7. Concentrations (mean ± S.E.M., 5 rats) of free glycine and free leucine in tissues of rats before and at intervals after injection of either glycylleucine (0.5 μmol/g body weight), equimolar mixture (0.5 μmol each/g body weight) of free glycine and free leucine, or saline solution. The differences in tissue concentrations of glycine or the differences in tissue concentrations of leucine in the dipeptide compared to amino-acid-injected rats were not statistically significant. # and * indicate that the concentrations of leucine or glycine in dipeptide- or amino acid-injected rats were significantly greater ($P < 0.01$ and < 0.05, respectively) than that in the paired saline-injected controls. (From Adibi *et al.* 1977.)

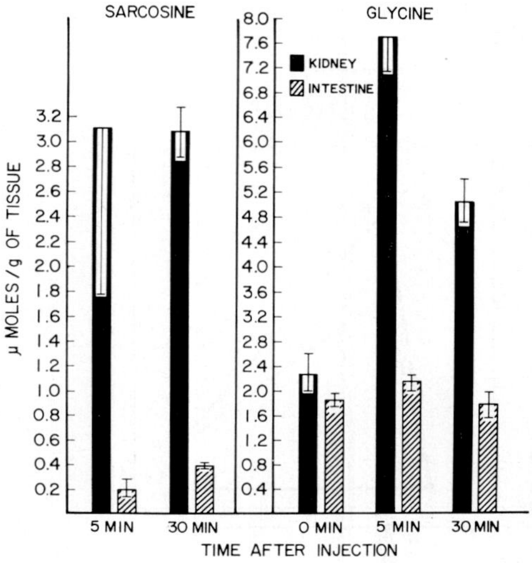

FIG. 8. Concentrations (mean ± S.E.M., 4–6 rats) of sarcosine and glycine before and 5 and 30 min after the injection of glycylsarcosine (0.5 μmol/g body weight) in the renal cortex and the intestinal mucosa. Sarcosine was not detected in tissues before the injection. Sarcosine was also not detected in plasma, liver or skeletal muscle after the injection of glycylsarcosine. (From Adibi *et al.* 1977.)

centrations in the renal cortex and the intestinal mucosa. The concentrations of sarcosine in the renal cortex were markedly greater than those in the intestinal mucosa. As compared to saline injection, there were no significant changes in glycine concentration in the liver, muscle and intestinal mucosa after the glycylsarcosine injections, but there were significant increases in glycine concentrations in the renal cortex (Fig. 8).

In addition to establishing the efficiency of dipeptides in enriching the amino acid pools in tissues, the above observations also suggested that this enrichment is mostly due to intracellular hydrolysis of dipeptides. This hypothesis is based on the fact that after glycylsarcosine injection sarcosine accumulated in the renal cortex and intestinal mucosa without any sarcosine appearing in plasma.

DIPEPTIDE STIMULATION OF HORMONE SECRETION

Among the many important metabolic functions of amino acids is their potential as stimulators of insulin and glucagon secretion and as substrates for gluconeogenesis. Insulin, glucagon and glucose may, in turn, change the plasma concentrations of amino acids (Adibi *et al.* 1975). Because of these

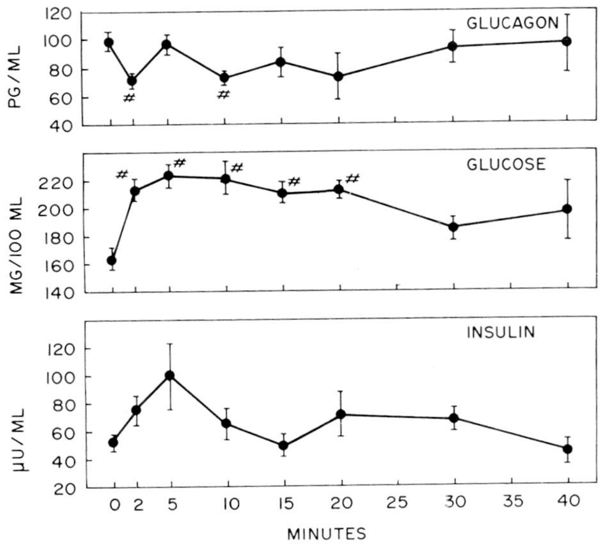

FIG. 9. Plasma concentrations (mean ± S.E.M., 6 rats) of insulin, glucose and glucagon before (0 min) and at intervals after intravenous injection of saline solution. # Indicates that at that interval the concentration of insulin, glucose, or glucagon in saline-injected rats was significantly different ($P < 0.01$) from that of non-injected rats. (From Adibi *et al.* 1977.)

metabolic interactions, we investigated the plasma concentration of insulin, glucose and glucagon in dipeptide-injected rats. We also did studies to determine whether alterations in plasma glucose, insulin and glucagon concentrations would differ if the same amino acids were injected in dipeptide or in free form. The same analyses were performed in saline-injected rats to determine which effects of amino acid and dipeptide injection could be attributed to the introduction of these substrates into the systemic circulation, and which effects were due to the experimental procedure.

Saline injection increased plasma glucose concentration without significantly altering plasma insulin and glucagon concentrations (Fig. 9). The injection of both glycylleucine and the mixture of free glycine and leucine raised the plasma glucose concentrations compared to the basal levels (Fig. 10). The pattern of increase in plasma glucose concentration was essentially similar in both groups of rats (Fig. 10) and did not significantly differ from that of saline-injected rats (Fig. 9). Although the amount of glucagon secreted after glycylleucine injection did not differ significantly from that secreted after the corresponding amino acid mixture, the dipeptide injection caused a greater

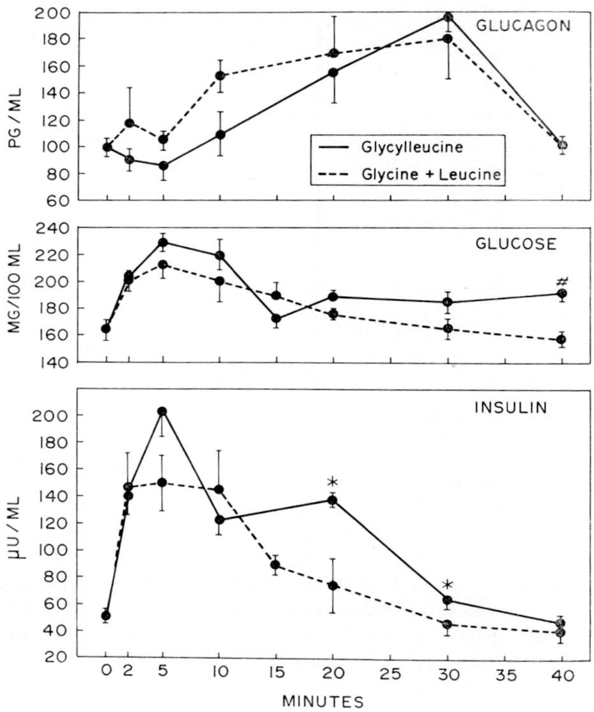

FIG. 10. Plasma concentrations (mean ± S.E.M., 5 rats) of insulin, glucose, and glucagon before (0 min) and at intervals after intravenous injection of either glycylleucine (0.5 μmol/g body weight) or glycine plus leucine (0.5 μmol each/g body weight). # and * indicate that at that interval the concentration of insulin or glucose was significantly different ($P < 0.01$ and $P < 0.05$, respectively) in amino acid- and dipeptide-injected rats. (From Adibi *et al.* 1977.)

insulin secretion than the corresponding amino acids (Fig. 10). Planimetric integration of the area under the curve describing insulin concentration as a function of time showed that the amount of insulin secreted after glycylleucine injection was 1.6 times the amount secreted after the mixture of glycine and leucine.

If the ability of dipeptides to stimulate greater insulin secretion can be more fully established by additional studies with other dipeptides, a physiological advantage of intravenously administered dipeptides over amino acids could be envisioned. Insulin stimulates protein synthesis (Manchester 1970) and inhibits protein degradation (Mortimore & Mondon 1970). These properties are of paramount importance in maintaining or achieving an optimal state of protein nutrition.

SPECULATION FOR FUTURE STUDIES

During the studies described above, we were motivated by the thought that the results might have clinical applications, especially in the field of parenteral nutrition. The amino acid solutions currently used for intravenous feeding are quite hypertonic, and the replacement of the amino acids with dipeptides or, better still, with tripeptides would markedly reduce the hypertonicity of these solutions. It is possible that a properly designed peptide solution might offer a variety of nutritional and metabolic advantages over amino acid solutions. Much work remains to be done in this area. The present studies relate to the metabolic fate of circulating dipeptides after a bolus injection. The use of dipeptides during continuous infusion, which may have more clinical and nutritional relevance than a bolus injection, remains to be investigated. Nevertheless, in contrast to a widely maintained belief that small peptides, when infused intravenously, are not efficiently utilized (Long *et al.* 1974), the results of the present experiments encourage additional metabolic and clinical studies on the use of dipeptides as substrates for parenteral nutrition.

ACKNOWLEDGEMENTS

This work was supported by a grant (AM-15855) from the National Institute of Arthritis, Metabolism and Digestive Diseases, National Institutes of Health, United States Public Health Service.

References

ABRAHAM, D., PISANO, J. J. & UDENFRIEND, S. (1964) Uptake of carnosine and homocarnosine by rat brain slices. *Arch. Biochem. Biophys. 104*, 160-165

ADDISON, J. M., BURSTON, D. & MATTHEWS, D. M. (1972) Evidence for active transport of the dipeptide glycylsarcosine by hamster jejunum *in vitro*. *Clin. Sci. (Oxf.) 43*, 907-911

ADIBI, S. A. (1971) Intestinal transport of dipeptides in man: relative importance of hydrolysis and intact absorption. *J. Clin. Invest. 50*, 2266-2275

ADIBI, S. A. (1975) Dipeptide absorption and hydrolysis in human small intestine, in *Peptide Transport in Protein Nutrition* (Matthews, D. M. & Payne, J. W., eds.), pp. 147-166, North-Holland, Amsterdam

ADIBI, S. A. (1976) Intestinal phase of protein assimilation in man. *Am. J. Clin. Nutr. 29*, 205-215

ADIBI, S. A. & KRZYSIK, B. A. (1977) Effect of nephrectomy and enterectomy on plasma clearance of intravenously administered dipeptides in rats *Clin. Sci. Mol. Med. 52*, 205-213

ADIBI, S. A. & SOLEIMANPOUR, M. R. (1974) Functional characterization of dipeptide transport system in human jejunum. *J. Clin. Invest. 53*, 1368-1374

ADIBI, S. A., MORSE, E. L. & AMIN, P. M. (1975) Role of insulin and glucose in the induction of hypoaminoacidemia in man: studies in normal, juvenile diabetic, and insuloma patients. *J. Lab. Clin. Med. 86*, 395-409

ADIBI, S. A., KRZYSIK, B. A. & DRASH, A. L. (1977) Metabolism of intravenously administered dipeptides in rats: effects on amino acid pools, glucose level, and insulin and glucagon secretion. *Clin. Sci. Mol. Med. 52*, 193-204

BOULLIN, D. J., CRAMPTON, R. F., HEADING, C. E. & PELLING, D. (1973) Intestinal absorption of dipeptides containing glycine, phenylalanine, proline, β-alanine or histidine in the rat. *Clin. Sci. Mol. Med. 45*, 849-858

FOGEL, M. R. & ADIBI, S. A. (1974) Assessment of the role of brush border peptide hydrolases in luminal disappearance of dipeptides in man. *J. Lab. Clin. Med. 84*, 327-333

GRAHL-NIELSEN, O., ØDEGAARD, P. & TRITSCH, L. (1974) Oligopeptides as source of indispensable amino acids for mammalian cells in culture. *In Vitro 9*, 414-420

KRZYSIK, B. A., PETERSON, J. A. & ADIBI, S. A. (1975) The potential of blood, liver, muscle and kidney for dipeptide hydrolysis. *Fed. Proc. 34*, 466 (abstr.)

LONG, C. L., ZIKRIA, B. A., KINNEY, J. M. & GEIGER, J. W. (1974) Comparison of fibrin hydrolysates and crystalline amino acid solutions in parenteral nutrition. *Am. J. Clin. Nutr. 27*, 163-174

MANCHESTER, K. L. (1970) Sites of hormonal regulation of protein metabolism, in *Mammalian Protein Metabolism*, vol. 4 (Munro, H. N., ed.), pp. 229-298, Academic Press, New York

MATTHEWS, D. M. & ADIBI, S. A. (1976) Peptide absorption. *Gastroenterology 71*, 151-161

MILNE, M. D. (1971) Transport of amino acids and peptides in the gut and the kidney, in *Scientific Basis of Medicine Annual Reviews* (Gilliland, I. & Francis, J., eds.), pp. 161-177, Athlone Press, London

MORTIMORE, G. E. & MONDON, C. E. (1970) Inhibition by insulin of valine turnover in liver. *J. Biol. Chem. 245*, 2375-2382

NÜTZENADEL, W. & SCRIVER, C. R. (1976) Uptake and metabolism of β-alanine and L-carnosine by rat tissues *in vitro:* role in nutrition. *Am. J. Physiol. 230*, 643-651

PERRY, T. L., HANSEN, S., TISCHLER, B., BUNTING, R. & BERRY, K. (1967) A new metabolic disorder associated with neurologic disease and mental defect. *N. Engl. J. Med. 277*, 1219-1227

PROCKOP, D. J., KEISER, H. R. & SJOERDSMA, A. (1962) Gastrointestinal absorption and renal excretion of hydroxyproline peptides. *Lancet 2*, 527-528

SMITH, E. L. (1951) The specificity of certain peptidases, in *Advances in Enzymology*, vol. 12 (Nord, F. F., ed.), pp. 191-257, Intersciences Publishers, New York

YAMAGUCHI, T., YAMAGUCHI, M. & LAJTHA, L. (1970) Inhibition of dipeptide transport in mouse brain slices. *J. Neurol. Sci. 10*, 323-329

Discussion

Nützenadel: Your results allow the interpretation that dipeptides in the circulation are transported into the tissue cells, Dr Adibi, and this extends our observations on dipeptide transport through cell membranes outside the gut epithelium. In kidney cortex slices and the hemidiaphragm of the rat, carnosine (β-alanyl-L-histidine) is taken up by an active transport system which is independent of the mechanism serving the entry of free amino acids (Nützenadel & Scriver 1976). The meaning of dipeptide transport through cell membranes which are thought not to be exposed to peptides remains to be clarified. In addition to unhydrolysed peptides from ingested protein reaching the circulation, other peptides released from intracellular pools are present in the blood-stream. Peptide uptake into tissue cells could prevent their loss in the urine. Hydrolases which cleave peptides are available in many tissues and in kidney cortex slices the oxidation of β-alanine derived from L-carnosine in the bathing medium is very rapid. Accordingly, peptides in the blood could serve as a source for intracellular metabolism.

De Laey: I was astonished by your results with glycylsarcosine. What are the figures for its clearance?

Adibi: Out of 150 μmol, 20 μmol appeared in the urine in 40 min.

De Laey: What was the clearance in ml/min?

Adibi: I haven't calculated that.

Radhakrishnan: With glycylglycine you found an immediate rise in free glycine in the plasma. With glycylsarcosine you didn't find any glycine in the plasma but at the same time you found hydrolysis in the kidney because you found free glycine and sarcosine. Does this mean that the kidney differentially prefers glycine to sarcosine?

Adibi: Among the tissues that we studied only kidney showed an apparent capacity for Gly-Sar hydrolysis. There was a trace hydrolase activity against this dipeptide in the intestine and none in the skeletal muscle or liver.

Radhakrishnan: So that means glycine doesn't come out?

Adibi: In comparison to Gly-Gly, Gly-Sar injection caused very little increase in blood glycine concentrations. This difference is probably due to (*a*) the amount of glycine injected as Gly-Sar being half the amount injected as Gly-Gly, and (*b*) there being either no or very little hydrolysis of Gly-Sar by tissues.

Radhakrishnan: Why doesn't the glycine come out when you use glycylsarcosine?

Baker: There must be a difference in utilization of glycine in the kidney and in other tissues with the same enzyme. Have you done this with glycylsarcosine with a ^{14}C label?

Adibi: Dr Silk suggested this to us recently and we are about to do that kind of study.

Silk: Can you tell us anything about levels of other non-injected amino acids in the plasma?

Adibi: There is a small rise and then a moderate fall for most of them. For example, in rat plasma valine would normally be around 200 μmol/l. It rises to around 275 μmol/l and then falls to about 75 μmol/l after 40 min. The explanation is that either insulin, glucagon, glucose, or all three have caused this.

Silk: Do the dipeptide and the equivalent free amino acid produce the same pattern?

Adibi: Yes.

Meister: What is the lowest concentration of glycylglycine you can detect with your method?

Adibi: We can estimate as low as 0.01 μmol/ml very precisely.

Meister: So when you said there was no glycylglycine remaining you mean there was less than 10 μmol/l. Doesn't that mean that the concentration of glycylglycine or a similar peptide in the blood plasma is less than 10 μmol/l?

Adibi: Yes; we could not detect it but it could be there below the concentration of 0.01 μmol/ml.

Meister: This raises the obvious question of why the kidney has so much peptidase activity when there is apparently virtually no substrate in the blood plasma. Do you have any ideas about this?

Adibi: I don't know of an exact answer to this question, but after meals small peptides could be concentrated in substantial concentrations in kidney and therefore be substrates for kidney peptides.

Josefsson: The kidney is metabolically a very active organ, so the turnover of protein may be much higher there than in other tissues. If there is a higher turnover there must be a much higher capacity for peptidase activity.

Adibi: Kidney is an active tissue but I don't think it is more active than the liver in terms of protein synthesis.

Meister: Another possibility is that peptidase activity was not put there for the purpose of splitting the peptide bond. It is what we happen to measure in the laboratory but it may not be a physiological activity. The kidney is metabolically very active but it is also very active in transport.

Rhodes: The small intestine is another active transport organ, but there is also a considerable turnover of cellular protein, in the face of a selectively short cell life of three to five days (Alpers 1973). Perhaps some cytoplasmic peptidases are involved in the synthetic and degradative processes which must be associated with this turnover.

Meister: I don't think there is any evidence that any of the components of the protein-synthesizing system are very active in cleaving peptide bonds.

Milne: The presence of tripeptidase in the kidney tissue is probably part of the unique ability of kidney to turn over plasma protein. The kidney has the highest protein turnover of plasma proteins per gram of tissue in the body. It delivers the protein back into the blood as free amino acids and this clearly needs peptidases. This doesn't surprise me at all.

Adibi: But what is the need for a high protein turnover in the kidney?

Milne: In plasma protein turnover the fraction degraded by renal tissue is quantitatively significant. One can show that a large part of the various fractions of plasma proteins are degraded in the kidney.

Allan: Isn't a small proportion of protein filtered through the glomerulus into the urine? Presumably the peptidase activity conserves amino acids from partially degraded proteins, which would otherwise be continuously lost from the body. Small proteins, with a molecular weight less than 50 000, are excreted in urine of subjects with damaged renal tubules (Harrison *et al.* 1968).

Kenny: Most estimations suggest that the protein concentration in the glomerular filtrate is extremely low (0.1–10 mg/ml). Over 24 h it would amount to 0.2–17 g protein in man and all the evidence indicates that this is taken up by pinocytosis in the proximal tubule. There is no reason to think that the brush border peptidases are involved in its hydrolysis

Gardner: What is the nitrogen balance in patients who have received Aminosol? This preparation for intravenous nutrition contains peptides and I had the impression some time ago that its use was accompanied by some loss of amino acid nitrogen in the urine (Lidström & Wretlind 1952). Professor Matthews suggested to me that this is due to a 'browning reaction', that is conjugation with sugar during heat sterilization.

Matthews: It has been shown that tissue utilization of small peptides is approximately as effective as that of free amino acids. There is an important historical feature here, which unfortunately produced a lingering impression that peptides were not well utilized. When solutions for intravenous feeding were being developed after World War II, partial hydrolysates of proteins containing small peptides were frequently autoclaved with glucose. This produced compounds which led to extremely heavy losses of peptides in the urine. About half of the material could be lost. When another method was used, with no autoclaving with glucose, peptide loss in the urine fell to less than 10%—little more than the loss after infusion of mixtures of free amino acids (Christensen *et al.* 1955; see also Matthews & Payne 1975*a*).

Kenny: Why does autoclaving the amino acids with glucose have that effect, Professor Matthews?

Matthews: Christensen and his colleagues suggested that it produces compounds which interfere with the transport or hydrolysis of peptides—the exact mechanism is unknown. [*Note added in proof:* Since this discussion took place, I have read a paper (Amaya-F. *et al.* 1976) which shows that a peptide, fructosylglycyl-L-leucine, produced from glycyl-L-leucine by heating with glucose (the Maillard reaction), is neither hydrolysed by leucine aminopeptidase nor absorbed from the small intestine of rats. If the Maillard reaction produces peptides which are neither hydrolysable nor transportable, the explanation of the heavy peptiduria occurring after intravenous infusion of solutions of small peptides heated with glucose is likely to be that the Maillard peptides formed by this procedure cannot be hydrolysed in the blood, and are neither taken up by the tissues nor reabsorbed by the renal tubules.]

Meister: The effect with glucose is very well known. It is apparently even more complicated than the subject being discussed here today. There are many papers on the chemical constitution of amino acid–carbohydrate complexes of one sort or another. That is why one doesn't normally autoclave glucose with an amino acid.

Gardner: This is also relevant for people who wish to analyse protein hydrolysates which contain sugar, such as plasma. It is a hazardous business.

Mellander: Bone tissue has very high phosphatase activity. Did you examine active bone tissue for peptidase activity, Dr Adibi?

Adibi: No. It took us several years to complete the study on kidney and intestine. I have not found bone tissue easy to work with.

Parsons: The amino acid turnover of the central nervous system might be quite interesting to examine after giving peptides. Presumably the peptides must get across the blood–brain barrier and also into the cerebrospinal fluid?

Adibi: It would be very interesting to study brain tissue.

Matthews: There are many dipeptides and tripeptides in brain. In many cases, nobody knows what they are doing there. Brain tissue also possesses a peptide uptake system (see Matthews & Payne 1975*b*).

Josefsson: Immunologically the brain dipeptidases are identical to other dipeptidases and there are many peptides that we know of within the brain tissue. Many of them are small peptides.

Radhakrishnan: A few more peptides have been derived from brain. One of them is enkephalin which may bind to opiate receptors. It is a naturally occurring peptide in the brain, with a Gly-Gly sequence (Simantov & Snyder 1976).

References

ALPERS, D. H. (1973) Regulation of small intestinal protein metabolism. *Gastroenterology 64*, 471-496

AMAYA-F., J., LEE, T.-C. & CHICHESTER, C. O. (1976) The intestinal absorption of fructosyl-glycine-L-leucine. *Nutr. Rep. Int. 14*, 229-235

CHRISTENSEN, H. N., WILBER, P. B., COYNE, B. A. & FISHER, J. H. (1955) Effects of simultaneous or prior infusion of sugars on the fate of infused protein hydrolysates. *J. Clin. Invest. 34*, 86-94

HARRISON, J. F., LUNT, G. S., SCOTT, P. & BLAINEY, J. D., (1968) Urinary lysozyme, ribonuclease and low molecular weight protein in renal disease. *Lancet 1*, 371

LIDSTRÖM, F. & WRETLIND, K. A. J. (1952) Effect of dialyzed casein hydrolysate. The effect of intravenous administration of a dialyzed enzymatic casein hydrolysate (aminosol) on the serum concentration and on the urinary excretion of amino acids, peptides and nitrogen. *Scand. J. Clin. Lab. Invest. 4*, 167-178

MATTHEWS, D. M. & PAYNE, J. W. (1975*a*) Peptides in the nutrition of microorganisms and peptides in relation to animal nutrition, in *Peptide Transport in Protein Nutrition* (Matthews, D. M. & Payne, J. W., eds.), pp. 1-60, North-Holland, Amsterdam

MATTHEWS, D. M. & PAYNE, J. W. (1975*b*) Occurrence and biological activities of peptides, in *Peptide Transport in Protein Nutrition* (Matthews, D. M. & Payne, J. W., eds.), pp. 392-463, North-Holland, Amsterdam

NÜTZENADEL, W. & SCRIVER, C. R. (1976) Uptake and metabolism of β-alanine and L-carnosine by rat tissues *in vitro*: role of nutrition. *Am. J. Physiol. 230*, 543-651

SIMANTOV, R. & SNYDER, S. H. (1976) Isolation and structure identification of a morphine-like peptide 'enkephalin' in bovine brain. *Life Sci. 18*, 781-788

Peptiduria in the Fanconi syndrome

A. M. ASATOOR, M. D. MILNE and J. M. WALSHE

Medical Unit, Westminster Medical School, London and Department of Medicine, University of Cambridge

Abstract Peptide excretion has been studied in 20 cases of Wilson's disease and in maleate-induced Fanconi syndrome in the rat, ligand-exchange column chromatography being used to separate peptides from free amino acids. There is a statistically significant increase in urinary peptides in both types of the Fanconi syndrome. In both man and the rat, a large fraction of the excreted peptides has been shown to contain hydroxyproline, and therefore to be derived from collagen degradation. In both groups there is a close correlation between the output of hydroxyproline and that of total peptide-bound amino acids. Arguments are advanced that the peptiduria is due to increased urinary clearance of plasma peptides rather than to a metabolic cause. Peptides excreted in patients with the Fanconi syndrome are shown to have a different proportionate amino acid composition to those in urine specimens from normal controls. The mean size of urinary peptides derived from collagen must be at least five or six amino acids per peptide chain. Maleate-induced Fanconi syndrome in the rat is thought to be a close analogue of the syndrome in man, and further results obtained in the animal model may well be directly applicable to human disease.

The functional defect of the Fanconi syndrome is a reduced power of re-absorption, and to a lesser extent of secretion, of solutes by the renal tubular epithelial cells. The defect usually involves increased output of at least three of six classes of substance: (*a*) amino acids; (*b*) low molecular weight proteins, particularly β-globulins; (*c*) glucose; (*d*) various electrolytes including Na^+, K^+, HCO_3^-, Cl^-, phosphate and Ca^{2+}; (*e*) urate; and (*f*) water (Scriver *et al.* 1976). Amino acids are reabsorbed by active transport, there being a higher concentration within the tubular cells than within both the tubular lumen and the peritubular capillary blood. Globulins of low molecular weight are filtered at the glomeruli, and are reabsorbed by a process of pinocytosis followed by considerable hydrolytic degradation to smaller peptides and free amino acids within the tubular cells. Corresponding abnormalities of renal excretion of nitrogenous compounds of molecular weights intermediate between those of amino acids and proteins, i.e. oligopeptides, have previously been neglected because of methodological difficulties. This paper describes an investigation of peptide excretion in two types of the Fanconi syndrome: Wilson's disease (hepatolenticular degeneration) in man, and the syndrome occurring after a single injection of sodium maleate in the rat.

The main difficulty in investigating peptide output in disease is that urine contains a large number of different peptides of widely divergent molecular size and amino acid composition. Most excreted peptides have the single common property of being relatively resistant to hydrolysis by tissue peptidases (Hanson & Ansorge 1967), as otherwise they would be rapidly hydrolysed to their constituent amino acids before being filtered at the glomeruli. Two main classes of urinary peptides with this characteristic have been described: peptides containing hydroxyproline and therefore derived from collagen degradation, and those containing a β-aspartyl or γ-glutamyl linkage. The size of excreted oligopeptides may vary from small dipeptides, e.g. prolylhydroxyproline and β-aspartylglycine (Meilman *et al.* 1963; Dorer *et al.* 1966), to large peptides of molecular weight about 5000 (Krane *et al.* 1970) and therefore containing 30–40 amino acids within the peptide chain.

Renal excretion of these peptides has not been investigated in detail. It is known that the average clearance of hydroxyproline-containing peptides in man averages 50% of the glomerular filtration rate (GFR) (Benoit & Watten 1968), although this may rise to values above the GFR at abnormally high plasma concentrations.

All analyses of urinary peptides were made on specimens deproteinized by 1% picric acid, with subsequent removal of the precipitant on ion-exchange columns (Stein & Moore 1954). Two separate methods were used throughout: (*a*) the difference method which involved analysis of amino acid content before and after hydrolysis by 6 N-HCl for 24 h at 110 °C, and (*b*) the direct method involving separation of amino acids from peptides by ligand-exchange column chromatography (Buist & O'Brien 1967). Both methods have their disadvantages. By the difference method amino acids in conjugates, e.g. hippuric acid, are included in the final analysis, and the error of a method dependent on two analyses is inevitably increased. Separation of amino acids and peptides by ligand-exchange columns is not absolutely perfect, as a proportion of the acidic amino acids appears in the effluent and some peptides containing basic amino acids may be retained. In addition, conjugates containing amino acid residues are not completely retained by the columns. Values obtained by the direct method are always somewhat lower than by the difference method, but there is an excellent correlation between the two estimates ($r = 0.90$, $P < 0.001$). The two methods combined probably put an upper and a lower limit on the true value of urinary peptides.

WILSON'S DISEASE

Abnormalities of urinary peptide output in the Fanconi syndrome are proportionately less than those involving free amino acids and therefore a statistical analysis of a substantial number of cases was necessary. Wilson's disease was selected because of the availability of sufficient patients in a clinic where therapy by the copper chelating agents, D-penicillamine and triethylene tetramine, was being assessed. Both treated and untreated cases had of necessity to be studied, but therapy does not appear to increase peptide output. There was no correlation between peptide excretion and the duration or dosage of the two drugs, and administration of penicillamine to rats at a dosage of 40 times that used in man on a body weight basis does not cause any rise in urinary hydroxyproline contained in peptides (Nimni *et al.* 1969).

Free amino acids in Wilson's disease were excreted in amounts averaging twice that in normal control subjects ($P < 0.01$). Values for individual amino acids were all above those in controls, but increases in basic and branched-chain amino acids were not significant statistically. Values for amino acids contained in oligopeptides were also higher (Asatoor *et al.* 1976), ranging from 1.75 times the values for controls by the direct method to 1.56 times by the difference method ($P < 0.01$), i.e. less than the increases in free amino acids. Individual peptide-bound amino acids were all higher in Wilson's disease, except methionine, leucine, tyrosine, phenylalanine and hydroxyproline. Hydroxyproline output has a very large standard deviation if adults and adolescents are considered together. When the adults were examined alone, hydroxyproline output was significantly higher than that of controls, and also the adolescent patients had a higher excretion rate when compared with values from a large group of normal subjects recorded by Wharton *et al.* (1972).

Since a large fraction of urinary peptides contains hydroxyproline and is derived from collagen degradation, a correlation should be expected between hydroxyproline output and total peptide-bound amino acids. This was found both in the patients and in the controls (Fig. 1). The regression line in Wilson's disease is significantly above that in controls, but the slopes are identical. For every mole increase in hydroxyproline output, there is a mean increase of 5 mol in peptide-bound amino acids, indicating that the mean chain length of the peptides containing hydroxyproline is at least five amino acids and possibly higher. An estimate of the output of peptides not containing hydroxyproline, but resistant to hydrolysis because of β-aspartyl and γ-glutamyl linkages, is given by the intercept of the two regression lines on the ordinate. This too is significantly higher in Wilson's disease than in controls ($P < 0.01$).

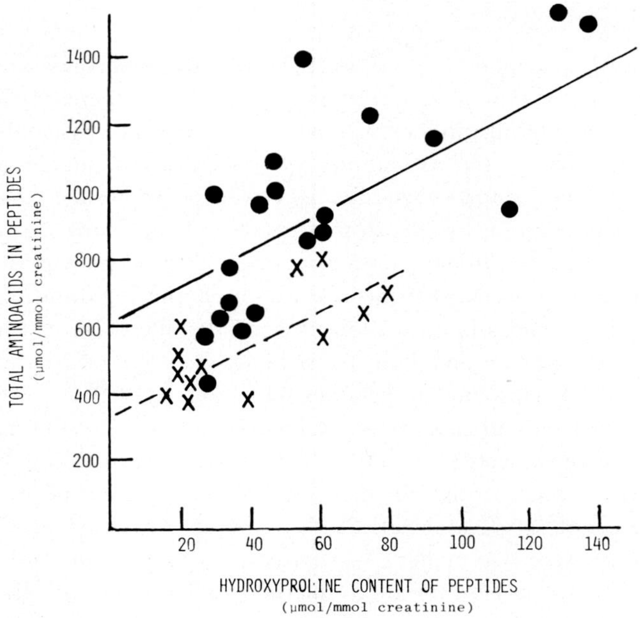

FIG. 1. Correlation between hydroxyproline content of urinary peptides and total output of peptide-bound amino acids (difference method) in 20 cases of Wilson's disease (closed circles) and in 13 normal controls (crosses). The two lines are the calculated regression lines. Similar results using the direct method of analysis have previously been published (Asatoor *et al.* 1976).

No evidence was found that excess peptide output accounts for the hypercupruria present in many cases of Wilson's disease, a view originally considered by Uzman (1953). The excess output of free amino acids accounted for 90% of the complexing power of the urinary copper both in the patients and in the controls, and no detectable complexing power could be detected for the peptides contained in the effluents from the ligand-exchange columns.

EXPERIMENTAL FANCONI SYNDROME IN THE RAT

Sodium maleate injected at a dosage of 3-4 mmol/kg body weight causes a reversible Fanconi syndrome in the rat (Harrison & Harrison 1954). The poison is also active *in vitro*, as shown by a reduced uptake of amino acids and other solutes by kidney slices (Rosenberg & Segal 1964). Within a few hours of a single injection of maleate, there is severe damage to the perimitochondrial membranes of renal tubular cells, and a rapid development of the

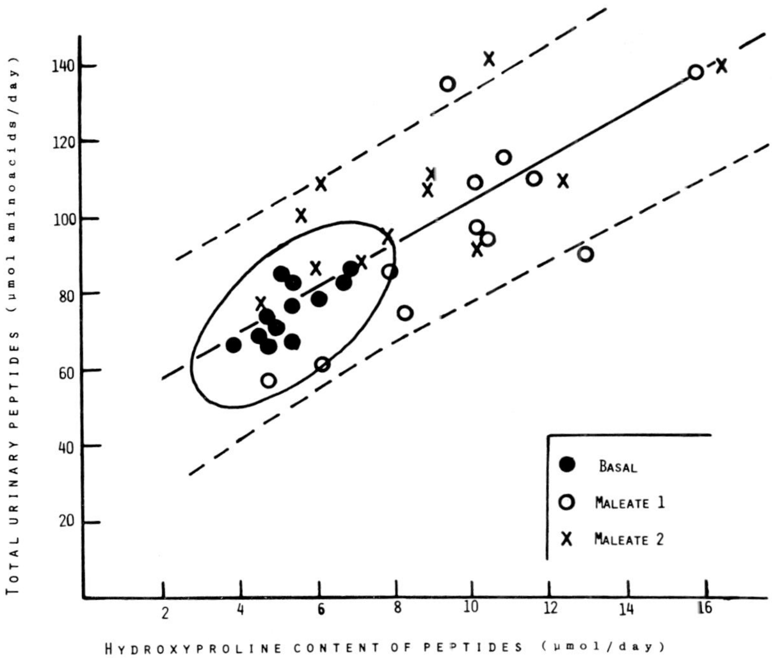

FIG. 2. Correlation between hydroxyproline content of urinary peptides and total output of peptide-bound amino acids (difference method) in rats before and after maleate injection. The ellipse gives the 95% fiducial limits in the controls. The continuous line gives the calculated linear regression between the two variables and the broken lines the 95% fiducial limits for all 36 observations.

functional abnormalities typical of the Fanconi syndrome. Both the structural and functional disorders recover about 96 h after the injection.

In studies of the functional renal abnormalities occurring on the first and second days after subcutaneous injection of sodium maleate, 3.5 mmol/kg body weight, in the rat (Asatoor *et al.* 1976), free amino acid output in the urine was found to be ten times control values on the first day, and nine times control on the second day. Peptide output also significantly increased, to 1.5 times the control on the first day and 1.8 times on the second day. Most of the individual peptide-bound amino acids were also higher on both days, and in particular hydroxyproline output was 1.9 times the control mean. A highly significant correlation was found between hydroxyproline output and total excretion of peptide-bound amino acids (Fig. 2), values for both being higher than in controls on both the first and second days after the injection. There is a mean increase of 6 mol in total bound amino acid for every 1 mol rise in urinary hydroxyproline. In view of the rapid rise in peptide output it seemed important to investigate possible changes in the composition of excreted peptides after the injection. The amino acid content did not differ significantly in ranking order from that in collagen itself. After the injection, however, there was a significant increase in the proportionate content of hydroxyproline, leucine and cystine, with a corresponding fall in glycine and lysine. As a comparison, urinary peptides in the patients with Wilson's disease contained a higher proportion of cystine and a lower glycine content than those of controls.

DISCUSSION

The available data are insufficient to allow us to decide with certainty whether the increased output of peptides in the two types of Fanconi syndrome is of metabolic or renal origin. Skeletal abnormalities are common in Wilson's disease (Strickland & Leu 1975), and in fact were present in 15 of the 20 patients in our study. There was no correlation between the severity of skeletal disease and the output of either peptides or free amino acids. Sodium maleate is almost a specific kidney poison owing to its concentration within renal parenchymal cells. Urinary peptides increased soon after the injection, at a time when there are known to be severe structural and functional abnormalities of renal tubules. It is extremely unlikely that maleate could cause a rapid increase in collagen breakdown, and therefore a renal cause for the peptiduria is much more probable. Direct determination of clearances of hydroxyproline peptides is technically difficult because of their low concentrations in plasma, with a relatively high content of free hydroxyproline. The evidence to date suggests that the peptiduria in both Wilson's disease and experimental Fanconi syndrome in the rat is of tubular origin and is comparable to the excess excretion and renal clearance of amino acids and small globulins characteristic of the syndrome.

There is current uncertainty about the precise nature of the renal tubular functional defect of the Fanconi syndrome. The classical view of the cause of increased amino acid clearance is that proximal tubular reabsorption of these compounds from the tubular lumen is reduced. Several recent studies cast considerable doubt on this simplified view of the disorder (Bergeron *et al.* 1976; Scriver *et al.* 1976). Clearance of amino acids in the Fanconi syndrome can considerably exceed the glomerular filtration rate (Scriver 1968). Normal competitive interactions between related molecules for active transport sites are preserved even in severe examples of the syndrome. After maleate injection in the rat, leucine is excreted at a rate indicative of net tubular flux, with a reduction in its intracellular concentration (Bergeron & Vadeboncoeur 1971). Maleate in the rat accelerates efflux of amino acids from renal cortical slices, as well as impairing uptake (Rosenberg & Segal 1964). The functional defect in the Fanconi syndrome can, therefore, probably be best explained by an enhancement of efflux on the luminal side of the tubular epithelial cells. This implies an equal or greater defect in the function of distal tubular cells as compared to those of the proximal tubule, since there is little, if any, influx of amino acids in the distal area. The structural abnormalities produced by maleate injections are equally prominent in both proximal and distal tubular cells.

If the peptiduria of the Fanconi syndrome is similarly due to increased transtubular flux, it would be expected that smaller peptide molecules would be preferentially excreted. Urinary peptides, both in Wilson's disease and after maleate in the rat, have been shown to have a different amino acid composition to those in control urine specimens, but there is no proof that they are of lower molecular weight. Smaller peptides derived from collagen degradation necessarily have a proportionately higher hydroxyproline content than larger peptides. Urinary peptides in both Wilson's disease and the experimental Fanconi syndrome have a higher proportion of hydroxyproline than peptides from controls, but this was statistically significant only in the rats.

These observations on the urinary output of peptides in the Fanconi syndrome can only be regarded as preliminary observations, and clearly much more detailed work is necessary before the implications can be adequately understood. It is, however, encouraging that the results of studies of Wilson's disease in man and maleate-induced Fanconi syndrome in the rat are comparable and have produced non-contradictory data. The ready availability of the rat model may therefore be valuable in the prediction of further details of the abnormalities of urinary peptides in the Fanconi syndrome in man.

References

Asatoor, A. M., Milne, M. D. & Walshe, J. M. (1976) Urinary excretion of peptides and of hydroxyproline in Wilson's disease. *Clin. Sci. Mol. Med. 51*, 369-378

Benoit, F. L. & Watten, R. H. (1968) Renal tubular transport of hydroxyproline peptides: evidence for reabsorption and secretion. *Metabolism 17*, 20-33

Bergeron, M. & Vadeboncoeur, M. (1971) Microinjections of L-leucine into tubules and peritubular capillaries of the rat. II. The maleic acid model. *Nephron 8*, 367-374

Bergeron, M., Dubord, L. & Hausser, C. (1976) Membrane permeability as a cause of transport defects in experimental Fanconi syndrome. A new hypothesis. *J. Clin. Invest. 57*, 1181-1189

Buist, N. R. M. & O'Brien, D. (1967) The separation of peptides from amino acids in urine by ligand exchange chromatography. *J. Chromatogr. 29*, 398-402

Dorer, F. E., Haley, E. E. & Buchanan, D. L. (1966) Quantitative studies of urinary β-aspartyl oligopeptides. *Biochemistry 5*, 3236-3245

Hanson, H. & Ansorge, S. (1967) Zur Ausscheidung von Peptiden im menschlichen Harn.

Das Verhalten isolierter Peptidfrakionen gegenüber proteolytischen Enzymen. *Hoppe-Seyler's Z. Physiol. Chem. 348*, 347-351

HARRISON, H. E. & HARRISON, H. C. (1954) Experimental production of renal glycosuria, phosphaturia, and aminoaciduria by injection of maleic acid. *Science (Wash. D.C.) 120*, 606-608

KRANE, S. M., MUNOZ, A. J. & HARRIS, E. D. (1970) Urinary polypeptides related to collagen synthesis. *J. Clin. Invest. 49*, 716-729

MEILMAN, E., URIVETZKY, M. M. & RAPOPORT, C. M. (1963) Urinary hydroxyproline peptides. *J. Clin. Invest. 42*, 40-50

NIMNI, M. E., DESHMUKH, K., GERTH, N. & BAVETTA, L. A. (1969) Changes in collagen metabolism associated with the administration of penicillamine and various amino and thiol compounds. *Biochem. Pharmacol. 18*, 707-714

ROSENBERG, L. E. & SEGAL, S. (1964) Maleic acid-induced inhibition of amino acid transport in rat kidney. *Biochem. J. 92*, 345-352

SCRIVER, C. R. (1968) Renal tubular transport of proline, hydroxyproline and glycine: III. Genetic basis for more than one mode of transport in human kidney. *J. Clin. Invest. 47*, 823-835

SCRIVER, C. R., CHESNEY, R. S. & MCINNES, R. R. (1976) Genetic aspects of renal tubular transport: diversity and topology of carriers. *Kidney Int. 9*, 149-171

STEIN, W. H. & MOORE, S. (1954) The free amino acids of human blood plasma. *J. Biol. Chem. 211*, 915-926

STRICKLAND, G. T. & LEU, M. L. (1975) Wilson's disease: clinical and laboratory manifestations in 40 patients. *Medicine (Baltimore) 54*, 113-137

UZMAN, L. L. (1953) On the relationship of urinary copper excretion to the aminoaciduria in Wilson's disease (hepatolenticular degeneration). *Am. J. Med. Sci. 226*, 645-652

WHARTON, B. A., GOUGH, G., WILLIAMS, A., KITTS, S. & PENNOCK, C. A. (1972) Urinary total hydroxyproline:creatinine ratio. Range of normal, and clinical application in British children. *Arch. Dis. Child. 47*, 74-79

Discussion

Mellander: I want to describe an experimental approach we used when we were interested in the biological properties of human milk as compared with other kinds of milk. When we compared the digestibility of caseins by proteolytic enzymes, human milk casein was found to be more resistant than cow's milk casein (Mellander 1947). Several other caseins were also investigated but no other casein is more resistant to proteolytic enzymes than the human casein. Casein from the milk of the blue whale is almost completely hydrolysed (Bennich *et al.* 1957). The enzyme-resistant residue after hydrolysis of casein consists of a mixture of peptides containing phosphorus and appearing as a calcium complex if prepared in the presence of calcium. The preparations usually contained about 10% Ca and 5–6% P. We had not known until then that phosphorus and calcium could be combined with amino acids in such a soluble form (50 g/100 ml solution) (Mellander 1955*a*).

We did a complete amino acid analysis of some phosphorylated peptides isolated from human casein (Strid 1961). Chemical synthesis of phosphorylated peptides was undertaken (Mellander & Fölsch 1972). Generally the peptide bonds close to phosphorylated serine residues were found to be enzyme-resistant. The metal-binding capacity of phosphorylated peptides has been investigated in our laboratory (Österberg 1966). The hypothesis that metal complexes of this type can act as physiological carriers for calcium and other metals has been discussed (Mellander 1955*b*).

In a few cases we gave phosphorylated Ca-peptides to infants and found a very high calcium utilization (Mellander 1950). In some cases with rickets, increasing ossification could be demonstrated before vitamin D was given. In tissue culture experiments it was demonstrated that if bone from young chickens was cultivated in a medium containing peptide-bound Ca, normal growth and calcification occurred (Edenö 1974).

We also investigated uptake in microorganisms (*Thermobacterium acidophilum*). Some of the amino acids in the phosphopeptide, for example serine, were utilized completely, while others such as aspartic acid were less well utilized (Løvtrup *et al.* 1957). At this stage of our research trail we were under the impression that most workers entirely denied the possibility of peptide absorption. Since we now understand that this is not true any more we have again taken up our peptide work. It has also been possible in this laboratory to isolate three different peptide fractions from human milk (A. Johnsson & A. Brohammer, unpublished). One of them has a carbohydrate content which suggests that it is the same substance as was once studied as the bifidus growth-promoting factor (György 1953).

In circulating blood, whatever load of peptide mixtures we gave, we could never demonstrate any peptides. We are now analysing portal blood in humans after test meals of peptides and amino acids.

Matthews: We (D. Burston & D. M. Matthews, unpublished) studied the phosphorylated peptide, Ser-P-Leu, that you gave us, with our usual preparation of rings of everted hamster jejunum *in vitro*, Professor Mellander. Its hydrolysis was indeed very much slower than that of the unphosphorylated peptide, as expected, but we found no intact peptide in the tissue. Nevertheless there was evidence that some of this peptide (Ser-P-Leu) at least was taken up by carrier-mediated transport, since uptake of Ser and Leu from it was inhibited by carnosine.

Meister: Professor Milne, could you elaborate on the interaction of maleate with amino groups?

Milne: That work was done on red cell ghosts, not renal tubules. Obaid *et al.* (1972) showed that this was a direct combination between the maleyl radicle and the amino groups in the proteins of the cell membrane. They showed this stoichiometrically. They measured the stability and determined the permeability characteristics of the membrane with what they call the maleyl-amino group attached. I take it this is a direct combination of the two carboxyl radicles to two adjacent amino groups.

Parsons: What were the concentrations of maleate in those experiments?

Milne: The dose was high, about 4 mmol/kg, but plasma concentrations were not measured. I would hesitate to do this on myself or other human beings even though I know that the effect is reversible in the experimental animal. Maleate *in vivo* is primarily nephrotoxic, being concentrated in the renal tubular cells.

Parsons: What is the concentration in the urine in these rats?

Milne: I regret we have not measured that. It is excreted, but of course the damage to the cell lasts much longer than the presence of maleate in the body. The renal functional defect is due to actual cell damage rather than to anything competitive with maleate still present. The concentration in the *in vitro* experiments is high, but the concentration *in vivo* must be high too. The trouble in the functional defect of the Fanconi syndrome is to explain these very diverse effects on cellular transport. There are too many solutes involved in the syndrome. If one has to think of active transport one must think of what is involved in all active transport, such as the ATPase system. The effect of maleate is certainly not proportional to any inhibition of ATPase but it is more proportional to the retention of the maleate radicle on cell membranes.

Meister: Of course maleate reacts readily with thiol groups. It can serve as an alkylating agent so one can picture a number of different chemical linkages which might affect a variety of systems. Tetracycline that has been carried around in a suitcase too long has been reported to produce the Fanconi syndrome in man.

Milne: Yes, it could do that. This cause of the syndrome has now disappeared. The manufacturers make the drug up differently and even in the hottest and most humid conditions no degraded tetracycline is produced. It would not be ethical to administer this in the human but it is a possible experimental tool in animals, although not so convenient as maleate.

Meister: You referred to the damaged mitochondria of the proximal renal tubule, yet later you spoke of distal tubular damage.

Milne: Bergeron & Laporte (1973) have shown that the damage is both proximal and distal, and it is equivalent in both sites. The only difference is that the damage proximally can be corrected by more distal active transport. The damage distally cannot be corrected and remains as a functional defect.

In relation to the thiol inhibition you mentioned, again there has not been the slightest correlation between the functional defect and inhibition of thiol groups in renal tubular enzymes. You are quite right in saying that maleate combines with thiol groups but there is plenty of evidence that this is not the reason for the functional defect in the Fanconi syndrome.

Meister: Is the defect seen in cystinosis of a similar sort?

Milne: The effect in cystinosis is a mystery. This is a lysosomal disease. Cystine is in the lysosomes and tissue culture has shown that the trapped cystine in the cell does not play any part in cell metabolism (Schneider *et al.* 1967). In a cystine-free medium the cells are as incapable of growing and dividing as other human cells such as fibroblasts, etc. They contain vast amounts of cystine but it is within an intracellular membrane, compartmentalized in the lysosome, and is not available for general cell metabolism. A tenable theory is that large amounts of cystine, just like large amounts of galactose-1-phosphate or fructose-1-phosphate, could be capable in the same way of producing a functional defect identical to the one produced by maleate.

Radhakrishnan: What is the mechanism for the increase in peptide-bound hydroxyproline with maleate? β-Aminopropionitrile increases a specific soluble collagen pool. So the increased excretion of hydroxyproline peptide there is understandable.

Milne: There is no evidence that within 24 h, or at any time, maleate has any effect on bone. However we haven't yet proved that there is no increase in plasma hydroxyproline peptides. This is technically difficult to do, because of the very low concentrations. But as the pharmacological effect of maleate in the rat is simply to produce the Fanconi syndrome, it is inconceivable to me that the tremendous differences could be due to a sudden breakdown, within 4 h, of bone collagen. That is a most unlikely hypothesis, but in cold science it must be excluded experimentally.

Radhakrishnan: If your assumption is right, will maleate bind the ε-amino group of lysine and behave just like β-aminopropionitrile?

Milne: As far as I know, maleate in the body is only present in high enough concentrations to damage the one tissue in which it is specifically concentrated about 10 times more than in any other tissue, that is the renal tubule. That is the accepted view of the pharmacology of maleate at 4 mmol/kg maximum.

Payne: If the reaction is with the ε-amino group it would be rather extraordinary. To obtain an ε-maleyl linkage routinely in protein chemistry, maleic anhydride is used. Furthermore, it is a labile linkage and comes off again readily at about pH 6. Thus, even if the substituent were formed, I doubt whether it would be stable for any significant length of time.

Milne: In fact the work on the red cell membrane was with maleyl anhydride, as I should have mentioned. Besides reporting cold facts we are of course engaged in certain speculations to attempt to explain our data. The evidence so far seems to be that maleate is affecting cell membranes rather than active transport processes. There is objective evidence of this from electron microscope studies of the mitochondrial membrane, while there is no evidence from the less sensitive boundary cell membranes. I take it that if this is disintegrated the renal tubule would be destroyed. It would show histological change under the light microscope, which it doesn't.

Parsons: Professor Smyth, have you ever exposed an intestine to maleic acid?

Smyth: No.

Semenza: At one time we used maleate buffer to study sugar uptake and found no inhibition by this buffer.

Milne: Wapnir *et al.* (1975) have recently shown that maleate in the rat intestine causes a decrease of Na^+ and an increase of K^+ transport, but without any effect on amino acid and glucose absorption.

References

BERGERON, M. & LAPORTE, P. (1973) Effet membranaire du maléate au niveau du néphron proximal et distal. *Rev. Can. Biol. 32*, 275-279

BENNICH, H., JOHANSSON, B. & MELLANDER, O. (1957) Enzymic hydrolysis of casein from different species. *Acta Soc. Med. Upsal. 62*, 67-76

EDENÖ, C. (1974) Effect of an enzyme-resistant phosphopeptide on calcification of embryo chicken bone *in vitro. Ups. J. Med. Sci. 79*, 18-20

GYÖRGY, P. A. (1953) A hitherto unrecognized biochemical difference between human milk and cow's milk. *Pediatrics 11*, 98-108

LØVTRUP, S., MELLANDER, O. & ROOS, K. (1957) The utilization of peptides by *Thermobacterium acidophilum. Arch. Mikrobiol. 26*, 83-88

MELLANDER, O. (1947) On chemical and nutritional differences between casein from human and from cow's milk. *Acta Soc. Med. Upsal. 52*, 107-198

MELLANDER, O. (1950) Phosphorylated peptides Ca absorption in infants. *Acta Soc. Med. Upsal. 55*, 247-255

MELLANDER, O. (1955*a*) The nutritional significance of some peptides. Reprint from '*Voeding*' [Nutrition] Vol. 16, pp. 363-366

MELLANDER, O. (1955*b*) Protein quality. *Nutr. Rev. 13*, 161-163

MELLANDER, O. & FÖLSCH, G. (1972) Enzyme resistance and metal binding of phosphorylated peptides, in *International Encyclopaedia of Food and Nutrition*, vol. 11 (Bigwood, E. J., ed.), pp. 569-579, Pergamon Press, Oxford

OBAID, A. L., REGA, A. F. & FARRAHAN, P. J. (1972) The effects of maleic anhydride on the ionic permeability of red cells. *J. Membr. Biol. 9*, 385-401

ÖSTERBERG, R. (1966) *Phosphorylated Peptides. Study of Primary Structure and Metal Complexity.* Thesis, University of Göteborg

SCHNEIDER, J. A., BRADLEY, K. & SEEGMILLER, J. E. (1967) Increased cystine in leucocytes from individuals homozygous and heterozygous for cystinosis. *Science (Wash. D.C.) 157*, 1321-1322

STRID, L. (1961) Phosphopeptides from a tryptic hydrolysate of human casein. *Acta Chem. Scand. 15*, 1423-1428

WAPNIR, R. A., EXENI, R. A., MCVICAR, M., DE ROSAS, F. J. & LIFSHITZ, F. (1975) Inhibition of sodium intestinal transport and mucosal (Na^+-K^+)-ATPase in experimental Fanconi syndrome. *Proc. Soc. Exp. Biol. Med. 150*, 517-520

General discussion II

COELIAC DISEASE

Fottrell: I wish to describe some results of ours on a human malabsorption condition, coeliac disease, where the patients are sensitive to a cereal protein, gliadin. The incidence of coeliac disease is very high in Ireland, with about 1 in 300 of the population suffering from it. The high incidence of intestinal cancer, both lymphoma and carcinoma, among patients with coeliac disease is also a very worrying aspect of the malabsorption condition. Information on the clinical, genetic and other aspects of this disease has been published (McNicholl *et al.* 1970, 1974). In the context of the present symposium the results I wish to describe are of interest because they suggest that (*a*) coeliac disease may *not* be attributable to the absence of a peptide hydrolase from the intestine and (*b*) the toxicity of gluten to coeliac patients may be due to a side-chain substituent (probably carbohydrate) on the polypeptide chains.

A preliminary account of our findings has been published (Phelan *et al.* 1974) and additional data are now available and will be published shortly (Phelan *et al.* 1977). In summary our findings are as follows:

(*a*) Gliadin, 70% alcohol-soluble fraction of gluten, was treated with carbohydrases from *Aspergillus niger*. This enzyme preparation removed all the carbohydrate covalently bound to the gliadin molecules. Gliadin treated in this manner was then fed to three coeliac patients who had been on gluten-free diets for several months, during which time the morphology of their mucosa improved.

(*b*) Several criteria such as xylose absorption, mucosal morphology and brush border enzymes were used to monitor the toxicity of carbohydrase-treated gliadin to the mucosa of the three coeliac patients in remission. Controls were coeliac patients who took untreated gliadin.

(*c*) On the basis of the criteria already mentioned, the gliadin treated with the carbohydrase preparation did not damage the intestinal mucosa of the patients with coeliac disease.

(*d*) The preparation of carbohydrases did not alter the primary structure of the gliadin proteins.

(*e*) These results therefore strongly suggest that a side-chain substituent, probably carbohydrate on the gliadin molecules, is involved in the toxicity of this protein mixture to patients with coeliac disease. They also explain why no purified proteolytic enzyme has removed the toxicity of gliadin to coeliac patients.

Semenza: Why didn't you use β-elimination? Is gliadin perhaps not soluble at alkaline pH or are the sugars bound via an *N*-asparagine bond?

Fottrell: Our evidence to date does not indicate that an *N*-asparagine bond is involved. Furthermore we did not detect any *N*-acetylglucosamine on hydrolysis.

Semenza: Everybody working with glycoprotein has been looking for an enzyme preparation which would chop off the sugars without damaging the protein. Could you say more about the carbohydrases you used?

Fottrell: At the moment it is a relatively crude preparation, containing several carbohydrases which are being characterized. We used several criteria to monitor any alterations in the protein, for example peptide maps of enzyme-treated and untreated gliadin and analysis of treated and untreated gliadin on ion-exchange resins, and we could detect no difference. Furthermore, when gliadin treated with this enzyme preparation was examined periodically for the appearance of amino groups, using 2, 4, 6-trinitrobenzene sulphonic acid, no proteolytic activity was detectable. Our enzyme preparation appears to remove all the sugars from the gliadin.

Adibi: Did you study intestinal alteration in the same individuals when they took treated and non-treated gliadin? Organ culture has been used for the investigation of the toxic effect of gluten on the intestine of coeliac patients. You could use this technique to further substantiate your observations.

Fottrell: One patient who took untreated gliadin showed the typical response of coeliac patients. Then after about nine months the same patient took treated gliadin and showed no ill effects. Dr Fiona Stevens in our group used organ cultures of human intestinal mucosa; but she finds that this system lacks specificity with regard to the type of peptides introduced into the medium. These findings are corroborated by others (T. J. Peters, personal communication).

Meister: Do your findings mean that these people congenitally lack a carbohydrase? If so, what happens when they get a little older? I understand that these patients frequently get better after some time.

Fottrell: It has not been the experience of our clinicians that coeliac patients improve with time; it is a life-long disease. We are not quite sure whether it is due to absence of carbohydrase or whether there is something abnormal on the membrane in the coeliac patient that allows gluten to bind to the surface of the mucosa where it would not normally bind. We have shown why gluten might be toxic to coeliacs; the next step will be to show how it is toxic.

Nützenadel: I would suggest that a young child is suitable for proving or disproving your hypothesis. In my experience morphological changes and clinical symptoms appear and disappear faster after gluten challenge or withdrawal in the very young child than in older children. Further, in my opinion eight days are very short for clear-cut evidence that your preparation is harmless.

Fottrell: This is an experiment we hope to try.

Josefsson: I am curious about the specificity of this carbohydrase. Is the carbohydrate moiety split directly from the peptide linkage or is part of the carbohydrate molecule left?

Fottrell: We do not know yet. Our preparation still contains a number of carbohydrases. Our own feeling is that, since we did not detect any *N*-acetylglucosamine, the hydroxyl group of serine or threonine may be involved. However we do not know the nature of the bond involved. We are now investigating the linkage between the gliadin peptides and the carbohydrate and determining whether there are one, two or three sugars, etc., attached at these linkages.

Josefsson: You haven't studied the specificity of this mixture of carbohydrases on synthetic substrates?

Fottrell: That is what we are doing now.

Silk: I would like to support Professor Fottrell. Studies by different independent groups have shown that morphological changes can occur in the mucosa very rapidly in response to gluten challenge. Professor Fottrell has offered us eminently satisfactory evidence that the carbohydrate-free gluten is not toxic. Actually there is little evidence that untreated coeliac disease ever regresses. We believe that adults have this mucosal lesion from childhood, though it may present very late. Our oldest patient was 75 on presentation.

Meister: Do you see an accumulation of glutamine-containing peptides in the blood in these patients, as I think was reported in this disease?

Fottrell: We have not investigated this.

Lloyd: I agree with what David Silk said about permanent gluten intolerance because that is the definition of coeliac disease. If there are situations, indistinguishable at the moment, where it is not permanent then that is something quite different. There does seem to be a very high incidence of coeliac disease in Ireland, and maybe the Irish intestine exposed to gliadin behaves differently in Ireland than in North America and other parts of the world. Professor Fottrell's experiments are, of course, excellent and the short-term experiments showing a reaction are absolutely clear-cut. However, there are studies showing that in some individuals, even after a long period of exposure, no abnormality can be detected by current techniques. Conclusions drawn from a few individuals studied for a short time may lead one into difficulties. If there is a negative response one has to be prepared to go on with further studies and this is what everybody is having to do at the moment.

Fottrell: We tried in these experiments to simulate the normal conditions as much as possible. If gluten is instilled into the mucosa, as Rubin *et al.* (1962) did many years ago, there is a response within six hours. We put our treated gliadin into gluten-free bread and gave 10 g of carbohydrase-treated gliadin in gluten-free loaves of bread (450 g) to each patient fives times in one week—that is, each patient took 50 g treated gliadin, which is a very large dose.

Lloyd: How long are you proposing to go on feeding your preparation in loaves of bread?

Fottrell: It would be useful to continue feeding the enzyme-treated gliadin for months but considerable effort is needed to prepare this material at present, so that is not practical yet, with our present technology.

Adibi: Are you giving this bread to young patients?

Fottrell: We will first have to ensure that the preparation is non-toxic in adults.

Adibi: How do you explain the observation that the treatment of gluten with hog mucosal homogenate removes the toxic effect? Is that observation artifactual?

Fottrell: No; I believe the observations of Frazer *et al.* (1958) on the abolition of toxicity by a crude homogenate of hog intestinal mucosa. However, the results may have been due to other enzymes to which gluten peptides might not be exposed during normal digestive processes. The ones I have in mind are enzymes such as lysosomal enzymes. The conclusions drawn from the experiment may not be valid. We have not attempted this experiment in our laboratory. Professor Meister asked earlier if a carbohydrase might be absent from the coeliac mucosa. This is a possibility.

Enari: Is this disease caused by cereals other than wheat?

Fottrell: Yes, by rye and perhaps by oats and barley as well.

Enari: What is the carbohydrate composition of the prolamines from these other cereals?

Fottrell: I don't know.

References

FRAZER, A. C., FLETCHER, F. R., ROSS, C., SHAW, B., SAMMONS, H. G. & SCHNEIDER, R. (1958) Gluten-induced enteropathy. The effect of partially digested gluten. *Lancet 2*, 252-255

MCNICHOLL, B., MITCHELL, B., DOLLY, J. O., DILLON, A. & FOTTRELL, P. F. (1970) Multiple forms of peptidases in intestinal mucosa from children with coeliac disease, in *Proc. Int. Conf. Coeliac Disease* (Booth, C. C. & Dowling, R. H., eds.), pp. 124-133, Churchill Livingstone, Edinburgh

MCNICHOLL, B., EGAN-MITCHELL, B. & FOTTRELL, P. F. (1974) Varying susceptibility in coeliac disease, in *Proc. 2nd Int. Conf. Coeliac Disease* (Hekkens, W. Th. J. M. & Pena, A. S., eds.), pp. 413-418, Stenfert Kroese, Leyden

PHELAN, J., STEVENS, F., MCCARTHY, C. F., MCNICHOLL, B. & FOTTRELL, P. F. (1974) Nature of gluten toxicity in coeliac disease. A new concept, in *Proc. 2nd Int. Conf. Coeliac Disease* (Hekkens, W. Th. J. M. & Pena, A. S., eds.), pp. 58-68, Stenfert Kroese, Leyden

PHELAN, J. J., STEVENS, F. M., MCNICHOLL, B., FOTTRELL, P. F. & MCCARTHY, C. (1977) Coeliac disease: the abolition of gliadin toxicity by enzymes from *Aspergillus niger*. *Clin. Sci. Mol. Med.* in press

RUBIN, C. E., BRANDBORG, L. L., FLICK, A. L., PHELPS, P., PARMENTIER, C. & VAN NIEL, S. (1962) Studies of celiac sprue III. The effect of repeated wheat instillation into the proximal ileum of patients on a gluten free diet. *Gastroenterology 43*, 631-641

Transport and hydrolysis of peptides by microorganisms

J. W. PAYNE

Department of Botany, University of Durham

Abstract The structural specificities of the dipeptide and oligopeptide permeases of *E. coli* are briefly reviewed and related to the requirements found for other microorganisms. New, quick, sensitive methods for studying peptide transport are described, based on the following: (*i*) peptide-dependent incorporation of free radioactive amino acid into newly synthesized protein by a double amino acid auxotroph, (*ii*) colorimetric assay of peptide-dependent enzyme synthesis by an amino acid auxotroph, (*iii*) dansyl fingerprint technique. These approaches provide information on peptide binding affinity to a permease and rates of peptide uptake and amino acid efflux. Among current and future research areas considered are: the influence of the p*K*b of the N-terminal amino group on transport, generality of peptide transport in microorganisms, energy coupling and regulation, involvement of binding proteins, and the 'smugglin' concept. Peptide hydrolysis, and nutritional utilization of peptides, by microorganisms are briefly discussed.

There is a large body of information on the transport and the hydrolysis of peptides by microorganisms, and it is clearly impossible to deal with it comprehensively here. Fortunately, a number of relevant reviews have appeared in the past years (Barak & Gilvarg 1975*a*; Holzer *et al.* 1975; Gilvarg 1972; Matthews & Payne 1975*a*; Miller 1975; Payne 1972*a*, 1975*a*, *b*, 1976; Payne & Gilvarg 1971, 1977; Simmonds 1970, 1972; Sussman & Gilvarg 1971). I shall therefore deal mostly with peptide transport and only to a lesser extent with peptide hydrolysis. Furthermore, I propose to treat transport from the standpoints of past, present and future studies; and in this way I hope to avoid repetition of much of the material to be found in the above reviews. Much of what is written here is speculative, and much of this may quickly be refuted by experiment and discussion; however, this meeting was conceived of as a platform for speculation and discussion, and one hopes it will act also as a catalyst for the appropriate experiments.

PEPTIDE TRANSPORT IN MICROORGANISMS: THE PAST

Most studies have used bacteria, in particular *Escherichia coli.* Recently, reports have appeared dealing with other bacterial species, and with yeast and fungi. To my knowledge peptide transport in algae has not been studied.

Defined radioactive peptides are not routinely available and consequently most of the information on peptide transport has been obtained using growth tests with amino acid auxotrophs. In the absence of extracellular peptidases, growth of an auxotroph on a peptide that contains the required amino acid may be taken as evidence for the transport of the peptide into the organism. Although recognized as an indirect approach, since growth response is far removed from actual transport, growth tests have provided much of the information on the general characteristics of the peptide transport systems. These studies have been complemented by use of toxic peptides that inhibit normal growth.

Distinct dipeptide and oligopeptide transport systems

The presence of separate transport systems for dipeptides and for oligopeptides in *E. coli* was first suspected from the poor competition between the two classes of peptide but was clearly revealed by the isolation of a mutant that was resistant to the toxic peptide triornithine (Payne 1968). The mutant had lost the ability to grow upon a variety of oligopeptides that supported growth of the parent strain whilst retaining normal ability to utilize dipeptides. Subsequently, similar oligopeptide permease mutants (*Opp*) were isolated from several different strains of *E. coli* and the responsible genetic locus was mapped (De Felice *et al.* 1973; Barak & Gilvarg 1974). Similar mutants have been isolated from *Salmonella typhimurium* (Ames *et al.* 1973; Jackson *et al.* 1976). An apparent oligopeptide-transport-deficient mutant of *Neurospora crassa* has been described (Wolfinbarger & Marzluf 1975*a*) but the situation was complicated by the inability of the parent to transport dipeptides and the presence of extracellular peptidases. Dipeptide-permease-deficient mutants (*Dpp*) have been isolated from *E. coli*, and the genetic defect has been shown to map separately from that for *Opp* (De Felice *et al.* 1973; Wasmuth & Umbarger 1974).

Structural requirements for peptide transport

The general results and conclusions have been recently summarized (Payne 1975*a*, 1976; Payne & Gilvarg 1977; Barak & Gilvarg 1975*a*).

N-terminal α-amino group. For dipeptide and oligopeptide transport the presence of a free, protonated N-terminal α-amino group is important. This is true for many bacterial species e.g. *Escherichia, Salmonella, Pseudomonas, Lactobacilli*, and (with some reservations) for the yeast *Saccharomyces cerevisiae* and the fungus *Neurospora crassa*. For *E. coli* it has been shown that substituents (e.g. α-*N*-acyl) that remove the positive charge on the amino group destroy the nutritional value of the peptide; mono-substituted peptides (e.g. α-*N*-alkyl) in which the positive charge is retained can still be transported and utilized (Payne 1971, 1974). Using the dansyl detection method (p. 316) (J. W. Payne, unpublished) it is found that α-*N*-methyl diglycine is taken up more slowly than diglycine but α-*N*-methyl triglycine is transported as rapidly as triglycine, in each case the tripeptide being transported faster than the corresponding dipeptide.

C-terminal α-carboxyl group. For those bacteria in which this feature has been studied (*Escherichia, Salmonella, Pseudomonas*) dipeptide transport via a dipeptide permease has a requirement for the free C-terminal α-carboxyl group. Removal or derivatization (e.g. ester, amide) of this group presumably prevents binding (no observed competitive ability) and consequently transport. Any ability for such dipeptides to be used nutritionally arises from their transport via an oligopeptide permease, related studies having shown that (in these three species of bacteria) the oligopeptide permease differs from the dipeptide system in having no requirement for the C-terminal α-carboxyl group: it is possible to derivatize the α-carboxyl group of oligopeptides in many ways and still retain transport (see the section Smugglins, p. 320). In yeast, requirement for this carboxyl group is still an open question (Naider *et al.* 1974), and no reports have appeared relating to fungi.

Stereospecificity of peptide transport. The transport systems show specificity for the L-stereoisomers; this general conclusion rests on the corroborative results of competition studies, growth tests, use of radioactive peptides and the dansyl detection technique (J. W. Payne, unpublished). It has been shown for strains of *Escherichia, Pseudomonas* and *Lactobacilli* (Payne & Gilvarg 1977). For oligopeptides, some evidence is available (Becker & Naider 1974) to support the suggestion (Payne & Gilvarg 1971) that the presence of a D-isomer at the C-terminus of a tripeptide should not prevent transport. No relevant reports have appeared for other bacterial species, yeast or fungi.

α-Peptide bond. Normal (protein-derived) α-linked peptides are the preferred substrates of the peptide permeases discussed here. A variety of β-, γ-, εlinked peptides have been examined but none showed affinity for the transport systems.

Peptide bond nitrogen atom. Study of this feature in *E. coli* (Payne 1972*b*) revealed that substitution (methylation) of the nitrogen atom did not prevent uptake via the dipeptide and oligopeptide permeases, although the substituted bonds were resistant to peptidase cleavage. Accumulation of these substrates can therefore provide good evidence for the separate nature of transport and hydrolysis. In addition, because they can be accumulated essentially intact they are extremely useful substrates with which to determine whether peptide transport is an active process, and for measuring its kinetic parameters. Use of the dansyl detection technique has recently shown that Gly-Sar, Gly-Sar-Sar and Gly-Sar-Sar-Sar (Sar is *N*-methylglycine) are transported by *E. coli* at about one-twentieth the rate of the corresponding unsubstituted peptides at pH 7, each at 50 μM (J. W. Payne, unpublished). Peptides of this type have also been used with much success in studies of intestinal peptide transport (Burston *et al.*, this volume, pp. 79-98) but have not yet been applied to other microorganisms (Payne & Gilvarg 1977).

Amino acid side chains. Most protein-derived peptides will be of mixed amino acid composition, and consequently for efficient peptide utilization it might be expected that the transport systems would not show specificity for the side chains. Results from studies with a variety of microorganisms support this view (Payne 1976; Payne & Gilvarg 1977). On the other hand, one can imagine that since different side chains significantly determine the conformation/topography/charge distribution, etc. of a peptide, and the structural specificities of (the binding component of) a peptide permease require a peptide to adopt a particular orientation for binding, then the affinity for binding (and hence transport) would be influenced by the nature of the side chains; competition studies have shown that different peptides have different transport binding affinities. Evidence that dipeptides and oligopeptides containing structurally very different amino acids can use general dipeptide and oligopeptide permeases in bacteria, yeast and fungi has come from competition studies, and from the isolation of transport mutants in which the ability to transport peptides of widely differing amino acid compositions is impaired. However, it should be stated that the exact natures of the mutations are uncharacterized, and that only a small fraction of the total possible peptides have so far been tested. The possibility of additional systems with particular specificities must therefore be left open.

In this connection, evidence has recently been presented for the existence of specific oligopeptide transport system(s) in addition to the general oligopeptide permease (*Opp*) (Barak & Gilvarg 1975*b*; Naider & Becker 1975; Jackson *et al.* 1976). I myself do not consider that the evidence so far presented justifies the conclusion for several additional *specific* oligopeptide transport systems, and I shall return to this point later (p. 315 *et seq.*).

Overall size of peptide. There is clearly an upper limit to the overall size of a dipeptide but this is not so for oligopeptides. It is likely, however, that a recognition/binding component of a microbial oligopeptide permease will be able to accommodate peptides only up to a certain overall size (analogous to the different substrate specificities of dipeptidases, tripeptidases and proteases). In addition, it seems likely that the peripheral layers of the microbial cell envelope overlying the permease in the cytoplasmic membrane will act as a barrier or sieve to the passage of large peptides (and other molecules). In fact, there is good evidence for such a size discrimination in oligopeptide transport by *E. coli* (Payne & Gilvarg 1968, 1977; Fickel 1973; Payne 1976; Matthews & Payne 1975*a*), *Pseudomonads* (Cascieri & Mallette 1976), yeast (Naider *et al.* 1974) and fungi (Wolfinbarger & Marzluf 1975*b*). The size at which the cut-off is observed is different for these various microorganisms; this accords with their differing cell envelope structures, for it appears in these cases that size discrimination is exercised here rather than at the membrane permease.

Summary of structural requirements (Fig. 1). The structural specificities for peptide transport in bacteria (*E. coli* and other microorganisms within the limits of the above discussion) are summarized in Fig. 1, which represents a dipeptide and the N-terminal portion of an oligopeptide. Numbers refer to the following features (see reviews cited on p. 305 for discussion of features not considered above): (1) the α-amino group should preferably be free but may be monosubstituted providing a positive charge is retained, e.g. monoalkyl; acyl

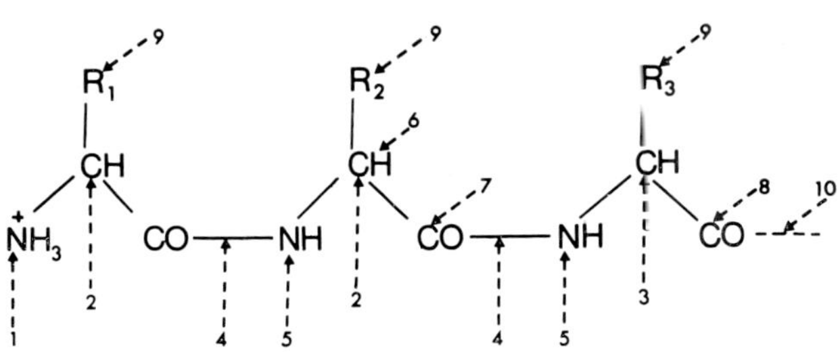

FIG. 1. Structural requirements for peptide transport in *E. coli*. Numbered features are discussed in the text. (Reproduced from Payne 1976, with permission.)

or dialkyl substituents prevent transport; (2) L-isomers specified; (3) in oligopeptides this residue (and perhaps others towards the C-terminus) may be D-isomer; (4) α-linkage specified; (5) nitrogen atom may be methylated; (6) α-hydrogen not specified; (7) free C-terminal-α-carboxyl specified for dipeptide transport but not (8) for oligopeptides in which it may be absent or substituted; (9) no absolute specificity for side chains, may be natural residues, synthetic or substituted, but their nature may affect transport affinity and/or rate; possibility of multiple transport systems; (10) large size limits oligopeptide transport.

Models and mechanisms for microbial peptide transport

Various theoretical models for peptide transport in microorganisms are shown in Fig. 2; detailed evidence in support of their occurrence has been discussed elsewhere (Payne & Gilvarg 1977). In Fig. 2A, peptides are hydrolysed by free peptidases either extracellularly or within the cell wall or periplasm and the liberated amino acids are transported by amino acid permeases. Here, competition could occur between a free exogenous amino acid and a related amino acid residue supplied in peptide form, as well as between residues from dipeptides and oligopeptides. Many microbial species utilize peptides in this way. In Fig. 2B, bound peptidases liberate cleavage products that can diffuse away and subsequently be absorbed through amino acid permeases; there is little evidence for the operation of this mode in microorganisms. Fig. 2C envisages a hydrolase component of a membrane permease releasing cleavage products vectorially to a separate carrier component, this carrier being inaccessible to the same substrates present in free solution. A simple form of this model might be a pore lined with the peptidase and the carrier in series. Fig. 2D is a simple group translocation mechanism in which hydrolysis and translocation are the sequentially coupled activities of a single permease entity that binds peptides from the outside and releases cleavage products to the interior. Fig. 2E envisages specific peptide carriers that deliver intact peptides to the cell interior where they are subsequently hydrolysed. This is 'true' peptide transport, and although conceptually simple it readily accounts for all the important characteristics found for *E. coli* and many other microorganisms. However, it is important to note that in many instances the available evidence does not permit us to exclude involvement of other mechanisms such as those shown in Fig. 2C and D.

PEPTIDE TRANSPORT IN MICROORGANISMS: THE PRESENT

Concerned with the inadequacies of the growth test, and recognizing the

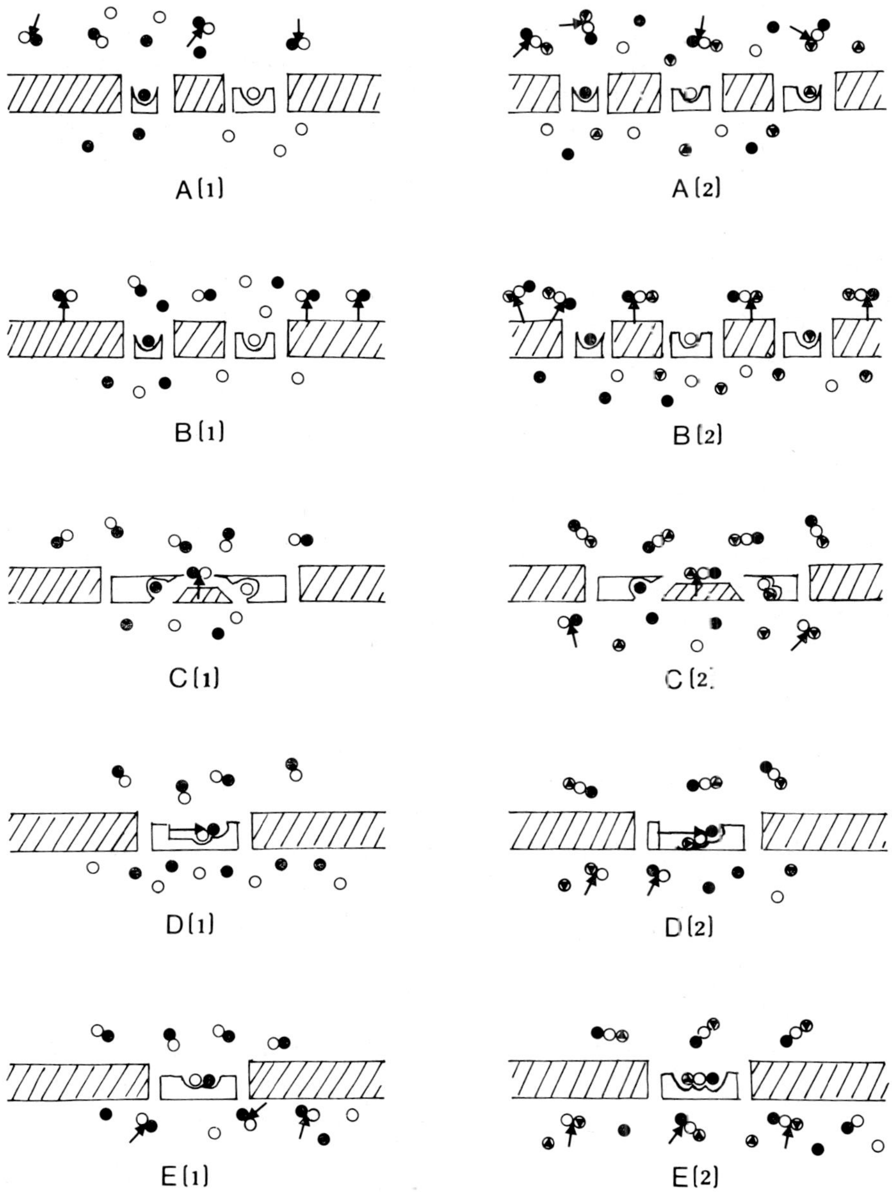

FIG. 2. Theoretical models for microbial peptide utilization. The cytoplasmic membrane (cross-hatched) contains various carriers (plain). Cell exterior, cell wall and periplasm are above the membrane, and cell interior is below. Arrows represent peptidase action. Transport and hydrolysis of dipeptides is on the left (A1, etc) and oligopeptides on the right (A2, etc). See text for discussion. (Reproduced from Payne & Gilvarg 1977, with permission.)

problems of obtaining radioactive peptides, we have recently been trying to develop new methods for studying peptide transport. These radioactive, enzyme and dansyl methods provide rapid and sensitive indications of whether a peptide can be transported and utilized, and in addition they can provide quantitative assessment of the affinity of peptides for the transport systems and of the rates of uptake (V_{max}).

Radioactive protein incorporation method

An auxotroph requiring at least two amino acids is needed. The transport of a peptide containing one of the required amino acids is studied by monitoring the peptide-dependent incorporation into newly synthesized protein of the other required amino acid which is supplied as the free form, radioactively labelled. The radioactive amino acid is supplied at a constant high concentration; the peptide is supplied at varied concentrations so that its transport into the cell (and hence the supply of its required amino acid) can become the rate-limiting step for protein synthesis. The method requires that intracellular peptide hydrolysis is not rate-limiting for protein synthesis. By measuring incorporated radioactivity one can tell within minutes, rather than after many hours as with growth tests, whether a peptide can be transported, hydrolysed and utilized for protein synthesis. Additionally, by relating the amount of peptide-dependent protein synthesis (i.e. counts incorporated) to the concentration of peptide one can get a measure of peptide binding affinity for the transport system. The method requires (say) 1 % of the peptide normally used in a growth test.

We have used (Payne & Bell 1977) the *E. coli* K12 mutant M148 × 478 *pro*$^-$, *lys*$^-$, *leu*$^-$, *met*$^-$, *trp*$^-$ (kindly provided by Dr M. Iaccarino: see De Felice *et al.* 1973). The strain is grown up into, for example, lysine starvation (about 0.04 μmol/ml); all other amino acids are supplied at about 0.2 μmol/ml, to an $E_{660} = 0.4$ (Spectronic 20), about 5×10^8 cells/ml. An appropriate limiting concentration for each amino acid is determined beforehand. When exponential growth stops (monitored by E_{660}) the culture is incubated further (about 1.5 h) so that lysine is completely depleted. (Although we have not tried, it is probable that non-starved cells could be used if they were efficiently washed to remove all exogenous and intracellular-pool amino acids; if these are not completely removed, protein synthesis cannot be made dependent on added peptide and high blank values of incorporated radioactivity are obtained.) The lysine-starved bacteria are collected by centrifugation and resuspended at 37 °C in fresh growth medium minus lysine and (say) proline. To aliquots (0.5 ml) of this suspension ($E_{660} = 0.1$) are added a mixture (0.5 ml) of a lysine peptide at

TABLE 1

Radioactive protein incorporation method for studying peptide transport

1. Starve lys⁻, pro⁻ auxotroph* for lysine;
   or, for use with non-starved organisms, completely remove lysine by efficient washing.
2. Add varied concentrations lysine peptide plus constant high concentration of radioactive proline.
3. Allow protein synthesis to proceed for 10 min.
4. Measure the extent of protein synthesis from the c.p.m. incorporated into trichloroacetic-acid-precipitated protein.
5. Determine the peptide concentration that limits the rate of protein synthesis and relate it to affinity of peptide for transport system.

* Lys and Pro are used only as examples.

varied *low* concentrations plus a constant amount of radioactive proline (tritiated: about 5×10^6 c.p.m./ml) at *high* concentration (about 0.015 μmol/ml). Controls are run with no added lysine and with a high concentration of free lysine so that the maximum extent of protein synthesis can be assessed. Protein synthesis starts immediately the required amino acids are supplied and is stopped after 10 min by the addition of 25% trichloroacetic acid (2 ml). It is important that the concentration of bacteria should be low enough and the peptide concentration high enough for the external peptide concentration to remain effectively constant during the 10 min period of protein synthesis. About 15 min later the samples are boiled (15 min) to ensure liberation of pool amino acids and tRNA; they are then filtered (Millipore, type HAWP, 0.45 μm) and, after the trichloroacetic-acid-precipitated protein has been washed, incorporated radioactive proline is determined with a scintillation counter. The overall procedure is summarized in Table 1 and should be applicable to most micro-organisms and to most amino acids/peptides.

Fig. 3A shows a plot of radioactivity (proline) incorporated into protein as a function of dilysine concentration; maximum protein synthesis under these conditions corresponded to about 3.5×10^4 c.p.m. This was equivalent to that obtained with high concentrations of lysine (Fig. 3A). The data are shown as a double reciprocal plot in Fig. 3B. The intercept on the abscissa gives a value for the external peptide concentration that gives half-maximal rate of protein synthesis and is a measure of transport 'binding affinity' or 'limit concentration' (Ames 1964): for dilysine it is 1×10^{-6} M. In Fig. 4A,B are shown double reciprocal plots for transport of leucylmethionine, with leucine- and methionine-starved cells respectively; radioactive proline was used in each case. Intercepts on the abscissa give a 'binding affinity' for leucylmethionine

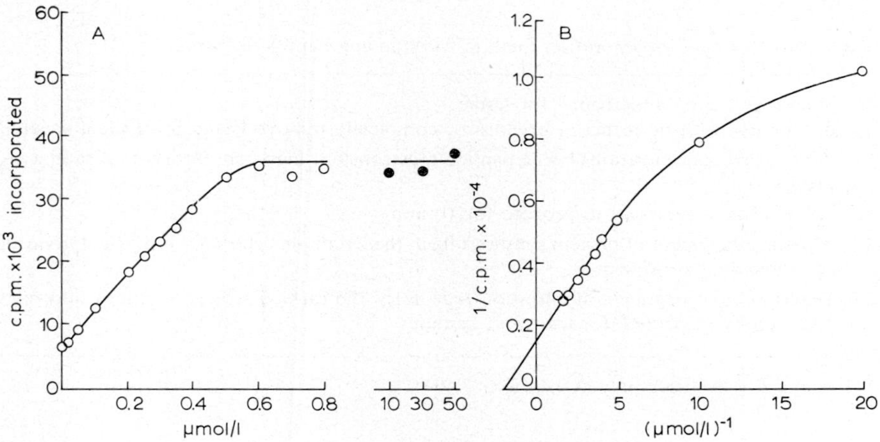

FIG. 3. Transport of dilysine in lysine-starved *E. coli* K12, *leu*⁻, *met*⁻, *pro*⁻, *lys*⁻, *trp*⁻ mutant monitored by incorporation of radioactive proline into protein. A: Proline incorporation as a function of dilysine concentration; (○) dilysine; (●) lysine; B: double reciprocal plot of dilysine results from A.

TABLE 2

Transport binding affinities (K_m) for peptides determined by radioactive incorporation method

Peptide	*Starved for*	$K_m \times 10^{-7}$ M[a]
Leu-Trp *	Trp	4.0-4.7
Leu-Trp *	Leu	3.0
Leu-Met *	Leu	7.5-10
Leu-Met *	Met	6.5
Leu-Met +	Leu	8.3-11.0
Met-Leu *	Leu	10.5
Met-Leu *	Met	10
Lys-Lys	Lys	10-11
Lys-Lys-Lys	Lys	1.1-1.7
Lys-Lys-Lys-Lys	Lys	4-5
Leu-Leu-Leu	Leu	10
Trp-Asp-Met-Phe-NH_2	Met	67-125
Trp-Asp-Met-Phe-NH_2	Trp	71
Lys-Lys	Lys	5-10
Lys-Lys-Lys	Lys	20
Leu-Leu-Leu	Leu	10

All 10 min incubations except for * 15 min, + 7 min.
[a] Values are for single experiment or range for several experiments. Lower part of table is for oligopeptide permease mutant (TOR) and upper part for parent strain.

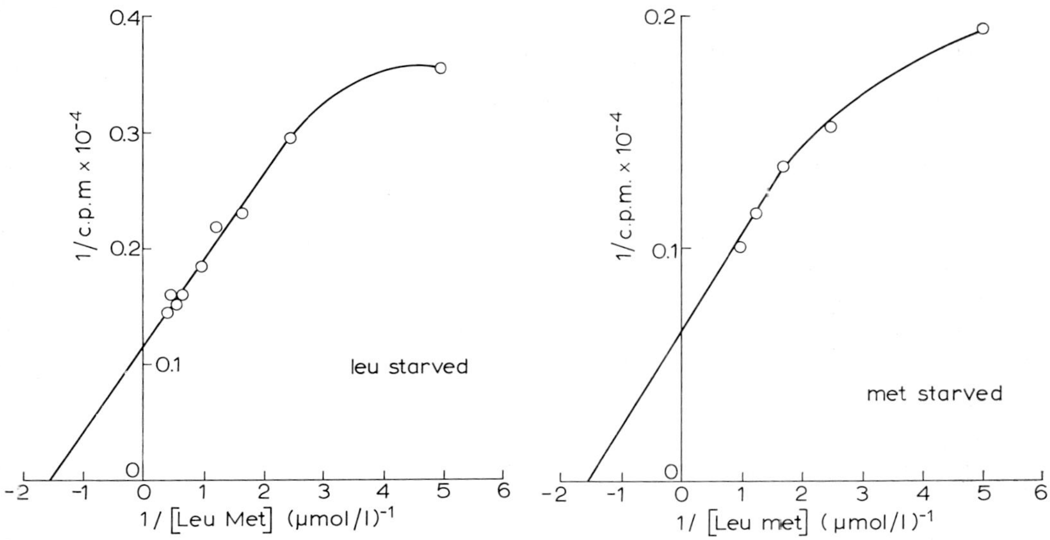

A B

FIG. 4. Transport of leucyl methionine by *E. coli* K12, *leu*$^-$, *met*$^-$, *pro*$^-$, *lys*$^-$, *trp*$^-$ mutant monitored by incorporation of radioactive proline into protein. A: Double reciprocal plot, Pro incorporation against Leu-Met concentration, Leu starved; B: Double reciprocal plot, Pro incorporation against Leu-Met concentration, Met starved.

of about 7.5×10^{-7} M (Fig. 4A) and 6.5×10^{-7} M (Fig. 4B). Some 'binding affinities' (K_m) for various peptides determined with differently starved organisms are given in Table 2. Results are independent of starvation condition. Trp-Asp-Met-Phe-amide is particularly interesting in having the lowest affinity of any of the tested peptides (could this be effect of amidation or size, or both?), and because similar results were obtained when organisms were starved for amino acid residues that probably require different peptidases for their intracellular release. For the parent and TOR mutant similar results were obtained for dipeptides but were markedly different for certain oligopeptides, e.g. $(Lys)_3$, but not for others, e.g. $(Leu)_3$. This TOR mutant is resistant to triornithine and trivaline, sensitive to dipeptides of valine, fails to grow on trilysine (at concentrations that support growth of the parent), and grows on trileucine. These results have been duplicated in a different strain of *E. coli* K12. It is clear that this new approach provides valuable information on the TOR mutation and the claims for additional peptide transport systems, and we are studying these aspects further.

Enzyme synthesis method

The enzyme synthesis method (Bell *et al.* 1977*a*) and the radioactive one are related. In this method, transport in amino acid auxotrophs of *E. coli* is monitored by following the peptide-dependent synthesis of an induced or derepressed enzyme in bacteria that have been starved for a required amino acid. We have studied induced β-galactosidase synthesis, for which a simple colorimetric assay is available. Affinity constants for peptide transport can be determined from the exogenous peptide concentrations that give half maximal rates of enzyme synthesis; these values are analogous to the 'limit concentrations' described by Ames (1964). The method (Bell *et al.* 1977*a*) was also used to study transport mutants and competition between peptides for uptake, and it endorsed earlier conclusions from growth tests about the structural specificities for peptide transport. The method is however quicker, more sensitive and more informative than growth tests. A related method has been described by Cascieri & Mallette (1974).

Dansyl detection method

We regard this method as the most important development for studying peptide transport in microorganisms. Thus, bacteria are incubated with defined peptides either alone or as mixtures, or with heterogenous peptone fractions, etc. Periodically, samples of the incubation mixtures are withdrawn with a syringe, and the bacteria are removed with a Millipore filter. Samples of the clear filtrate (containing residual non-transported peptides) are dansylated by standard techniques and then fractionated on polyamide sheets in two dimensions, using three solvent systems. Estimates of peptide transport are then made from direct visual assessment of fluorescent intensities of the peptides on the chromatograms. The method allows peptide transport to be studied under various conditions of temperature, pH, etc., and the influence of bacterial physiology (e.g. exponential phase, stationary phase, starved), and of various additions to the medium (e.g. energy source, inhibitors, cations, etc.) can also be assessed. Auxotrophic strains are not, of course, required. The method is extremely sensitive (0.1–1.0 nmol peptide visually detected); it can also be made objectively quantitative by labelling with radioactive dansyl chloride and using an automatic radioactive scanner on the polyamide sheets.

With this method, one can assess the different rates at which peptides are taken up (either alone or from mixtures) and relate this to peptide structure, net charge, p*K* of N-terminal amino group, etc. Uptake into spheroplasts and vesicles can also be studied. Additionally, one can study specificities in efflux of amino acids, for we have observed that this process occurs *simultaneously* with peptide influx. Clearly, this also provides much information on peptidase activities *in vivo*. (We have studied the above aspects using *E. coli*: J. W. Payne *et al.*, manuscripts in preparation.)

The method is also potentially valuable in studying transport of substances other than peptides. Thus, mixtures of amino acids can be used and their relative influx/efflux rates measured. This has many advantages over the normal methods using ^{3}H- and ^{14}C-labelled substrates in which usually one, and at most two, differently labelled substrates, are monitored.

PEPTIDE TRANSPORT IN MICROORGANISMS: THE FUTURE

In addition to those aspects already discussed, it seems that the topics in this section deserve investigation. Studies should make use of all possible approaches, e.g. growth tests, radioactive peptides, and the 'radioactive', 'enzymic' and 'dansyl' methods.

Charge variation on the N-terminal α-amino group

Evidence was presented earlier (p. 307) for the importance of the N-terminal α-amino group in transport. However, I believe insufficient attention has been directed to the possible importance of the p*K*b of this group in peptide transport. Thus, with a p*K*b of (say) 7.0, a peptide at pH 7 effectively exists as an equal mixture of the protonated ($NH^+{}_3$) and unprotonated (NH_2) form. It seems possible that if separate dipeptide and oligopeptide permeases occur with differential specificity for the protonated and unprotonated amino group this would provide for efficient complementarity of function. With many peptides the ratio of the ($NH^+{}_3$) form to the (NH_2) form could change significantly over the pH range 7.2–5.8, a pH range typical of that over which bacteria like *E. coli* will grow. It should also be noted that during typical batch culture of such bacteria the pH of the medium can change over this range. As the proportion of the $NH^+{}_3$ to the NH_2 form may vary for different peptides at a particular pH, it is a feature to keep in mind in competition experiments; it may be particularly relevant in cases of 'reciprocal competition' in which peptide A can completely inhibit uptake of peptide B, but peptide B can only partially inhibit uptake of A, etc. For any particular peptide the p*K*b of its α-N-amino group will be mainly dependent upon the electrostatic, and inductive and mesomeric effects of its N-terminal side-chain residue, but the distance away of the C-terminal carboxylate and other side-chain residues will be influential and the p*K*b will change within a homologous peptide series (a feature not to be ignored when 'size effects' only are being considered). Monoalkylation of the α-amino group (Payne 1974) will make it more basic, i.e. lower its p*K*b, through an inductive effect (ethyl > methyl); dialkylation will lower the p*K*b slightly further still. However, sequential removal of the hydrogen atoms leads to lower stabilization of the cation species by hydration, which tends to decrease basicity. It is possible therefore to make predictions from peptide structure about likely p*K*b values, and these can of course be determined directly by titration. Experiments relating such p*K*b values to peptide transport, and on the influence of pH on uptake, might prove rewarding. However, it should always be kept in mind that the microenvironment of the peptide permease may be considerably different from that in bulk solution and the presence of a high-charge density layer at the membrane surface may profoundly influence charged species; these membrane effects could vary significantly for different microorganisms.

Peptide transport in various microbial species

The generality of peptide transport systems in microorganisms remains to be established, and it is not yet known whether one or more dipeptide, oligopeptide permeases occur commonly. The structural requirements of such systems should be established. Although it appears reasonable that for a permease to handle the vast array of peptides that can arise from enzymic cleavage of proteins its 'recognition component' will have requirements only for those structural features common to all α-linked peptides, it is also possible that certain microorganisms occupy niches of highly specific protease activity that will generate peptides with particular N- or C-terminal residues; this could lead to the evolution of permeases with specificity for these features. If apparently different sequence-specificity for peptides is found using growth tests with different species, it should be kept in mind that this may reflect differences in the specificity of peptidases (Watson 1976; Hirshfield & Price 1975) rather than peptide permeases. Due attention should also be paid to the way in which the varied cultural conditions required for different species may affect the form in which the peptide exists, e.g. pH differences giving different charged species.

Energy coupling for peptide transport

Although peptide transport in microorganisms has been shown to be an active process (Payne 1976; Payne & Gilvarg 1977), little is known about the energy-coupling mechanism. Cowell (1974) showed that phosphate bond energy is required for transport of diglycine in *E. coli* via a system sensitive to osmotic shock.

However, with the growing acceptance of the chemiosmotic hypothesis (Mitchell 1966), it is apt to consider this hypothesis in relation to peptide transport. The chemiosmotic hypothesis states that during bacterial respiration, protons are pumped out of the cell to give an electrochemical gradient across the cell membrane. This proton-motive force has electrical and chemical components and is thought able to drive ATP synthesis through the membrane-bound proton-translocating ATPase, and the transport of neutral and ionic nutrients through a series of proton symport and uniport carriers (Hamilton 1975; Mitchell 1970). Peptides provide a particularly interesting study because, depending on their primary sequence, they can be neutral, cationic or anionic, and the ionic species may be singly or multiply charged. This clearly will require varied stoichiometry between coupled proton movements (Collins *et al.* 1976) and peptide transport, with all the implications that may have for 'different' peptide transport systems. Anionic species (especially multiply charged) present particular problems, arising from the fact that co-transport with a single proton is non-electrogenic and driven only by the pH gradient which normally would seem unable to support a high accumulation ratio. As well as considering energy coupling in relation to the *net* charge on a peptide, one should also consider the demonstrated involvement of the α-amino group in transport and the implications of its existence in an NH^+_3 and NH_2 form. A consideration compatible with the published properties of TOR mutants is that these mutants are defective in an energy-coupling component but can still carry out facilitated diffusion; under these circumstances different oligopeptides may show varied rates of facilitated diffusion depending on their structures and charge, some—e.g. $(Leu)_3$, $(Met)_3$—being fast enough to support growth but others—$(Lys)_3$, $(Orn)_3$, $(Val)_3$—being taken up or hydrolysed (or both) too slowly to support or inhibit growth, as the case may be. This aspect is currently being studied with the dansyl method.

Finally, when considering theoretical mechanisms for peptide transport that involve coupled peptidase activity in the membrane (Fig. 1), one should consider the fact that hydrolysis of the peptide bond (at about neutral pH) generates a carboxyl and an amino function, the different pK values of which lead to 'asymmetric' proton binding and effective acidification of the immediate environment. In this context it is essential to know how water molecules gain access to the membrane-bound peptidase.

Involvement of binding proteins in peptide transport

Although osmotic shock, which is used generally to release binding proteins, impairs peptide uptake (Cowell 1974; Payne & Gilvarg 1977), there is no positive evidence for the involvement of binding proteins in microbial peptide transport and it would be of great value to clarify this situation. Possible problems of peptidase action during tests for the presence of peptide-binding proteins in osmotic-shock fluids should be alleviated by use of 'peptidase-resistant' peptides with N-substituted peptide bonds. In attempts to reconstitute peptide transport function in osmotically shocked cells, the possibility should be investigated that a free or peripheral binding protein can attach to an integral membrane carrier system only after a conformational change in the binding protein that results from the peptide binding to it.

Regulation of peptide transport

This important feature should be investigated alongside that of amino acid uptake; the likely interrelations of the two have been discussed (Payne 1976; Payne & Gilvarg 1977). Recent studies with the dansyl detection method (J. W. Payne, unpublished) indicate that unequal accumulation of amino acid residues can occur from a peptide as a result of selective active efflux of particular residues. The relevance of this process to the regulation of peptide transport is being studied. The involvement of tRNA synthetases in the regulation of amino acid transport (Quay *et al.* 1975) and the effects of amino acids and peptides on the properties of these enzymes (Hirshfield *et al.* 1975, 1976) is clearly also relevant.

'Smugglins'

At the previous Ciba Foundation symposium on peptide transport, and elsewhere, the nature of 'smugglins' has been discussed (Payne 1972*a*, 1976; Matthews & Payne 1975*b*). When an impermeant substance is chemically linked to a peptide, and the resultant complex can be transported via a peptide permease, the impermeant moiety can thus be 'smuggled' across the cell membrane and the complex is accordingly called a 'smugglin'. This principle has been elegantly demonstrated by Fickel & Gilvarg (1973), and by Ames *et al.* (1973). The potential this approach offers for introducing antimicrobial agents into the cell should provide the stimulus for much more research in this area. The possibility of rationalizing the structural basis for relative peptide binding affinities and rates of uptake offers the hope that 'smugglins' of maximal peptide permease affinity can be designed. The principle could be valuable in helping to unambiguously demonstrate peptide transport in the intestine, where surface hydrolysis of peptides plus uptake of liberated amino acids can occur; accumulation of a non-transportable amino acid analogue when it was coupled to a peptide would provide unambiguous evidence for peptide transport *per se*.

PEPTIDE HYDROLYSIS IN MICROORGANISMS

Generalizations about peptide hydrolysis by microorganisms are much more difficult to make than about peptide transport. Peptidases may occur as extracellular, cell-bound, or intracellular forms and different microorganisms may have one, or more types (Holzer *et al.* 1975; Miller 1975; Payne 1975*b*, 1976). This feature largely determines whether peptides are likely to be hydrolysed before, during or after transport. With respect to regulation, there are reports indicating that in certain species induced peptidases may be secreted in response to the addition of peptides to the medium. There is no good evidence that intracellular peptidases are induced in response to added peptides. There is still much debate over the regulation of intracellular peptidase activity in bacteria and whether this exists mainly in a latent form. The dansyl method has recently provided unambiguous evidence that *E. coli* peptidases exist *in vivo* in a predominantly active form (J. W. Payne, unpublished). Total intracellular peptidase activity in *E. coli* appears to be vastly in excess of that required simply for the nutritional utilization of peptides (Payne 1972*c*; Bell *et al.* 1977*a*), and there has been much speculation about their additional functions (see reviews above). In bacteria there are commonly many peptidases able to cleave a particular peptide bond, and this seeming redundancy helps to explain why mutants in which several peptidases are missing are still viable (Miller 1975). Particularly important with respect to substrate specificity are recent studies showing how specificity for particular peptide bonds can be dramatically changed by the addition of different cations (Patterson *et al.* 1975; Simmonds *et al.* 1976), or by different associations between enzyme subunits (Stoll *et al.* 1973).

Only a few peptidases have been purified and their characteristics and substrate specificities determined. In this regard I would like to make a plea that only *true* peptides be used as substrates in studies of peptidase activity, and that the use of artificial substrates such as aminoacyl-β-naphthylamides, aminoacyl-*p*-nitroanilides, etc. be curtailed. Recently, a shudder went through the world of bacterial peptide transport when reports appeared of the occurrence of an inducible periplasmic aminopeptidase in *E. coli* (Lazdunski *et al.* 1975; Murgier *et al.* 1976). However, it turned out that although the enzyme apparently cleaved *p*-nitroanilide- and β-naphthylamide-derivatives it actually had zero activity against *true* peptides. True peptide hydrolysis is readily monitored by loss of peptide-bond absorbance (210-220 nm) and by differential trinitrobenzene sulphonate determination of amino acid and peptide α-amino groups (Payne 1972*c*), etc.

PEPTIDE TRANSPORT AND HYDROLYSIS IN MICROBIAL NUTRITION

Although treated separately here, transport and hydrolysis act in concert during the nutritional utilization of peptides by microorganisms. It is apparently energetically advantageous to an organism to absorb nutrients from its environment rather than synthesizing them, and it is not uncommon for about 30% of its energy to be used for transport. It would seem to be a particular advantage for an organism to be able to absorb peptides, which in most cases probably takes less energy than is needed to transport or to synthesize the component amino acids. With some species there is the added advantage that cell yield is greater when amino acids are supplied in peptide rather than in the free form; this occurs because of a 'sparing effect' with the peptide (Matthews & Payne 1975*a*; Payne 1976). Whether this occurs seems to depend on the intracellular peptidase activity and the level of metabolic regulation of the organism. In *E. coli*, which has high peptidase activity, it appears that growth yield on peptide-bound and free amino acids is almost always identical. Using amino acid auxotrophs of *E. coli* this feature has been made the basis for a microbiological assay of amino acids in protein foods (Bell *et al.* 1977*b*) and has thereby provided a novel application of the results of academic studies of peptide transport and hydrolysis in microorganisms.

ACKNOWLEDGEMENTS

I am indebted to those who have collaborated in this work, in particular G. Bell and G. M. Payne. Financial support from the Science Research Council, the Royal Society, Smith, Kline and French Foundation, and Lawson Tait Scientific and Medical Research Fund is gratefully acknowledged.

References

AMES, B. N., AMES, G. F., YOUNG, J. D., ISUCHIYA, D. & LECOCQ, J. (1973) Illicit transport, the oligopeptide permease. *Proc. Natl. Acad. Sci. U.S.A. 70*, 456-458

AMES, G. F. (1964) Uptake of amino acids by *Salmonella typhimurium. Arch. Biochem. Biophys. 104*, 1-18

BARAK, Z. & GILVARG, C. (1974) Triornithine-resultant strains of *Escherichia coli:* isolation, definition and genetic studies. *J. Biol. Chem. 249*, 143-148

BARAK, Z. & GILVARG, C. (1975*a*) Peptide transport, in *Biomembranes*, vol. 7 (Eisenberg, H., Katchalski-Katzir, E. & Manson, L. A., eds.), pp 167-218, Plenum, New York

BARAK, Z. & GILVARG, C. (1975*b*) Specialised peptide transport system in *Escherichia coli*. *J. Bacteriol. 122*, 1200-1207

BECKER, J. M. & NAIDER, F. (1974) Stereospecificity of tripeptide utilization in a methionine auxotroph of *Escherichia coli* K-12. *J. Bacteriol. 120*, 191-196

BELL, G., PAYNE, G. M. & PAYNE, J. W. (1977*a*) Monitoring enzyme synthesis as a means of studying peptide transport and utilization in *Escherichia coli*. *J. Gen. Microbiol. 98*, 485-491

BELL, G., HIGGINS, C. F., PAYNE, G. M. & PAYNE, J. W. (1977*b*) Use of *Escherichia coli* to determine available lysine in plant proteins, in *Proceedings of the Meeting on Nutritional Evaluation of Cereal Mutants:* Vienna 1976, FAO/IAEA, Rome, In press

BURSTON, D., MARRS, T. C., SLEISENGER, M. H., SOPANEN, T. & MATTHEWS, D. M. (1977) Mechanisms of peptide transport, in this volume, pp. 79-98

CASCIERI, T. & MALLETTE, M. F. (1974) New method for study of peptide transport in bacteria. *Appl. Microbiol. 27*, 457-463

CASCIERI, T. & MALLETTE, M. F. (1976) Peptide utilization by *Pseudomonas putida* and *Pseudomonas maltophilia*. *J. Gen. Microbiol. 92*, 283-285

COLLINS, S. H., JARVIS, A. W., LINDSAY, R. J. & HAMILTON, W. A. (1976) Proton movements coupled to lactate and alanine transport in *Escherichia coli:* isolation of mutant with altered stoichiometry. *J. Bacteriol. 126*, 1232-1244

COWELL, J. L. (1974) Energetics of glycylglycine transport in *Escherichia coli*. *J. Bacteriol. 120*, 139-146

DE FELICE, M., GUARDIOLA, J., LAMBERTI, A., & IACCARINO, M. (1973) *Escherichia coli* K-12 mutants altered in the transport systems for oligo- and dipeptides. *J. Bacteriol. 116*, 751-756

FICKEL, T. E. (1973) *The Oligopeptide Permease of* E. coli *as a Vehicle for the Transport of Impermeant Substances and its Accessibility to Large Oligopeptides*. Ph. D. thesis, Princeton University

FICKEL, T. E. & GILVARG, C. (1973) Transport of impermeant substances in *E. coli* by way of oligopeptide permease. *Nat. New Biol. 241*, 161-163

GILVARG, C. (1972) Peptide transport in bacteria, in *Peptide Transport in Bacteria & Mammalian Gut (Ciba Found. Symp. 4)*, pp. 11-16, Associated Scientific Publishers, Amsterdam

HAMILTON, W. A. (1975) Energy coupling in microbial transport. *Adv. Microb. Physiol. 12*, 1-53

HIRSHFIELD, I. N. & PRICE, M. B. (1975) Utilization of selected leucine amides by *Escherichia coli*. *J. Bacteriol. 122*, 966-975

HIRSHFIELD, I. N., YEH, F. M. & SAWYER, L. E. (1975) Metabolites influence control of lysine transfer ribonucleic acid synthetase formation in *Escherichia coli* K-12. *Proc. Natl. Acad. Sci. U.S.A. 72*, 1364-1367

HIRSHFIELD, I. N., YEH, F. M. & ZAMECNIK, P. C. (1976) An *in vivo* effect of the metabolites L-alanine and glycyl-L-leucine on the properties of the lysyl-tRNA synthetase from *Escherichia coli* K-12. *Biochim. Biophys. Acta 435*, 290-305

HOLZER, H., BETZ, H. & EBNER, E. (1975) Intracellular proteinases in microorganisms, in *Current Topics in Cellular Regulation*, vol. 9 (Horecker, B. L., ed.), pp. 103-156, Academic Press, New York

JACKSON, M. B., BECKER, J. M., STEINFELD, A. S. & NAIDER, F. (1976) Oligopeptide transport in proline peptidase mutants of *Salmonella typhimurium*. *J. Biol. Chem. 251*, 5300-5309

LAZDUNSKI, A., MURGIER, M. & LAZDUNSKI, C. (1975) Evidence for an amino endopeptidase localized near the cell surface of *Escherichia coli*. *Eur. J. Biochem. 60*, 349-355

MATTHEWS, D. M. & PAYNE, J. W. (1975*a*) Peptides in the nutrition of microorganisms and peptides in relation to animal nutrition, in *Peptide Transport in Protein Nutrition* (Matthews, D. M. & Payne, J. W., eds.), pp. 1-60, North-Holland, Amsterdam

MATTHEWS, D. M. & PAYNE, J. W. (1975*b*) Occurrence and biological activities of peptides, in *Peptide Transport in Protein Nutrition*, (Matthews, D. M. & Payne, J. W., eds.), pp. 392-463, North-Holland, Amsterdam

MILLER, C. G. (1975) Peptidases and proteases of *Escherichia coli* and *Salmonella typhimurium*. *Annu. Rev. Microbiol. 29*, 485-504

MITCHELL, P. (1966) Chemiosmotic coupling in oxidative and photosynthetic phosphorylation. *Biol. Rev. (Camb.) 41*, 445-502

MITCHELL, P. (1970) Membranes of cells and organelles: morphology, transport and metabolism, in *Organization and Control in Prokaryotic & Eukaryotic Cells (Soc. Gen. Microbiol. Symp. 20)*, pp. 121-166, Cambridge University Press, London

MURGIER, M., LAZDUNSKI, A., & LAZDUNSKI, C. (1976) Control of aminopeptidase and alkaline phosphatase synthesis in *Escherichia coli:* evidence for a regulation by inorganic phosphate independent of pho R and pho B phosphatase regulatory genes. *FEBS (Fed. Eur. Biochem. Soc.) Lett. 64*, 130-134

NAIDER, F. & BECKER, J. M. (1975) Multiplicity of oligopeptide transport systems in *Escherichia coli. J. Bacteriol. 122*, 1208-1215

NAIDER, F., BECKER, J. M. & KATZIR-KATCHALSKI, E. (1974) Utilization of methionine containing peptides and their derivatives by a methionine-requiring auxotroph of *Saccharomyces cerevisiae. J. Biol. Chem. 249*, 9-20

PATTERSON, E. K., GATMAITAN, J. S. & HAYMAN, S. (1975) The effect of Mn^{2+} on the activities of a zinc metallo-dipeptidase from a mouse ascites tumor. *Biochemistry 14*, 4261-4266

PAYNE, J. W. (1968) Oligopeptide transport in *Escherichia coli*: specificity with respect to side chain and distinction from dipeptide transport. *J. Biol. Chem. 243*, 3395-3403

PAYNE, J. W. (1971) The requirement for the protonated α-amino group for the transport of peptides in *Escherichia coli. Biochem. J. 123*, 245-253

PAYNE, J. W. (1972*a*) Mechanisms of bacterial peptide transport, in *Peptide Transport in Bacteria & Mammalian Gut (Ciba Found. Symp. 4)* pp. 17-32, Associated Scientific Publishers, Amsterdam

PAYNE, J. W. (1972*b*) Effects of N-methyl peptide bonds on peptide utilization by *Escherichia coli. J. Gen. Microbiol. 71*, 259-265

PAYNE, J. W. (1972*c*) Variations in the peptidase activities of *Escherichia coli* in response to environmental changes. *J. Gen. Microbiol. 71*, 281-289

PAYNE, J. W. (1974) Peptide transport in *Escherichia coli:* Permease specificity towards terminal amino group substituents. *J. Gen. Microbiol. 80*, 269-276

PAYNE, J. W. (1975*a*) Transport of peptides in microorganisms, in *Peptide Transport in Protein Nutrition* (Matthews, D. M. & Payne, J. W., eds.), pp. 283-364, North-Holland, Amsterdam

PAYNE, J. W. (1975*b*) Microbial peptidohydrolases, in *Peptide Transport in Protein Nutrition* (Matthews, D. M. & Payne, J. W. eds.) pp. 365-391, North-Holland, Amsterdam

PAYNE, J. W. (1976) Peptides & microorganisms, *Adv. Microb. Physiol. 13*, 55-113

PAYNE, J. W. & BELL, G. (1977) A radioisotope incorporation method for studying the transport and utilization of peptides by *Escherichia coli. FEMS (Fed. Eur. Microbiol. Soc.) Lett. 1*, 91-94

PAYNE, J. W. & GILVARG, C. (1968) Size restriction on peptide utilization in *Escherichia coli. J. Biol. Chem. 243*, 6291-6299

PAYNE, J. W. & GILVARG, C. (1971) Peptide transport. *Adv. Enzymol. Relat. Areas Mol. Biol. 35*, 187-244

PAYNE, J. W. & GILVARG, C. (1977) Active transport of peptides, in *Bacterial Transport* (Rosen, B. P., ed.), Marcel Dekker, New York, In press

QUAY, S. C., KLINE, E. L. & OXENDER, D. L. (1975) Role of leucyl-tRNA synthetase in regulation of branched-chain amino acid transport. *Proc. Natl. Acad. Sci. U.S.A. 72*, 3921-3924

SIMMONDS, S. (1970) Peptidase activity and peptide metabolism in *Escherichia coli* K-12. *Biochemistry 9*, 1-9

SIMMONDS, S. (1972) Peptidase activity and peptide metabolism, in *Peptide Transport in Bacteria & Mammalian Gut (Ciba Found. Symp. 4)*, pp. 43-53, Associated Scientific Publishers, Amsterdam

SIMMONDS, S., SZETO, K. W. & FLETTERICK, C. G. (1976) Soluble tri- and dipeptidases in *Escherichia coli* K-12. *Biochemistry 15*, 261-270

STOLL, E., ERICSSON, L. H. & ZUBER, H. (1973) The function of the two subunits of thermophilic aminopeptidase I. *Proc. Natl. Acad. Sci. U.S.A. 70*, 3781-3784

SUSSMAN, A. J. & GILVARG, C. (1971) Peptide transport & metabolism in bacteria. *Annu. Rev. Biochem. 40*, 397-408

WASMUTH, J. J. & UMBARGER, H. E. (1974) Role for free isoleucine or glycylleucine in the repression of threonine deaminase in *Escherichia coli*. *J. Bacteriol. 117*, 29-39

WATSON, R. R. (1976) Substrate specificities of aminopeptidases: a specific method for microbial differentiation, in *Methods in Microbiology*, vol. 9 (Norris, J. R. & Ribbons, D. W., eds.), pp. 1-15, Academic Press, London

WOLFINBARGER, L. & MARZLUF, G. A. (1975*a*) Specificity and regulation of peptide transport in *Neurospora crassa*. *Arch. Biochem. Biophys. 171*, 637-644

WOLFINBARGER, L. & MARZLUF, G. A. (1975*b*) Size restriction on utilization of peptides by amino acid auxotrophs of *Neurospora crassa*. *J. Bacteriol. 122*, 949-956

Discussion

Radhakrishnan: Can pentalysine be hydrolysed in your system?

Payne: Yes. That was shown by Gilvarg & Katchalski (1965).

Radhakrishnan: Many of your reciprocal plots, especially those with dilysine, were non-linear. Would you take that as evidence for the possible existence of two separate systems with different affinities?

Payne: No. I think that is an experimental error arising from the very low peptide concentrations used for these experimental points. The external concentration has probably changed during the 5–10 min incubation, and if we could accurately measure it over, say, 1 min the deviation should not occur. The assumption of the method (Payne & Bell 1977) is that the external concentration does not significantly change during the time of the experiment, but at very low peptide concentrations I am not sure that is always true; one can of course calculate the percentage of the peptide that is taken in. We only make use of the results on the linear portion of the plot.

Meister: The *E. coli* must certainly have evolved or adapted in a manner similar to that of the enterocyte. They live in the same environment and eat the same food. This might be one of the best arguments, aside from the fact that you unequivocally demonstrated peptide transport, that it does occur.

You said there are no peptidases on the cell membrane but did you say that because of the overall phenomenon you have observed, namely obvious lack of peptidase activity? Or have you actually taken the membrane apart to see whether any peptidase activity can be detected?

Payne: Considerable effort has gone into attempts to detect periplasmic or membrane peptidases in *E. coli* but all the reports are negative (see, e.g., Simmonds 1972; Miller 1975). Of course with all negative results one is aware of the problem that in the isolation procedure one might have inactivated the enzymes, or washed away a co-factor or something. There was recently a report (Lazdunski *et al.* 1975) that *E. coli* has a periplasmic aminopeptidase that is induced under phosphate limitation. Of course that appeared as a bombshell to all of us who work on peptide transport in *E. coli*—not however that we normally work in phosphate limitation. It turned out that the substrates used in those experiments were natural amides and *p*-nitroanilides, and fortunately when true peptides were examined the enzyme showed zero activity, even though it was referred to at times as an aminopeptidase.

Meister: So the cell membrane can act on a nitroanilide?

Payne: Apparently so. Under the particular conditions of phosphate limitation a number of periplasmic proteins were induced, one of which was a well-recognized enzyme, namely alkaline phosphatase, and another was this so-called aminopeptidase which actually showed amidase and protease activity.

Meister: I was pleased to hear you use the word permease, which was originally intended to indicate that transport is mediated by an enzyme or enzyme-like activity. Are you prepared to postulate what kind of enzyme or enzyme-like mechanism might be involved in the transport of peptides?

Payne: I think we need more information on the nature of the components of the peptide permeases before we speculate on a mechanism. I use the word permease to indicate that a multicomponent system rather than a single enzyme mediates peptide transportation. There is evidence that binding proteins may be involved in peptide transport; these could constitute the recognition component to which a peptide binds initially, perhaps before binding to an intrinsic membrane component. There must also be an energy-coupling component in this system.

Adibi: I am intrigued with the way you assayed the affinity for transport by looking at protein synthesis. Riggs & Walker (1963) and others tried to tie up the mammalian system of amino acid transport with protein synthesis. The data indicated that the processes of transport and incorporation of amino acids into protein are independent of one another. Are amino acids or peptides uniquely used for protein synthesis or do they serve other purposes when inside bacteria?

Secondly, high affinity for transport may not be advantageous to protein synthesis since amino acids with high affinity reach saturation in transport at concentrations which are smaller than those reached by amino acids with low affinity. Therefore, one may expect a lower rate of protein synthesis with amino acids with high affinity. How do these concepts help in the bacterial system?

Payne: In the experiments on enzyme synthesis and radioisotope incorporation, probably all the amino acid goes to protein; certainly in *E. coli* the protein yield is the same for free amino acids as for peptide-bound residues, and there doesn't seem to be preferential degradation of amino acid residues.

The question of whether high affinity is advantageous for protein synthesis cannot be answered. Clearly V_{max} is important here and I think that as soon as the peptides get in they are cleaved and the residues used.

Kim: Is the size limitation for transport due to the specific structure of cell wall or periplasmic space or cell membrane?

Payne: The evidence is compatible with it being external to the permease, but the exact location remains to be proved (Payne & Gilvarg 1968; Fickel 1973).

Kim: It is interesting when one considers the possible biological role of glycocalyx in mammalian cells. Would it be possible to selectively remove the cell wall, periplasmic space or cell membranes and do transport studies?

Payne: If one does that, the resultant spheroplasts do not grow, so we couldn't do the experiment using growth tests. Radioactive peptides such as pentalysine are not available either. However, it may be possible to do those types of experiment now, using the dansyl method.

Rhodes: In the cells without the external rigid wall can you show uptake incorporation by enzymic methods? Since you can't show it by a growth method can you show an increase in enzymes within the cell, particularly in a starved cell?

Payne: Conceivably one could, because one can certainly get protein synthesis; however, it is technically difficult and we obtained inconclusive results when this was first tried (J. W. Payne & C. Gilvarg, unpublished results).

Matthews: If your suggestion about protonation of the terminal amino groups is correct, Dr Payne, the implication is that in looking for multiple transport systems one is likely to get different results at different pH values. The people working on intestine will have to take this possibility seriously. What is a physiological pH in, say, the jejunum? I believe the intraluminal pH in man is in the region of 6.0.

Parsons: The pH of the microclimate at the surface of the membrane where these sorts of reactions go on must be less than that in the bulk phase in the intestinal lumen.

Matthews: It might be less. But that difference is likely to persist *in vitro*, so to work at pH 6 *in vitro* might be more reasonable than pH 7.

Dr Payne, you showed a slide of the growth response of *Escherichia coli* to a series of lysine peptides, dilysine, trilysine, tetralysine etc. Was this from equimolar solutions?

Payne: No. The solutions were adjusted to contain the same equivalents of lysine residues, so trilysine would be one-third of the molarity of lysine but would have the same number of equivalents.

Matthews: Then it looks as though, in this particular instance, the transport of a given number of amino acid residues in the form of peptide is slightly slower with a larger peptide, tetralysine, than with the free amino acid of dilysine or trilysine.

Payne: The slide I showed (Gilvarg & Katchalski 1965 [not printed here]) is compatible with that. Uptake of tetralysine is conceivably rate-limiting for growth, though I very much doubt it; peptide cleavage might also in certain cases be the step that limits rate of supply of required amino acid. It may be relevant to this discussion to mention that the pK of the α-amino groups would vary along that homologous series.

Kenny: Within what limits does it vary?

Payne: I don't know. We would have to do the titration. No figures have been reported to my knowledge.

Kenny: The value of 5 for the pK of an α-amino group seems very low.

Payne: I agree, but given the right side chains one might get unusually low pK values, but maybe pK 6 is more realistic.

Kenny: I think the pK of the α-amino group of a dipeptide is about 8.

Payne: It would depend on the nature of the side chains and the proximity of the C-terminal carboxyl. I was not actually suggesting that peptides in free solution were likely to have pKb values of 5. I merely used 5 to illustrate how changing the pH would affect the proportion of protonated species. It seems that it could be important to measure the pKb value and to be aware of it when doing an experiment, for if the pH changes, the ratio of protonated to unprotonated species will change accordingly.

Smyth: Is there any way of distinguishing between the capacity to enter the cell and the capacity to stimulate growth? These could be quite different things. How do you know pentalysine can't enter the cell?

Payne: If it did enter the cell, excessive intracellular peptidase activity would cleave it and it would be utilized for growth. Secondly, it doesn't compete with peptides that are known to enter.

Milne: You showed a very poor entry of diglycine but a good entry of all the higher glycines. Is there any explanation for that?

Payne: That particular strain was deficient in dipeptide uptake.

Meister: Does the peptide go in at the C-terminal or the N-terminal end?

Payne: That is not known. It is only known that the N-terminus requirement implies that this end must become oriented with respect to the permease.

Mellander: Could you extrapolate your results to organisms depending on a photosynthetic process, for example plankton occurring at different depths?

Payne: I am sure microorganisms can adapt in this way. That kind of evolution has been shown to occur in the test tube. Thus, by using chemostats and exposing organisms to very low concentrations of a required nutrient Rigby *et al.* (1974) demonstrated a mutational change that affects the K_m of a particular enzyme that acts on the nutrient. Plankton, for example, could simply have two uptake systems with different K_m values. This situation is commonly found in bacteria which may, for example, have low and high affinity transport systems for amino acids, the former perhaps involved in absorbing large amounts as nitrogen sources, the latter acting as a scavenger system for low concentrations.

Josefsson: How much do you know about the enzymes taking care of the different peptides? Where diglycine was a less good substrate there might have been significant dipeptidase activity which was not working as well as the aminopeptidase. So the rate of hydrolysis might not reflect the absorption of the dipeptides. Have you any idea about the proteolysis inside the bacteria?

Payne: We have looked at the peptidase activity. If we get a presumptive transport mutant we always check that the peptidase activity of the mutant hasn't changed. That particular strain is a transport-negative strain. With some of these oligopeptides, for example tetralysine, they are split into two dilysines, which would then have to be processed by the same route as an original dipeptide.

Joseffson: Is there any extra bacterial proteolysis in that system?

Payne: No.

Kim: What are the specificities of the cytoplasmic peptidases in terms of the type of peptides and their chain length?

Payne: Few of them have been characterized. Sussman & Gilvarg (1970) looked at the enzymes in *E. coli* that cleave lysine peptides. Several distinct peptidases are involved in cleaving Lys-Lys bonds; there are activities specific for tetrapeptides, tripeptides and dipeptides and there seems to be an aminopeptidase activity. There are certainly four different specificities for chain length. Miller (1975) has recently reviewed this topic.

Radhakrishnan: Enzyme hydrolysis might be rate-limiting in a homologous series because one would decrease the actual concentration of these species. For instance with pentalysine the concentration will be much less than with dilysine on a molar basis if the amino acid is taken as the index.

Payne: I don't think this would become rate-limiting. All the evidence is that there is an enormous surplus of peptidase activity. With the enzymes we have looked at there seems often to be about 1000 times the activity required to supply the amino acid residues from peptide linkage at a rate sufficient for normal protein synthesis in an auxotroph.

Parsons: Would you like to amplify the different ways in which peptides can get across the membrane of the bacteria? You pointed out that there was one route for the dipeptides but you didn't say what happens to higher oligopeptides.

Payne: Oligopeptides are taken in by a system distinct from the dipeptide transport system. For amplification, I imagine you are referring to the figure on theoretical models for transport (Fig. 2, p. 311). It is just that with oligopeptides, transport coupled to membrane hydrolase activity is complex. Thus, if a tripeptide is cleaved, initially this generates an amino acid and a dipeptide, and one somehow has to accommodate the dipeptide into the scheme for subsequent transmembrane flow. It seems to me that the hydrolase system would only work with oligopeptides if an aminopeptidase or carboxypeptidase was used which would transport residues sequentially. When one tries to draw hydrolase schemes for tripeptides, tetrapeptides, etc., similar to the common dipeptide scheme, these complications appear.

Parsons: So there is perhaps a transport system for higher oligopeptides.

Adibi: We have shown (Adibi & Morse 1976) that the human enterocyte is not able to transport a tetrapeptide such as tetraglycine, clearly indicating a cut-off point in the size of glycine peptide that can be transported. How do you account for the difference between bacterial and enterocyte membranes in this respect? I understand that bacteria can even transport hexapeptides.

Payne: I don't know. Maybe tetraglycine is a particularly bad choice. It may be a pH effect; perhaps tetraglycine uptake could be demonstrated at a different pH.

Adibi: In vitro one should be able to test the effect of manipulating the pH, though *in vivo* in the human intestine it would be more difficult. But I doubt whether pH is the whole answer.

Payne: I think it is premature to say that we are absolutely sure that tetrapeptides cannot get in.

Adibi: I always wonder why the membrane is so radically different in bacteria and enterocytes. Have electron microscope studies been done to compare these two membranes?

Payne: I don't think the membranes can be too different structurally, although they may show varied properties. One explanation is that there are peptidases outside the enterocytes. A very large concentration flux might be needed for peptides to run the gauntlet of peptidases and to arrive at the peptide transport systems. The possibility of high concentrations of tetrapeptides ever getting to the transport sites may be very small, which might make transport specificity for them rather unnecessary.

Matthews: Professor J. S. Fruton suggested to me that the methylation of the peptide bonds in the series Gly-Sar, Gly-Sar-Sar and Gly-Sar-Sar-Sar might limit the conformation of these peptides in such a way that their behaviour was not typical of ordinary peptides. The failure of the gut to transport Gly-Sar-Sar-Sar does not necessarily prove that other tetrapeptides will not be transported.

Adibi: I realize that. I use tetraglycine, which is more soluble and more likely to be taken up intact, but I found no evidence of uptake. I am willing to conclude that there is no tetrapeptide uptake in appreciable amounts by enterocytes.

Kim: We really haven't examined that in detail. We studied the transport of one tetrapeptide, Leu-Gly-Gly-Gly, and it seems to be absorbed intact to a significant degree.

Silk: When extrapolating results of a very few experiments with tetrapeptides to the whole group, we must be very cautious, as there are perhaps 160 000 possible tetrapeptides. So far between us we have studied absorption of only four or five. Whereas I would agree that there appears to be no intact uptake of Gly-Sar-Sar-Sar in a recent study, Professor Kim, Dr Chung and I already have some preliminary data to support intact uptake of at least proportions of a further tetrapeptide, Leu-Gly-Gly-Gly. Evidence for intact uptake of this tetrapeptide correlates well with the presence of aminopeptidase activity within the cytoplasm against this peptide.

Adibi: Logically, if there is no uptake of tetraglycine, it is going to be difficult to envision uptake of other tetrapeptides.

Silk: As mentioned, we have preliminary evidence, derived from competition experiments, that a component of this peptide is absorbed intact. In the context of interpreting such competition experiments, Dr Payne raised the important point that transport by a peptide carrier, or surface hydrolysis, may be determined by protonation of free amino groups of the peptide molecule. Thus, results of some competition experiments may have been overinterpreted if the addition of further peptides to the test solutions merely alters the conditions of the milieu at the surface of the cell and thus merely alters the physicochemical properties of the test peptide molecule.

Matthews: It is apparently possible for animal cells to take up much larger peptides than tripeptides and tetrapeptides. Grahl-Nielsen and colleagues (1974) have described the ability of cells in culture to take up a series of homologous peptides. These cells were neoplastic cells derived from a carcinoma of the small intestine of the hamster. In culture the cells took up a series of homopeptides of Lys, up to deca-Lys. So although they were neoplastic cells and in culture, some animal cells can do this.

Adibi: Do they take up appreciable quantities?

Matthews: Towards the decapeptide it got progressively less, but it was measurable with the decapeptide, as estimated by growth.

Adibi: The entire protein can be taken up by the cells. The question is the quantity.

Matthews: The quantities of peptide were nutritionally significant.

Smyth: Many years ago we raised the question of interaction between hexoses and amino acids. Is there any interaction between hexoses and peptides?

Matthews: We found that glucose failed to inhibit uptake of one of our unhydrolysable peptides, Gly-Sar-Sar (Addison *et al.* 1975), but it was just one set of concentrations (peptide 5 mM, glucose 11 mM) and this needs exploring further.

Smyth: Did you try galactose?

Matthews: No.

Adibi: There have been a number of publications reporting on the effect of hexose on amino acid transport (see Adibi 1976). The conditions have to be defined, since the results are conflicting.

Smyth: The effect of hexoses depends on a number of conditions. In the hamster the effect of galactose on amino acid transfer may be a mixed effect of an allosteric interaction in the membrane and a competition for energy: in the rat the effect of galactose can best be explained by a competition for energy. Has such a competition been looked for in connection with peptide transfer?

Matthews: Not properly.

Radhakrishnan: We are probably oversimplifying the problem of the cut-off point. In the homologous series, diglycine is rare, triglycine would be even rarer and tetraglycine extremely rare. To extrapolate the results to dietary peptides would be an oversimplification of the problem.

I endorse what Dr Silk has said. It seems that we don't know what peptide to use at the moment. Probably we should think in terms of using a natural digest with pancreatic enzymes.

Adibi: With pancreatic enzymes, you can't be sure what peptide you are studying.

Radhakrishnan: We shall have to isolate them and look at these peptides.

Adibi: That has been the problem—how to isolate a single pure peptide.

Smyth: Many years ago we asked (Newey & Smyth 1960) whether glycylglycine was a physiological peptide. It happened to be a convenient one to study but is it really formed during protein digestion?

Radhakrishnan: I have seen one Gly-Gly sequence in a peptide from brain (Simantov & Snyder 1976) but rather strangely it doesn't occur in a protein like collagen in which about one-third is glycine.

Smyth: I ask this because we found, and many people have since found, that there is a slow rate of hydrolysis of glycylglycine in the intestine.

Payne: The affinities that we found for peptide binding suggest that glycine peptides have the lowest affinity. I would question whether, from putting this structure down on paper, one can actually predict anything about whether it is the most likely tetrapeptide to be transported. Much bigger things appear to get in much faster in bacteria.

The sugar linkage worried us a little as well. In our system we use glucose for bacterial nutrition, and it can give rise to Schiff base formation with amino groups. Addition of borohydride can inhibit peptide uptake.

Smyth: Have you used galactose?

Payne: No.

Rubino: In rabbit intestine, glycylproline influx is not affected by the addition of glucose to the incubation medium or by the tissue being preloaded with glucose. We have not checked any other monosaccharide.

Adibi: Using an *in vivo* perfusion technique, Cook (1972) found that glucose inhibited glycine absorption. This inhibition was reduced if glycine was presented as diglycine.

Rhodes: Perhaps we are over emphasizing the di-, tri-, and tetra- aspects. Dr Peters said (1972) that molecular size was a factor, and Try-Gly is not equivalent to Try-Leu or Try-Met.

Parsons: Diglycine has a smaller molecular weight than tryptophan and indeed so has alanylalanine.

Josefsson: Dr Payne, what is known of the molecular properties and specificities of the *E. coli* peptidases?

Payne: A review (Miller 1975) gives details of the peptidases of *E. coli.*

Kim: Do bacteria use peptides for protein synthesis?

Payne: Yes, subsequent to cleavage to amino acids.

Kim: I thought there was tRNA for peptides in bacteria.

Meister: You are right in one sense. In certain bacteria, dipeptides such as D-alanyl-D-alanine can be used for the synthesis of cell wall peptides.

Payne: This is non-ribosomal synthesis, of course. The cross-linking reaction in mucopeptide synthesis is an enzyme-coupled reaction. In addition, these are frequently non-protein amino acids such as D-alanine.

Rhodes: Is prolylglycylglycine taken up by bacteria, and if so by what method?

Payne: I think it is taken up by the same mechanism as other oligopeptides.

Rhodes: What does that do to your theory about a free N-terminal amino group?

Payne: It is not too bad, because it is an imino group, which is much the same as α-N-methyl. It will take up α-N-alkyls, although not α-N-acyls; seemingly if one preserves the positive charge it needn't be a primary amino group and it can be a secondary group (Payne 1971, 1974).

References

ADDISON, J. M., BURSTON, D., DALRYMPLE, J. A., MATTHEWS, D. M., PAYNE, J. W., SLEISENGER, M. H. & WILKINSON, S. (1975) A common mechanism for transport of di- and tripeptides by hamster jejunum *in vitro*. *Clin. Sci. Mol. Med. 49*, 313-322

ADIBI, S. A. (1976) Intestinal phase of protein assimilation in man. *Am. J. Clin. Nutr. 29*, 205-215

ADIBI, S. A. & MORSE, E. L. (1976) The size of glycine oligopeptide which limits its intact absorption in human jejunum. *Clin. Res. 24*, 431

COOK, G. C. (1972) Comparison of intestinal absorption rates of glycine and glycylglycine in man and the effect of glucose in the perfusing fluid. *Clin. Sci. (Oxf.) 43*, 443-453

FICKEL, T. E. (1973) *The Oligopeptide Permease of* E. coli *as a Vehicle for the Transport of Impermeant Substances and its Accessibility to Large Oligopeptides*, Ph.D. Thesis, Princeton University

GILVARG, C. & KATCHALSKI, E. (1965) Peptide utilization in *Escherichia coli. J. Biol. Chem. 240*, 3093-3098

GRAHL-NIELSEN, O., ØDEGAARD, P. & TRITSCH, G. L. (1974) Oligopeptides as sources of indispensable amino acids for mammalian cells in culture. *In Vitro, 9*, 414-420

LAZDUNSKI, A., MURGIER, A. & LAZDUNSKI, C. (1975) Evidence for an amino-endopeptidase localized near the cell surface of *Escherichia coli. Eur. J. Biochem. 60*, 349-355

MILLER, C. G. (1975) Peptidases and proteases of *Escherichia coli* and *Salmonella typhimurium. Annu. Rev. Microbiol. 29*, 485-504

NEWEY, H. & SMYTH, D. H. (1960) Intracellular hydrolysis of dipeptides during intestinal absorption. *J. Physiol. (Lond.) 152*, 367-380

PAYNE, J. W. (1971) The utilization of prolyl peptides by *Escherichia coli. Biochem. J. 123*, 255-260

PAYNE, J. W. (1974) Peptide transport in *Escherichia coli.* Permease specificity towards terminal amino group substituents. *J. Gen. Microbiol. 80*, 269-276

PAYNE, J. W. & BELL, G. (1977) A radioisotope incorporation method for studying the transport and utilization of peptides by *Escherichia coli. FEMS (Fed. Eur. Microbiol. Soc.) Lett. 1*, 91-94

PAYNE, J. W. & GILVARG, C. (1968) Size restriction on peptide utilization in *Escherichia coli. J. Biol. Chem. 243*, 6291-6299

PETERS, T. J. (1972) Subcelluar fractionation of the enterocyte with special reference to peptide hydrolases, in *Peptide Transport in Bacteria and Mammalian Gut (Ciba Found. Symp. 4)*, pp. 107-114, Associated Scientific Publishers, Amsterdam

RIGBY, P. W. J., BURLEIGH, B. D. & HARTLEY, B. S. (1974) Gene duplication in experimental enzyme evolution. *Nature (Lond.) 251*, 200

RIGGS, T. R. & WALKER, L. M. (1963) Some relations between active transport of free amino acids into cells and their incorporation into protein. *J. Biol. Chem. 238*, 2663-2668

SIMANTOV, R. & SNYDER, S. H. (1976) Isolation and structure identification of a morphine-like peptide 'Enkephalin' in bovine brain. *Life Sci. 18*, 781-788

SIMMONDS, S. (1972) Peptidase activity and peptide metabolism, in *Peptide Transport in Bacteria and Mammalian Gut* (*Ciba Found. Symp. 4)*, pp. 45-53, Associated Scientific Publishers, Amsterdam

SUSSMAN, A. J. & GILVARG, C. (1970) Peptidases in *Escherichia coli* K-12 capable of cleaving lysine homopeptides. *J. Biol. Chem. 245*, 6518-6524

Peptidases in germinating barley grain: properties, localization and possible functions

T-M. ENARI and J. MIKOLA

Biotechnical Laboratory, Technical Research Centre of Finland, Helsinki and Department of Biology, University of Jyväskylä, Jyväskulä, Finland

Abstract Barley grain contains about 10% insoluble reserve proteins. When the grain germinates the reserve proteins are hydrolysed to amino acids and transported to the growing tissues of the seedling.

In the resting grain most of the reserve proteins are 'packed' into the non-living storage tissue, the starchy endosperm. During germination the internal pH of the starchy endosperm is about 5, and it contains high activities of proteinases (secreted by the living aleurone cells) and carboxypeptidases, all with pH optima between 4 and 6. As a whole the starchy endosperm of a germinating grain resembles a giant secondary lysosome.

Adjacent to the starchy endosperm is a specialized absorptive and processing tissue, the scutellum. This organ contains very high activities of the 'acid carboxypeptidases' and also two 'alkaline peptidases': a leucine aminopeptidase and a dipeptidase, both with pH optima at 8 to 10. The high peptidase activities in the scutellum suggest that the hydrolysis products of the reserve proteins are absorbed from the starchy endosperm as a mixture of amino acids and small peptides, which are hydrolysed to amino acids in the scutellum before transport to the growing seedling tissues.

A barley grain is composed of two main parts. There is a small diploid embryo, which after germination develops to a new barley plant, and there is a much more extensive reserve tissue, the triploid endosperm, which provides energy and nutrients for the growing seedling until it is able to begin autotrophic growth. Both the embryo and the endosperm contain two separate tissues (or organs) with different physiological functions (Fig. 1). In the embryo, there is an 'embryo proper' with initials of the first leaves and roots, and another tissue called the scutellum. During germination the scutellum absorbs nutrients from the endosperm and processes them into suitable form for transport to the growing tissues of the seedling. There is no cell division in the scutellum during germination, and it degenerates after the reserves in the endosperm have been used up. In the endosperms there is a large starchy endosperm which contains

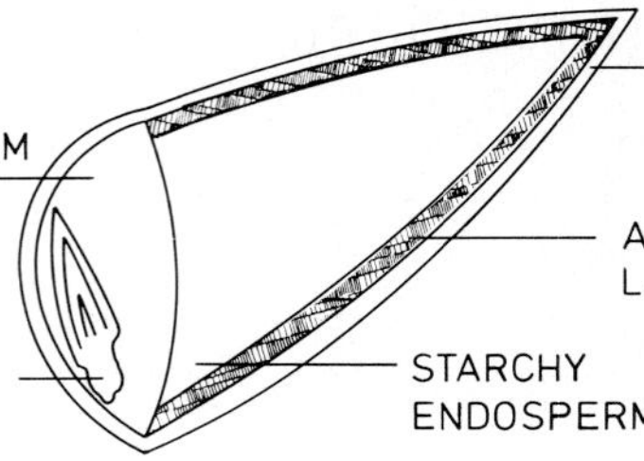

FIG. 1. Structure of resting barley grain.

mainly starch grains and insoluble storage proteins. It is a non-living tissue in the sense that it does not respire or synthesize proteins during germination. The starchy endosperm is surrounded by a living aleurone layer from two to four cells in thickness.

The whole grain has a protein content of 10 to 15%, the bulk of which consists of insoluble storage proteins (globulins, prolamins, glutelins); about two-thirds of these are localized in the starchy endosperm and the remaining third in the aleurone layer.

When the grain germinates the reserve materials disappear from the endosperm and appear in different forms (cellulose, seedling protein) in the seedling (Folkes & Yemm 1958). This mobilization of storage materials is based on a coordinated interaction between the four organs of the grain (Fig. 2). First the embryo synthesizes a plant hormone, gibberellin (or a mixture of closely related gibberellins), which diffuses into the aleurone layer and induces a massive *de novo* synthesis of hydrolytic enzymes and their active secretion into the starchy endosperm (e.g. Chrispeels & Varner 1967). Starch, cell walls and reserve proteins are hydrolysed, and the main reaction products are glucose, amino acids and probably a mixture of dipeptides and oligopeptides. The hydrolysis products are absorbed by the scutellum. Higher plants, however, are not able to transport glucose and many amino acids. Therefore, these are converted to sucrose and glutamine in the scutellum before the long-distance transport to the growing tissues.

As a whole, there is a certain analogy between the digestion and use of food by mammals and by the barley seedling. The starchy endosperm is a non-living digestive compartment corresponding to the alimentary tract, while the aleurone layer is an enzyme-secreting organ corresponding to the pancreas. The scutellum is an absorptive compartment which corresponds to intestinal mucosa. The scutellum, however, also functions as a processing compartment, thus having some of the functions of the mammalian liver as well.

In the mammalian digestive system, proteins are hydrolysed in the digestive

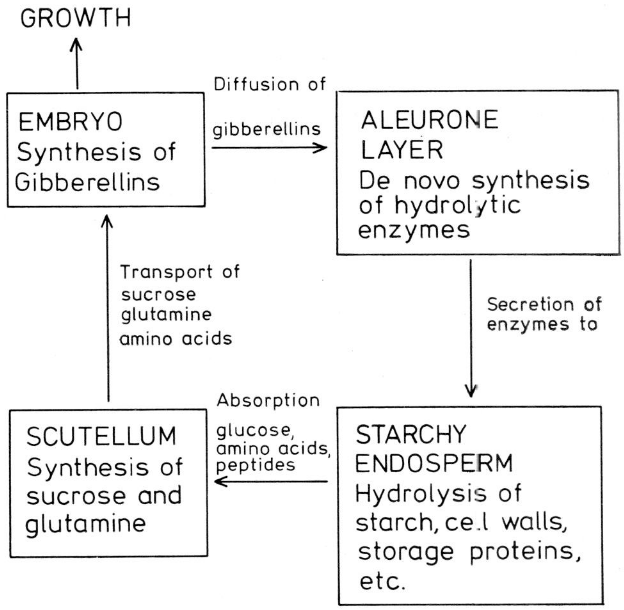

FIG. 2. Physiological processes in germinating barley grain.

compartment by a set of extracellular proteinases and peptidases which do not occur in other tissues of the organism. The hydrolysis products are absorbed by the absorptive compartment as a mixture of amino acids and dipeptides and oligopeptides (Matthews 1975). The latter are then hydrolysed to amino acids by peptidases which seem essentially similar to the intracellular peptidases of other tissues (e.g. Norén 1974).

We can now ask: what kinds of proteinases and peptidases are present in the starchy endosperm of germinating barley grain? In which form are the hydrolysis products absorbed by the scutellum? And, what kinds of peptidases, if any, are present in the scutellum?

NATURE AND LOCALIZATION OF BARLEY PROTEINASES

Ungerminated barley contains several proteinases (Enari *et al.* 1963). They fall into two main groups: SH enzymes with pH optima between 3.9 and 6.0, and metal-activated enzymes with pH optima between 5.5 and 8.5 (Enari & Mikola 1967).

The proteinases have been studied extensively, but because of the great number and lability of these enzymes our knowledge is still fragmentary (cf.

Ryan 1973). The following results, however, are relevant in this context. First, no proteinases resembling trypsin, chymotrypsin or pepsin have been detected in barley. Resting grains contain low proteolytic activity (pH 5.4), and there is an increase of about 20-fold in activity during germination (Mikola & Enari 1970). Isolated barley aleurone layers are induced by gibberellic acid (a gibberellin) to synthesize and release proteinases; the synthesis is *de novo* (Jacobsen & Varner 1967), and several different enzymes are released (Sundblom & Mikola 1972), the two most abundant being thiol proteinases with pH optima at pH 3.9 and 5.5. As these enzymes are released by the aleurone layer, their apparent localization in the germinating grain is the starchy endosperm, where the internal pH is between 4.8 and 5.2 during germination. The combination of high activities and suitable pH optima suggests that these proteinases catalyse the initial breakdown of insoluble storage proteins to soluble peptides.

PROPERTIES OF BARLEY PEPTIDASES

Compared to the proteinases, knowledge concerning barley peptidases is fairly exhaustive. The five enzymes with the highest activities have been purified and characterized, four other enzymes have been characterized to some extent, and it seems probable that if additional enzymes are present their activities must be low. The properties of the nine known peptidases are summarized in Table 1, pH and activity curves for representative substrates are shown in Fig. 3, and the substrate specificities of four enzymes are presented in Tables 2 to 5. (Abbreviations: -β-NA, -β-naphthylamide; Z-, *N*-carbobenzoxy-; U, enzyme unit, corresponds to the hydrolysis of 1 μmole of substrate per minute at 30 °C.)

Firstly, there are three carboxypeptidases. The first enzyme was purified to homogeneity with Z-Phe-Ala as a marker substrate (Visuri *et al.* 1969) and the second was purified to enzymic homogeneity (Moeller *et al.* 1970) with casein substrate (the enzyme liberates C-terminal tryptophan from α-casein). Later Yabuuchi *et al.* (1973) purified both enzymes with different procedures and demonstrated that the relatively high activity of unfractionated barley extracts in the hydrolysis of Z-Phe-Phe was due to the enzyme studied by Moeller *et al.* The enzyme acting on Z-Pro-Trp is labile and low in activity (Mikola *et al.* 1972); no purification has been reported.

Physiologically the important properties of barley carboxypeptidases are pH optima around pH 5, high activity in the liberation of C-terminal amino acids from N-substituted peptides of different chain length (Z-dipeptides, four residues from Phe-Phe-Val-Ala-Pro-Phe-Pro-Glu-Val-Phe-Gly-Lys, α-casein), and inability to hydrolyse dipeptides and tripeptides.

TABLE 1

Peptidases present in germinating barley grain

Enzyme and marker substrate	*pH optimum*	*Inhibitors*	*References*
Carboxypeptidases			
I Z-Phe-Ala	5.2	Diisopropylfluorophosphate	Visuri *et al.* 1969
II Z-Phe-Phe	5.6	*p*-Hydroxymercuribenzoate	Moeller *et al.* 1970
III Z-Pro-Trp	4.8		Yabuuchi *et al.* 1973
Neutral aminopeptidases			
I Phe-β-NA	7.2	*p*-Hydroxymercuribenzoate	Kolehmainen & Mikola 1971
II Arg-β-NA	7.2		Hejgaard & Bøg-Hansen 1974
III Leu-β-NA	7.2		
IV Leu-β-NA	7		
Leucine aminopeptidase Leu-Tyr, Leu-NH_2	8-10		Sopanen & Mikola 1975
Dipeptidase Ala-Gly	8.8	*o*-Phenanthroline *p*-Hydroxymercuribenzoate	Sopanen 1976

Barley carboxypeptidases differ completely from mammalian carboxypeptidases A and B (EC 3.4.12.2 and EC 3.4.12.3) but they resemble the 'acid carboxypeptidases' of other higher plants (for references see Salmia & Mikola 1976), yeast carboxypeptidase Y (Hayashi *et al.* 1973) and the mammalian cathepsin A. The common properties include pH optima around 5, complete inactivation by diisopropylfluorophosphate, lack of inhibition by chelating agents, and ability to hydrolyse esters of N-substituted amino acids (Mikola & Pietilä 1972). Cathepsin A is a lysosomal enzyme, carboxypeptidase Y is localized in the small vacuoles of yeast cells (Matern *et al.* 1974) and the 'acid carboxypeptidases' of the algae *Nitella* and *Chara* are located in the large central vacuoles of the cells (Doi *et al.* 1975). The vacuoles of yeasts and higher plants seem homologous with the mammalian lysosomes. Therefore, we can expect the barley carboxypeptidases also to be localized in some lysosome-like compartments.

In addition to the carboxypeptidases there is a group of four enzymes which can be called neutral aminopeptidases. Their most characteristic property is the ability to hydrolyse the β-naphthylamides of various amino acids; therefore, they are frequently called naphthylamidases or arylamidases (especially in cases where the hydrolysis of peptides has not been directly demonstrated). We observed that Phe-β-naphthylamide was the substrate most rapidly broken

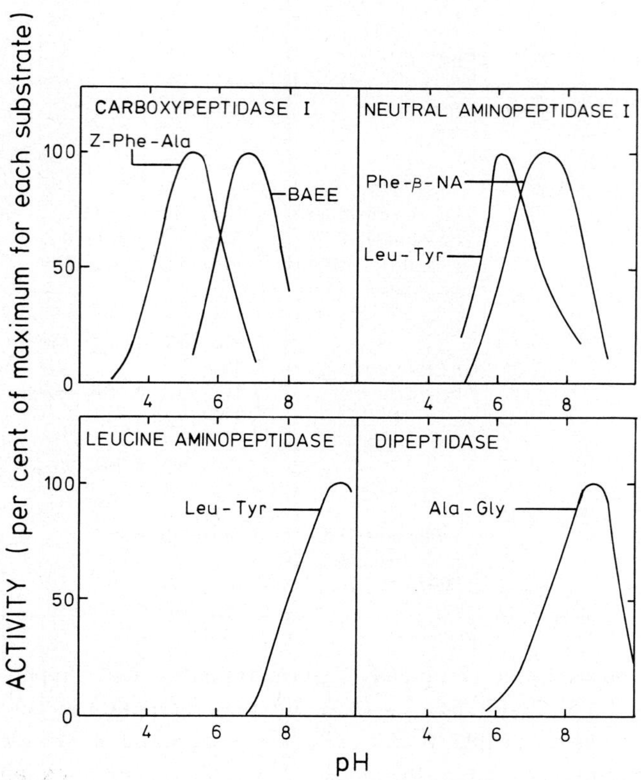

FIG. 3. Effects of pH on the hydrolysis of representative substrates by purified barley peptidases (redrawn from references cited in Table 1). BAEE denotes benzoyl-L-arginine ethyl ester.

down by barley extracts and we purified the enzyme to enzymic homogeneity (Kolehmainen & Mikola 1971). The purified enzyme hydrolysed the β-naphthylamides of amino acids with hydrophobic side-chains optimally at pH 7.2 and it was also highly active in the hydrolysis of dipeptides, tripeptides and dipeptidylamides at pH 5 to 7 (Fig. 3, Table 3). Disc electrophoretic separations followed by activity staining showed that unfractionated barley extracts contained in addition to the purified enzyme at least two other 'naphthylamidases' with different substrate specificity. Later Hejgaard & Bøg-Hansen (1974) demonstrated with crossed immunoelectrophoresis that the actual number of enzymes was four; all of these enzymes were also able to hydrolyse peptides (J. Hejgaard, personal communication).

Finally, there are two peptidases with relatively high pH optima. These

TABLE 2

Substrate specificity of barley carboxypeptidase I (Visuri *et al.* 1969, Mikola & Pietilä 1972)

Substrate	*Relative reaction rate at pH 5.2*
Z-Phe-Ala	100
Z-Phe-Ser	10.6
Z-Phe-Phe	8.3
Z-Gly-Ala	1.5
Z-Gly-Leu	0.9
Benzoyl-DL-arginine ethyl ester	40 (pH 7.0)
Benzoyl-L-arginine methyl ester	20 (pH 7.5)
Acetyl-L-phenylalanine ethyl ester	30 (pH 6.8)
No hydrolysis observed:	
Z-Gly-Pro	
Z-Pro-Trp	
Phe-Ala	
Leu-Gly-Gly	

TABLE 3

Substrate specificity of barley neutral aminopeptidase I (Kolehmainen & Mikola 1971)

Substrate	*Relative reaction rate*	
Phe-β-NA	100	
Leu-β-NA	36	
Met-β-NA	34	pH 7.2
Ala-β-NA	3.0	
Leu-*p*-nitroanilide	23	
Phe-Ala	26	
Leu-Tyr	24	
Leu-Ala	10	
Ala-Gly	3.3	
Pro-Gly	2.3	pH 6.5
Leu-Gly-Gly	2.6	
Phe-Leu-NH_2	++	
Ala-Gly-NH_2	+	
Pro-Gly-NH_2	+	
No hydrolysis observed:		
Gly-β-NA, Glu-β-NA, Arg.-β-NA, Lys-β-NA		
Gly-Gly		
Insulin B chain		

TABLE 4

Substrate specificity of barley leucine aminopeptidase (Sopanen & Mikola 1975)

Substrate	*Relative reaction rate at pH 9.2*
Leu-Gly-Gly	168
Met-Leu-Gly	146
Leu-Tyr	104
Leu-Gly	87
Ala-Gly	1.5
Leu-NH_2	27
Leu-β-NA	0.10
Insulin B chain	++

TABLE 5

Substrate specificity of barley dipeptidase (Sopanen 1976)

Substrate	*Relative reaction rate at pH 8.8*
Ala-Gly	100
Ala-Ala	63
Ala-Leu	102
Val-Gly	32
Leu-Tyr	8.3
Pro-Gly	0.9
Gly-Gly	0.7
No hydrolysis observed:	
Leu-NH_2	
Ala-Gly-Gly, Ala-Gly-NH_2	
Ala-β-NA	

enzymes differ from each other in most chemical and enzymic properties, but their 'physiological behaviour' in the germinating seed is similar. Therefore, we have found it convenient to call them collectively the alkaline peptidases of barley. Both enzymes have been purified to '90% homogeneity' and extensively characterized (Sopanen & Mikola 1975; Sopanen 1976).

The barley leucine aminopeptidase (cytosol aminopeptidase, EC 3.4.11.1) liberates N-terminal amino acids from peptides at pH 8 to 10 without regard to chain length (Leu-NH_2 $\rightarrow$ insulin B chain), but it has low activity on Ala-Gly and minimal activity on Leu-β-naphthylamide (Table 4), both marker substrates of the other peptidases. The dipeptidase, on the other hand, seems to have absolute specificity for neutral dipeptides (Table 5). Both these enzymes

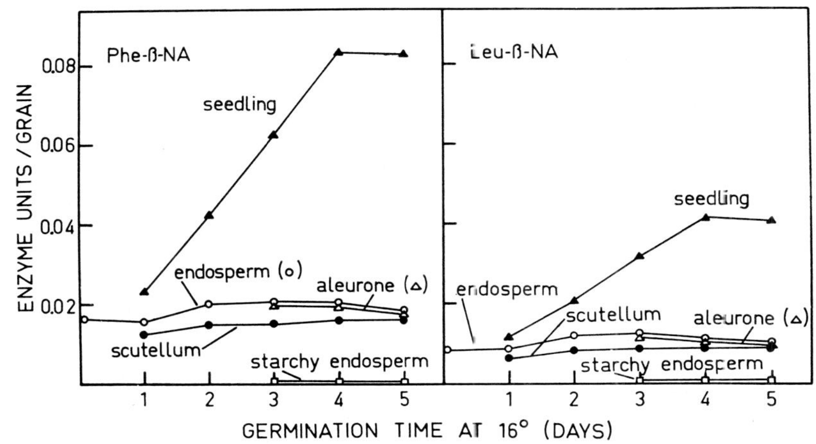

FIG. 4. Activities of the neutral aminopeptidases of barley in different tissues of germinating grains. (From Mikola & Kolehmainen 1972.)

share a number of common properties with the best-known mammalian and microbial leucine aminopeptidases and dipeptidases, respectively.

LOCALIZATION OF THE PEPTIDASES IN THE GERMINATING SEED

The activities of the various peptidases were assayed from the four organs of the grain at different stages of germination (Mikola & Kolehmainen 1972). The neutral aminopeptidases (Fig. 4) showed relatively small and constant activities in the aleurone layer and the scutellum, no activity was detected in the starchy endosperm, and higher and progressively increasing activities were present in the seedling. For the alkaline peptidases the results were similar (Fig. 5), except that the activities were higher, and the activities increased progressively also in the scutellum. The carboxypeptidases showed completely different behaviour (Fig. 6). Very high activities were present in the starchy endosperm, relatively high and apparently constant activities were found in the aleurone layer, high and progressively increasing activities were present in the scutellum, and the lowest activities were found in the seedlings.

CONCLUSIONS

The peptidase activities in the two most interesting compartments are summed up in Table 6.

The starchy endosperm (digestive compartment) contains high activities

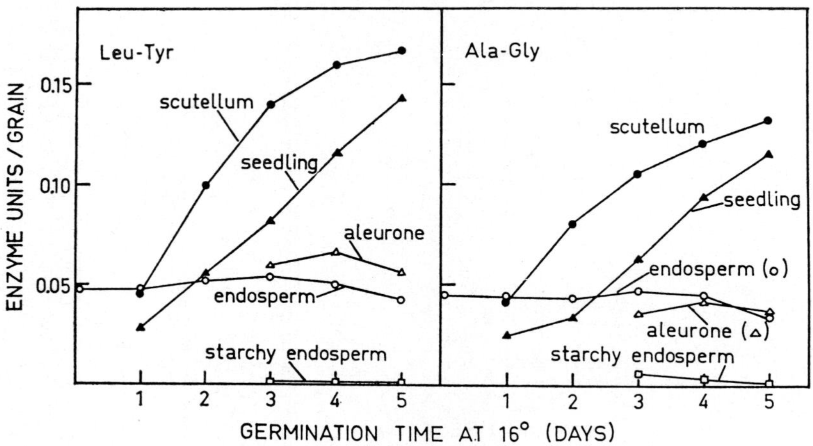

FIG. 5. Activities of barley leucine aminopeptidase (Leu-Tyr) and dipeptidase (Ala-Gly) in different tissues of germinating grains. (From Mikola & Kolehmainen 1972.)

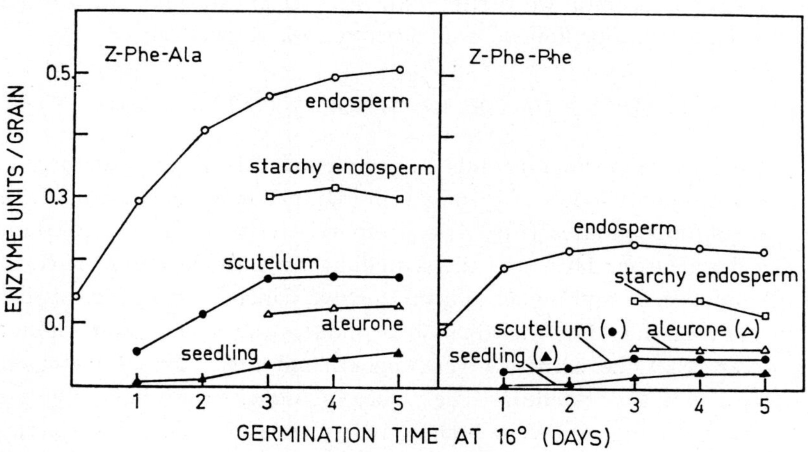

FIG. 6. Activities of barley carboxypeptidases in different tissues of germinating grains. (From Mikola & Kolehmainen 1972.)

of acid carboxypeptidases but no neutral or alkaline peptidases. As already described, the starchy endosperm also contains high activities of proteinases with pH optima at 4 to 6 and its internal pH is around 5; moreover, the barley carboxypeptidases seem related to the lysosomal carboxypeptidases of mammals and yeast. Therefore, the starchy endosperm of germinating barley grain

TABLE 6

Peptidase activities in two tissues of barley grains after five days germination at 16 °C (data from Mikola & Kolehmainen 1972)

Substrate	*Activity, μmol substrate hydrolysed/(min grain) at 30 °C at the respective pH optima*	
	Starchy endosperm	*Scutellum*
Phe-β-NA	–	0.02
Leu-β-NA	–	0.01
Leu-Tyr	–	0.15
Ala-Gly	–	0.13
Z-Phe-Ala	0.30	0.17
Z-Phe-Phe	0.15	0.05
Total	0.45	0.53

resembles a giant secondary lysosome. Apparently, during the course of evolution the barley plant has acquired a lysosomal enzyme battery for the hydrolysis of the protein reserves in the starchy endosperm.

The scutellum, on the other hand, contains high activities of both the alkaline peptidases and the acid carboxypeptidases. The high overall activity suggests that some hydrolysis products are absorbed as peptides, which are hydrolysed to amino acids in the scutellum. Consequently, there is a certain resemblance between the absorptive compartments of the mammalian intestine and the germinating barley grain.

References

CHRISPEELS, M. J. & VARNER, J. E. (1967) Gibberellic acid-enhanced synthesis and release of α-amylase and ribonuclease by isolated barley aleurone layers. *Plant Physiol. 42*, 398-406

DOI, E., OHTSURU, C. & MATOBA, T. (1975) Lysosomal nature of plant vacuoles II. Acid hydrolases in the central vacuole of internodal cells of Charophyta. *Plant Cell Physiol. 16*, 581-588

ENARI, T-M. & MIKOLA, J. (1967) Characterization of the soluble proteolytic enzymes of green malt, in *Eur. Brewery Conv. Proc. 11 Congr. Madrid*, pp. 9-16, Elsevier, Amsterdam

ENARI, T-M., PUPUTTI, E. & MIKOLA, J. (1963) Fractionation of the proteolytic enzymes of barley and malt, in *Eur. Brewery Conv. Proc. 9 Congr. Brussels*, pp. 39-44, Elsevier, Amsterdam

FOLKES, B. F. & YEMM, E. W. (1958) The respiration of barley plants. X Respiration and the metabolism of amino acids and proteins in germinating grain. *New Phytol. 57*, 106-131

HAYASHI, R., MOORE, S. & STEIN, W. H. (1973) Serine at the active center of yeast carboxypeptidase. *J. Biol. Chem. 248*, 8366-8369

HEJGAARD, J. & BØG-HANSEN, T. C. (1974) Quantitative immunoelectrophoresis of barley and malt proteins. *J. Inst. Brew. 80*, 436-442
JACOBSEN, J. V. & VARNER, J. E. (1967) Gibberellic acid-induced synthesis of protease by isolated aleurone layers of barley. *Plant Physiol. 42*, 1596-1600
KOLEHMAINEN, L. & MIKOLA, J. (1971) Partial purification and enzymatic properties of an aminopeptidase from barley. *Arch. Biochem. Biophys. 145*, 633-642
MATERN, H., BETZ, H. & HOLZER, H. (1974) Compartmentation of inhibitors of proteinases A and B and carboxypeptidase Y in yeast. *Biochem. Biophys. Res. Commun. 60*, 1051-1057
MATTHEWS, D. M. (1975) Intestinal absorption of peptides. *Physiol. Rev. 55*, 537-608
MIKOLA, J. & ENARI, T-M. (1970) Changes in the contents of barley proteolytic inhibitors during malting and mashing. *J. Inst. Brew. 76*, 182-188
MIKOLA, J. & KOLEHMAINEN, L. (1972) Localization and activity of various peptidases in germinating barley. *Planta (Berl.) 104*, 167-177
MIKOLA, J. & PIETILÄ, K. (1972) Hydrolysis of ester substrates of trypsin and chymotrypsin by barley carboxypeptidase. *Phytochemistry 11*, 2977-2980
MIKOLA, J., PIETILÄ, K. & ENARI, T-M. (1972) Inactivation of malt peptidases during mashing. *J. Inst. Brew. 78*, 384-388
MOELLER, M., ROBBINS, G. S., BURGER, W. C. & PRENTICE, N. (1970) A carboxypeptidase from germinated barley and its action on casein. *J. Agric. Food Chem. 18*, 886-890
NORÉN, O. (1974) Studies on a soluble dipeptidase from pig intestinal mucosa. Enzymatic properties. *Acta Chem. Scand. B 28*, 711-716
RYAN, C. A. (1973) Proteolytic enzymes and their inhibitors in plants. *Annu. Rev. Plant Physiol. 24*, 173-196
SALMIA, M. A. & MIKOLA, J. (1976) Localization and activity of a carboxypeptidase in germinating seeds of Scots pine, *Pinus sylvestris. Physiol. Plant 36*, 388-392
SOPANEN, T. (1976) Purification and partial characterization of a dipeptidase from barley. *Plant Physiol. 57*, 867-871
SOPANEN, T. & MIKOLA, J. (1975) Purification and partial characterization of barley leucine aminopeptidase. *Plant Physiol. 55*, 809-814
SUNDBLOM, N.-O. & MIKOLA, J. (1972) On the nature of the proteinases secreted by the aleurone layer of barley grain. *Physiol. Plant. 27*, 281-284
VISURI, K., MIKOLA, J. & ENARI, T-M. (1969) Isolation and partial characterization of a carboxypeptidase from barley. *Eur. J. Biochem. 7*, 193-199
YABUUCHI, S., DOI, E. & HATA, T. (1973) The mode of action of a carboxypeptidase from malt. *Agric. Biol. Chem. 37*, 687-688

Discussion

Kenny: Your carboxypeptidase I seems rather like cathepsin A (or lysosomal carboxypeptidase A) which is also a serine peptidase. I was confused that your enzyme was sensitive to both diisopropyl phosphorofluoridate (DipF) and to *p*-chloromercuribenzoate (*p*CMB). Should it be classed as a serine or a thiol enzyme? Do you think the *p*CMB effect is significant?

Enari: Yes, *p*CMB at the normal concentrations used in this sort of work completely inhibits carboxypeptidase I.

Kenny: I would like to think that the active site is a serine rather than a cysteine residue, since DipF is the more specific reagent. Is it really a carboxypeptidase? Its esterase activity clearly doesn't require a free carboxyl group.

Enari: It splits off C-terminal amino acid from carbobenzoxy peptides and also from α-casein, for instance, and it is not active on dipeptides or tripeptides.

Kenny: Do you think it has any endopeptidase activity?

Enari: No, I don't think so.

Kenny: Were the molecular weights you quoted for subunits?

Enari: We haven't studied the subunit molecular weight. The molecular weights I gave were determined by Sephadex chromatography and ultracentrifugation.

Matthews: If we assume that organisms are not silly, why is it that peptides are transported into the scutellum? One can see no difficulty in the necessary peptides being secreted into the food store. Why not reduce the food store completely to free amino acids? The situation suggests that there must be some advantage in having peptides transported across the membrane. I wonder what that advantage is?

Silk: The pH of the alkaline dipeptidases was 8 whereas the pH of the starchy endosperm was pH 5, so has peptide transport evolved purely on the basis of pH? The dipeptidases have an optimal pH that is alkaline and clearly the endosperm is not an optimal environment for these enzymes.

Rhodes: In a test tube, biochemically, one selects those conditions where enzymes work maximally. This may have no relation to the rate of hydrolysis *in situ* in an intact biological system.

Smyth: Professor Matthews might also have asked why the liver stores glycogen and not glucose, since living tissues usually store polymers.

Matthews: Yes, but are we dealing with storage?

Smyth: Aren't we discussing why storage in the plant is not in the form of amino acids?

Matthews: No, I don't think so. If storage is in the form of monomers there will be osmotic problems, at least in animal cells.

Rhodes: Since we believe that glucose, at least for most systems, is quite adequate, why does barley make sucrose?

Payne: This aspect of peptide versus amino acid uptake is presumably associated with energy-coupling, and is a case of efficiency. Many cells devote up to 30% of their energy to transport, so it is clearly efficient for the cell to spend energy that way rather than to synthesize things inside. Transporting individual amino acids or transporting the same things linked in peptide form may well take the same energy per molecule, making the latter more efficient. In addition, the charge can be neutralized if, say, an Asp is linked to a Lys, and in terms of chemiosmotic coupling this feature should make it even more efficient to transport peptides than to transport the free amino acids.

Parsons: Generally speaking, cells have a high concentration of amino acids inside them, so presumably the cells would have to do more work getting an amino acid in from the environment whether it carried a charge or not. If the amino acid carries a negative charge, that makes it worse, whereas a peptide could go into the cell down its electrochemical gradient, as you say.

Josefsson: In animal tissues peptidases have to be in different places because there are a lot of proteinases which are dangerous for the peptidases. If all the peptidases were in the intestinal lumen the pancreatic endopeptidases would destroy their activity. The same may happen in the plant. One reason why they are located in different compartments in both plants and animal tissues might be so that they will not destroy each other.

Parsons: Transport across animal cell membranes surely need not always be coupled to an electrochemical gradient of, say, a cation such as Na^+ or H^+. Do you know of any cases where there is a flux of amino acid across a membrane that is not dependent on some energy supply, such as sodium moving down a gradient?

Semenza: The uptake of some amino acids in the small intestine depends very little on Na^+ (e.g. leucine; Curran *et al.* 1967).

Matthews: According to Schultz there is a substantial Na^+-independent component in lysine uptake by rabbit ileum (see Curran & Schultz 1970).

Semenza: Yes, but lysine has a positive charge, so it may be driven by the membrane potential.

Gardner: I am worried about the automatic assumption that transport down an electrochemical gradient does not require energy. It frequently does require energy. Cell membranes, at least in the intestine, appear to be impermeable enough for cellular energy to be required if, for instance, sodium transport down an electrochemical gradient is to occur (Gardner 1975, 1977). This would go rather against the argument which Professor Matthews put forward (pp. 10-11) about the energetic advantages of peptides as the currency by which amino acid residues enter cells.

Payne: Peptides differ from many other classes of substance because their charges can be variable. One has to invoke varied stoichiometric ratios between sodium and different peptides, and that could present problems for the transport systems. Peptides with a certain number of net negative charges may perhaps not be transported because not enough sodium ions can be coupled to them. These peptides may be ones that require prior cleavage by membrane peptidase activity. Other peptides will actually possess net positive charges. So when one says that sodium coupling has been demonstrated, the picture is perhaps more complicated than is suggested by a mere requirement for sodium for certain peptides.

Fottrell: Storage proteins are normally like a gluten-type protein. They contain a lot of glutamic acid and proline. Professor Enari, how does the composition of the storage protein in barley compare to that of the storage protein of wheat gluten or oat gluten?

Enari: The main storage protein in barley is a prolamine similar to gliadin in wheat. There is a high content of proline and glutamine.

Fottrell: It makes sense to me to see the peptides being transported. As I understood you, the gibberellic acid seems to stimulate the proteolytic enzymes which are a special stage in the development of the endosperm itself.

Enari: Gibberellic acid induces *de novo* synthesis of both proteolytic enzymes and 'amylolytic' enzymes. We have not studied the peptides formed in the endosperm.

Josefsson: As I understand it, there is a lot of proline in all the storage proteins. But your peptidases do not split proline peptides very specifically. Are there other peptidases or proteinases which are known to have this specificity, or what is the explanation?

Enari: We have been looking for a peptidase which acts on the proline peptides but we cannot find one. It is a missing enzyme.

Josefsson: It must be there.

Rhodes: Is there any evidence that dipeptides are incorporated as such in protein synthesis in plants?

Enari: No. The amino acids are transformed into glutamine in the scutellum and transported as glutamine into the growing seedling. That is another transport problem.

Sopanen: Professor Josefsson told us that the peptidases in the intestinal mucosa are present in other tissues too. Also in plants all the peptidase activities found in germinating seeds seem to be present in all the other living tissues. So it is evident that these enzymes have other functions in addition to hydrolysing the storage proteins of seeds. Probably they take part in the catabolism of proteins which goes on in all the living cells.

Meister: The analogy between plant and mammalian systems is interesting. About 20% of the amino acids in human blood plasma can be accounted for as glutamine, and quite a bit of glutamate and NH_3 are transported as glutamine in other species.

Burston: Concerning the transformation of amino acids into glutamine, didn't you tell us, Dr Payne, that Gly-Sar which is not hydrolysed in the scutellum is transported into the shoot?

Payne: In studies where the scutellum rests on a filter paper so that it alone is in contact with a solution of Gly-Sar, and the shoots and roots are therefore spatially removed from the Gly-Sar, Gly-Sar accumulates in the shoots and roots as well as being present in the scutellum. It must therefore be translocated into those organs after being transported across the scutellum membrane. The concentration can be similar in all three organs and concentrated 10- to 20-fold higher than in the external medium. Incidentally, we find significant quantities of alanine, as well as other amino acids, and not just glutamine as Professor Enari mentioned.

Enari: Compared with animals the plant has a special problem. It has no urine in which to excrete any compounds it does not want. So what should a plant do with this glycylsarcosine?

Payne: I don't know but it is apparently moved by long-distance translocation mechanisms; whether it is being taken up into vacuoles or something else remains to be seen.

Rhodes: Is glutamyl transpeptidase present in barley?

Enari: No one has looked for it.

Rhodes: So the germinating barley plant uses glutamine and then resynthesizes all the amino acids to support its growth from glutamine?

Enari: Yes; when the plant becomes autotrophic it synthesizes all the amino acids from inorganic nitrogen compounds.

Matthews: Dr Payne mentioned the translocation of glycylsarcosine from the scutellum to the tissues of the growing barley plant. In this connection, it is known that in some seeds and bulbs there are high concentrations of various γ-glutamyl peptides, and it has been suggested that these peptides may be translocated from one part of the plant to another (Thompson *et al.* 1962). Perhaps this is a different aspect of peptide transport—translocation, not simply membrane transport.

Parsons: Can you tell us something about the methods used in the experiments where peptide transport occurs in the scutellum?

Sopanen: The procedure is simple. We first let barley grains germinate, then after three days we dissect out the scutella. They are very small, about 1 mm in diameter. We wash them, put them in a solution of the peptide, incubate for 1 h, then extract the peptide and estimate it (see Burston *et al.*, this volume, pp. 79-98).

Parsons: So you measure the accumulation by comparing the concentration in the scutellum with the concentration in the medium?

Sopanen: Yes.

Matthews: The incubation is at 30 °C and initially it was done in water. Monovalent cations are reported to inhibit the transport of amino acids by the scutellum (Stewart 1971) and we didn't want to have the complications of adding a buffer. We are investigating this question now.

Payne: Although we have come to the same conclusions as Professor Matthews, we worked in a different way. We soaked barley grains for 2–8 h, then separated the scutellum from the storage organ and looked at transport into the scutellum in that system. We looked both in aqueous solution, so that penetration could be from any direction, so to speak, and in an oriented system on filter paper. Because of our ignorance about the effect of sodium and potassium upon amino acid transport we used a sodium–potassium-phosphate buffer and found that it worked, although it is possible that greater uptake might be observed in their absence.

Baker: Have you looked at both sides of the scutellum?

Payne: No, only in the endosperm to the embryo direction.

Matthews: In the scutellum uptake of glycylsarcosine seems to be linear with time, at least up to 2 h. Prolonged linear uptake of glutamine has also been reported in corn scutella (Stewart 1971). This is very unfamiliar for people who study the intestine *in vitro*, where uptake by the wall nearly always reaches a plateau in 20–30 min. It is almost as though we have an influx which goes on over very prolonged periods. This is hard to understand.

Josefsson: Perhaps it is easier to keep the scutellum active experimentally than it is to keep the intestine active.

Smyth: I think there are two reasons for the rate of uptake by the *in vitro* intestine falling off. In the first place the intestine *in vivo* has a blood-stream, which carries away the products of transfer, and this is not present *in vitro*. The accumulation of these products could slow down the rate of transfer. A second reason is that the *in vitro* intestine is gradually deteriorating, I believe noticeably so after 15–20 min.

Sopanen: In general, plant organs are much more independent than animal organs. Also most plant cells differ from typical animal cells in having a large central vacuole, which takes up most of the space in the cell. Many substances including amino acids are concentrated in this vacuole. It may be that the unhydrolysed dipeptides go there and that there is no efflux of the kind seen in animal cells.

Rhodes: Have you compared glycylsarcosine transport into the germinating and the non-germinating scutellum in the absence of the peptidases or without the 20-fold increase?

Sopanen: Not yet.

References

Curran, P. F. & Schultz, S. G. (1970) Coupled transport of sodium and organic solutes. *Physiol. Rev. 50*, 637-718

Curran, P. F., Schultz, S. G., Chez, R. A. & Fuisz, R. E. (1967) Kinetic relations of the Na-amino acid interaction at the mucosal border of intestine. *J. Gen. Physiol. 50*, 1261-1286

GARDNER, M. L. G. (1975) Inhibition by iodoacetate of 'downhill' transport of sodium ions across isolated rat small intestine. *Biochem. Soc. Trans. 3*, 255-256

GARDNER, M. L. G. (1977) Metabolic energy dependence of glucose, water and sodium absorption in the presence and absence of 'downhill' sodium gradients across isolated rat small intestine. *J. Physiol. (Lond.) 265*, 231-247

STEWART, C. R. (1971) Some characteristics of the uptake of glutamine by corn scutellum. *Plant Physiol. 47*, 157-161

THOMPSON, J. F., MORRIS, C. J., ARNOLD, W. N. & TURNER, D. H. (1962) γ-Glutamyl peptides in plants, in *Amino Acid Pools* (Holden, J. T., ed.), pp. 54-64, Elsevier, Amsterdam

Final discussion

SITE OF PEPTIDE HYDROLYSIS

Parsons: One topic on which we might try to reach agreement now is the classification of the site of peptide hydrolysis in tissues such as the small intestine. As discussed earlier (pp. 145-146), confusion has arisen partly because the terminology was based either on where the substrate is when it has been hydrolysed, or on where the product goes to after the hydrolysis, or on the location of the enzyme. Professor Matthews published a classification in *Physiological Reviews* last year which we might consider now.

Matthews: This scheme (Fig. 1) (Matthews 1975) applies to the intestine. It is relatively simple and I think it should already be possible to distinguish all these different sites of hydrolysis experimentally. However, the scheme doesn't cover every possibility. For example, it doesn't cover the situation with a peptide such as carnosine which is to some extent absorbed intact from the gut into the blood and hydrolysed in various organs of the body as well as excreted in the urine.

There are two main divisions in the scheme. Firstly, *intralumen hydrolysis,*

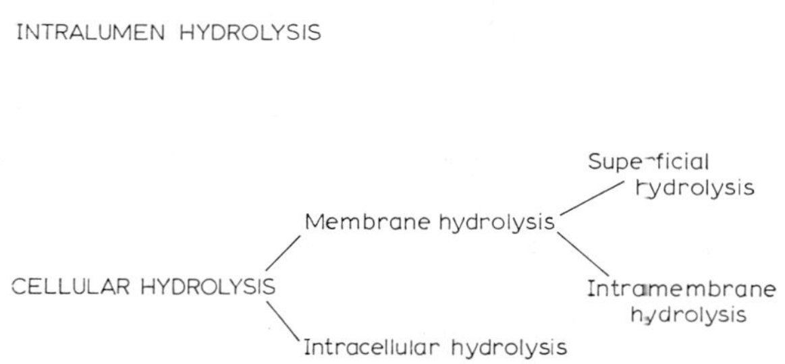

FIG. 1 (Matthews). Terms used in discussion of intestinal hydrolysis of peptides. (From Matthews 1975).

defined as hydrolysis within the fluid in the intestinal lumen, with release of amino acids in free solution. For *in vitro* preparations that could be called intra-medium hydrolysis. The second major division is *cellular hydrolysis*. This applies when hydrolysis is not in the lumen. It is associated with the intestinal mucosa but nothing is known about its subcellular site.

Hydrolysis can take place in association with peptidases of the plasma membrane of the brush border or by the action of proteases adsorbed in the brush border region. I call this *membrane hydrolysis*. When membrane hydrolysis appears to be occurring with release of amino acids in free solution so that they behave just as they would if you introduced them into the lumen in free solution, that is what I call *superficial hydrolysis*. If they should behave as though they are not released in free solution, as in Ugolev's scheme for example, I should call this *intramembrane hydrolysis*.

Finally, when hydrolysis appears to occur definitely deep to the plasma membrane of the absorptive cells, either in the cytosol or in some subcellular organelle within the cell, I term this *intracellular hydrolysis*. Such hydrolysis is, of course, literally intracellular in the most obvious sense.

Smyth: I think the situation is more complicated. Intraluminal digestion can occur at different sites. There are soluble peptidases in the intestinal lumen, which can take part in hydrolysis. But there are also peptides in the shed epithelial cells and, as Starling (1906) showed many years ago, most of the peptidase activity of the succus entericus depends on the presence of epithelial debris.

Matthews: Experimental results on this question have been variable. In the published version of my scheme I included hydrolysis by cells free in the lumen under intralumen hydrolysis.

Smyth: This role of the epithelial debris was one of the reasons why we thought of intracellular digestion (Smyth 1963). If the peptides could enter the shed epithelial cells and be hydrolysed, the possibility must be considered that they could enter the epithelial cells that were lining the intestine. Another question that arises is the place of glycocalyx in luminal digestion. Where does it fit into the scheme?

Parsons: It could be considered as one of the sites where superficial hydrolysis occurs; this will include hydrolysis by adsorbed enzymes of exogenous origin (see Parsons 1972).

Smyth: Does Professor Ugolev accept that?

De Laey: It is what Professor Matthews would call superficial and I would accept that definition as being within Professor Ugolev's membrane concept.

Baker: Is superficial hydrolysis in the stirred or the unstirred layer? Where does the lumen end?

Adibi: It varies.

Matthews: If one knows that the hydrolases responsible are in the brush border region, either adsorbed pancreatic proteases in the glycocalyx or membrane-bound hydrolases on the plasma membrane, the lumen has then stopped and it is membrane hydrolysis. One has to sort out whether it is superficial hydrolysis with release in free solution or intramembrane hydrolysis coupled to transport of amino acids.

Baker: In other words the scheme is based on where the enzyme is rather than where the hydrolysis occurs.

Silk: Intralumen hydrolysis has to be defined according to the experimental techniques used to measure it. As Professor Matthews pointed out, it is a measure of the activity of those enzymes which are free in the mucosal medium in *in vitro* preparations and in the bulk luminal phase medium in *in vivo* preparations.

Parsons: It seems important to remember that there can be two types of superficial hydrolysis: one due to enzymes which are an integral part of the membrane and one due to other enzymes adsorbed from some other origin such as the pancreas.

Smyth: If a peptide can be hydrolysed either in the brush border or in the cytosol, how do you know which is the site of hydrolysis physiologically? This would be possible only if the particular enzymes are absent from one or other of these sites and in such a case it might be useful to use the term cytosol hydrolysis or brush border hydrolysis, depending on the location of the enzyme.

Matthews: One can go on subdividing possible sites of hydrolysis.

Josefsson: Intracellular hydrolysis can be located either in the cytosol or in subcellular organelles, just as you have subdivided membrane hydrolysis. If you can show the enzyme to be soluble or localized to the cytosol, you can refer to intracellular hydrolysis.

Nicholson: In that case, since there are membranes inside the cell, perhaps we ought to talk about brush border membrane hydrolysis, because if a hydrolytic enzyme is found in the the endoplasmic reticulum, that would also be membrane hydrolysis and would give rise to confusion.

Josefsson: You cannot call it 'brush border' membrane hydrolysis because that refers only to the enterocyte. 'Plasma membrane hydrolysis' would be a better term.

Radhakrishnan: We are trying to provide definitions which will be useful for future work and help our own understanding. Professor Matthews' scheme is quite simple and we should not complicate it any further. I would only suggest that instead of 'superficial' we should use 'surface' hydrolysis. This can then be distinguished from the bulk phase reaction which occurs in the lumen. By 'intramembrane' we really mean hydrolysis during transport, for instance amino acids arising from peptides are not in true equilibrium with the luminal amino acids.

Josefsson: We must not be too simple. Our terminology must express where the functions are going on, if possible. We should try to say whether an action is located in the cytosol, or the mitochondria, or the membrane. In other cases we are just saying that it is hydrolysis without relating it to anything.

Smyth: I would support that very strongly. If we know where an enzyme is, let's say so.

Radhakrishnan: If you really want to pinpoint the location of an enzyme you get into problems again. You have to be very sure of the location. It is easy in pure cytosol but location of the enzyme after the cell is disrupted is not so easy; for example there could be a redistribution of the enzymes after cell disruption.

Parsons: We are trying to look ahead. But there are other tissues than the enterocyte and I would like to know what Dr Payne thinks of our attempts to develop a nomenclature.

Payne: We don't have the same problems in *E. coli*! I think we might use the following terms to describe membrane locations: intrinsic or integral for firmly bound or embedded elements, and extrinsic or peripheral for loosely attached entities, and periplasmic to describe a region adjacent to the membrane and contiguous to the cell wall. Thus, intraluminal would become periplasmic for us. Superficial hydrolysis would be extrinsic or peripheral hydrolysis, and intramembrane hydrolysis would be intrinsic or integral hydrolysis. Intracellular hydrolysis would be the same, or perhaps cytoplasmic hydrolysis.

Smyth: The difference between your cells and the enterocytes is that all sides of bacterial cells are in contact with the same fluid. The enterocytes have a definite polarity with one side of the cell in contact with the intestinal contents, and the other side in contact with a quite different fluid.

Silk: Professor Matthews, could you briefly explain again the difference between superficial and intramembrane hydrolysis?

Matthews: In superficial hydrolysis, hydrolysis occurs in the region of the membrane but the amino acids (or, to put it more generally, the monomers) behave as though released in free solution. In intramembrane hydrolysis—and I am still not convinced that this has been demonstrated in the gut in the case of peptides, though it may occur with disaccharides—the amino acids (monomers) do not behave as if released in solution.

Silk: This really means that you are leaving it to people to interpret their own data according to whether it is superficial or intramembrane. If we are trying to define the situation clearly we should avoid personal interpretation of data.

Maroux: Is hydrolysis by aminopeptidase superficial hydrolysis or intramembrane hydrolysis?

Matthews: It depends on how the amino acids that are released behave subsequently. Any classification will have loopholes, but one classification would be better than three or four. It is confusing if what one person calls intracellular another calls extracellular!

Parsons: I am against saying where the products go to. I would much rather say where the enzymes are, or where we think they are from our knowledge of membrane structure and shape of the proteins, etc.

Smyth: The function of classification is not to solve problems but to enable us to discuss problems.

Matthews: As far as we know, an enzyme which is in exactly the same subcellular location could do one of two things, or possibly both. It might be situated on the lumen side of the membrane, and there it could release amino acids which behave as if they were in free solution. On the other hand, if Professor Ugolev is right it could also engage them in transport mechanisms, in which case they would not behave like that.

Parsons: If, from the molecular architecture of the membrane, you knew where the active site of the enzyme was, you would then be able to say that the active site of some particular enzyme is superficial, therefore it ought to be called a superficial hydrolase.

Radhakrishnan: I don't think that would help. All you know is the location of the enzyme. That doesn't mean that during transport the binding of the substrate would not alter the conformation of the enzyme.

De Laey: I would like to support Professor Matthews' system. We are dealing at the moment with the functional topography of the cell, that is with enzyme activities and with their products of hydrolysis. We can try to put Professor Ugolev's concept of intramembrane digestion into the scheme for dipeptidases as well as for disaccharidases. Ugolev's concept of membrane digestion, on the other hand, would be the superficial phenomenon of absorbed enzymes and it happens in the glycocalyx. So membrane hydrolysis would be very much within the scope of Ugolev's work.

PEPTIDE TRANSPORT THROUGH MEMBRANES

Parsons: We might next consider the evidence for membrane transport of peptides, by which I mean movement of the unchanged peptide right through the membrane.

Adibi: There are many indirect but just two direct proofs for membrane transport of intact peptides: (*a*) they are found inside the cell; (*b*) they appear in the fluid bathing that cell (plasma or serosal fluid).

Smyth: And there is the fact that peptides can disappear from the mucosal or luminal fluid without being hydrolysed.

Matthews: I think there are one or two pieces of good evidence for membranes being traversed by peptides. Possibly the most important is the demonstration of the appearance of Gly-Gly, carnosine and other peptides in the blood during their absorption. The experiments on accumulation of Gly-Sar and similar peptides by small intestine *in vitro* (see Matthews 1975) are very suggestive indeed. It hasn't been proved to the hilt that they really cross the membrane into the absorptive cells but it is extremely unlikely that the accumulation can be attributed to adsorption at the surface.

Radhakrishnan: I think hydroxyproline peptides arising from dietary sources are found in the urine and I think there is good evidence for intact transport of the peptides.

Matthews: Yes, one could give at least a dozen examples (see Matthews 1975).

Nützenadel: In the kidney tubule cell the intact dipeptide can be found after exposure to extracellular carnosine and the concentration is higher intracellularly than extracellularly. Although this preparation is open to criticism since two different membranes (luminal or basal) could facilitate the entry, active dipeptide uptake through non-luminal cell membranes is evident from Dr Adibi's results.

Parsons: Professor Mellander, you spoke about phosphopeptides in milk (see p. 294). Would any peptides be transported from the interior of the mammary gland cells to the lumen of the duct, and does any hydrolysis occur within the duct?

Mellander: The free peptides are probably transported through the mammary gland. Casein peptides are synthesized in the mammary gland.

Adibi: This would be peptide secretion. There is evidence for renal secretion of peptides too.

Josefsson: There is clear evidence that peptide hormones are synthesized in, for example, various cells of the central nervous system and transported through the axons. Their transport can be followed all the way.

MEMBRANE DIGESTION OF PEPTIDES

Parsons: The next question is whether, within the luminal-facing plasma membrane of the intestine and the kidney, there is membrane digestion of peptides.

Josefsson: There is no unequivocal evidence that digestion is occurring within the membrane.

Silk: So membrane digestion is intramembrane digestion?

Josefsson: Yes. We are not talking about superficial hydrolysis.

Adibi: Whether enzyme activity measured *in vitro* has any meaning *in vivo* has to be established by physiological studies. There is *in vivo* evidence that free amino acids appear in the lumen when peptides are infused into intestine. Whether they are coming from superficial, intramembrane, or intracellular hydrolysis of peptides has yet to be fully established.

Radhakrishnan: I think this was taken care of in the definition (Fig. 1, p. 353). In superficial hydrolysis the products are in true equilibrium with the luminal amino acids. In intramembrane hydrolysis they are not in equilibrium.

Adibi: At the moment the fact is that when some dipeptides are put in the lumen, free amino acids are recovered, some in small amounts and others perhaps a little more.

Radhakrishnan: There may be an efflux phenomenon.

Adibi: The question is, what is the origin of these hydrolytic products? There are three choices and I am not certain which is the correct one—superficial, intramembrane, or intracellular hydrolysis.

Smyth: The only way to separate these would be to find a peptide which was hydrolysed and for which there was no appropriate enzyme in the cytosol.

De Laey: I thought I had presented some evidence (Ugolev *et al.*, this volume, pp. 221-239) which convinced you that membrane digestion occurred.

Josefsson: We have no real experimental evidence yet that there is digestion within the membrane.

Adibi: We are all looking for an inhibitor which will selectively inhibit either the hydrolytic or the transport system.

Josefsson: All we know about intramembrane-bound enzymes comes from the type of enzyme which has most of its protein parts outside the membrane. The parts within the membrane have little chance of having an active site. That means that the active site of that enzyme is certainly on the outside. Until we know that an enzyme which is completely hidden within the membrane has a certain activity we can't really be sure that hydrolysis occurs within the membrane.

Meister: If there is no water in the membrane, there may be no digestion.

Baker: We are getting mixed up in our definitions. As I understand the definition proposed earlier, intramembrane digestion doesn't necessarily mean that the active site of the enzyme is within the cell membrane. Professor Matthews is saying that the active site of the enzyme may still be sticking out from the surface but there is some special mechanism by which the products of digestion get through within the cells. The products of digestion are not in equilibrium with the free amino acids which may happen to be in the lumen. It is not so much a question of where the active site is as of what happens to the product when it gets in.

Josefsson: We don't know yet of any certain examples where we can say there is an acceptance of products.

Silk: Professor Matthews has defined superficial hydrolysis as being characterized by the production of amino acids which move into free solution and behave as though they are in the free form. By definition, using experimental techniques that we have described in this symposium, it would therefore be impossible to locate an intramembrane site of hydrolysis because hydrolysis is only indicated when hydrolytic products are detected in free solution. I thus agree entirely that we cannot exclude intramembrane digestion, but we have presented no evidence in favour of it.

Meister: Why not define membrane digestion in terms of where the water molecule comes from in hydrolysis? Does the water come from inside the cell or outside?

Silk: I think the question that Dr Parsons is posing is: what evidence have we presented in this symposium to favour intramembrane digestion? To my way of thinking, nobody has produced sufficient evidence to support the existence of intramembrane digestion, because the techniques so far used are not sophisticated enough.

Parsons: Could Dr Kenny and Dr Maroux tell us what is known about the aminopeptidases that are in the membrane? In the kidney, I think 10% of the brush border protein is an aminopeptidase.

Kenny: In the kidney, 4% of the microvillus protein can be attributed to aminopeptidase M. We haven't any data on the other aminopeptidase (A) which exhibits a specificity for N-terminal Asp- and Glu- residues.

Maroux: In the pig brush border, 8% of the protein is aminopeptidase.

Kenny: Another 4% of the microvillus protein is attributable to the neutral endopeptidase. This enzyme differs from the other microvillus peptidases in its resistance to release by papain—either because it is mostly within the lipid bilayer of the membrane or because no papain-sensitive bond is exposed on the external portion. In this respect it resembles alkaline phosphatase.

Parsons: Are these enzymes glycoproteins?

Kenny: Yes. Aminopeptidases M and A, endopeptidase and dipeptidyl peptidase IV, which also amounts to about 4% of the protein, are all glycoproteins. Aminopeptidase M is the major glycoprotein in the kidney microvillus membrane.

Parsons: Dr Kenny, have you any evidence on whether your enzymes are transmembrane proteins?

Kenny: Not yet. We are working on the same lines as Dr Maroux with dipeptidyl peptidase IV, but we have no data from labelling experiments at present.

Parsons: What about subunit size?

Kenny: Many seem to be dimeric when purified. This is true of aminopeptidase M and dipeptidyl peptidase IV as well as alkaline phosphatase and phosphodiesterase I. Whether this is an artifact caused by the treatment needed to release the enzymes or whether they exist as dimers in the membrane is an important question.

Parsons: What are the molecular weights?

Kenny: There seems to be some similarity in that they are all rather large. The monomer of aminopeptidase M is 160 000 in the native state and 140 000 when released by proteinases. The monomer of dipeptidyl peptidase IV is about 130 000, both in the native and the purified forms. The neutral endopeptidase is unusual in being monomeric when purified and in being smaller (about 95 000). We haven't purified aminopeptidase A, but in gel filtration in non-dissociating conditions it is not resolved from aminopeptidase M.

Maroux: The presence of three subunits in our aminopeptidase preparation might be an artifact, because the sum of the molecular weights of the two smaller subunits (95 000 and 50 000) is equal to the molecular weight of the heavier subunit (135 000). A peptide bond particularly sensitive to trypsin could be split even on the membrane-bound enzyme.

Norén: Wacker *et al.* (1976) studied the Triton X-100-solubilized form of aminopeptidase M and found that it was composed of two subunits of similar molecular weight. Is there any evidence that these two subunits are identical? If they are identical, one would expect two hydrophobic anchors per molecule but if they are not there may only be one per molecule.

Maroux: I have no evidence on whether there are two or only one. In rabbits, aminopeptidase is a monomer with a molecular weight of 140 000.

Kenny: You are talking about enzymes that have been released by one means or another from the membrane. What we need to know is their native form in the membrane.

Parsons: These are oligoaminopeptidases: they will split dipeptides but they are not true dipeptidases according to Dr Josefsson's definition.

Norén: They are not dipeptidases according to the definition of the Enzyme Commission!

Fottrell: Dr Kenny, regarding the use of papain where the enzyme is outside the membrane, since these enzymes are glycoproteins there may be steric hindrance to the activity of papain. I am not sure about the details of the technique but have others used mixtures of enzymes such as pronase and papain to produce a situation where the bond will be cleaved?

Kenny: We tried many enzymes in our early attempts to solubilize the neutral endopeptidase. The toluene–trypsin method was the only one that gave reasonable yields, although recently we have been successful with detergents.

De Laey: Dr Parsons, are you saying that there is membrane transport of peptides and no membrane digestion of peptides? Do all those peptides cross the membrane as peptides?

Parsons: We have no evidence of that.

Josefsson: When we are talking about membrane hydrolysis we do not include hydrolysis outside or inside the membrane. What we doubt, and what we have no evidence for, is that digestion of peptides takes place within the membrane. Everyone agrees there is superficial membrane digestion.

Matthews: I want to return to the question of the influence of pH on transport and hydrolysis and to describe some puzzling results suggestive of entry of peptides into the portal blood.

With hamster jejunum *in vitro*, uptake of Glu from Glu-Glu is maximal at pH 6.0, and so, as judged by the medium appearance of free Glu, is hydrolysis (see Fig. 5, p. 89). In similar experiments with Gly-Leu the situation is different. Total uptake of amino acids with this peptide is maximal at pH 7.0, and so is hydrolysis (Fig. 6, p. 90). The difference in the effect of pH on hydrolysis is compatible with hydrolysis of these substrates by two different enzymes. With some mixed peptides, especially *in vitro*, uptake of different amino acids from a peptide can be grossly unequal. There are two possible explanations (see Matthews 1975). One is that there are different rates of exit of the amino acids from the cells. The other is that a large component in total uptake is hydrolysis at the brush border or in the medium, followed by entry of free amino acids at different rates. If, with hamster small intestine *in vitro*, pH is reduced from the standard 7.0 to 5.0, uptake from Gly-Leu of the two amino acids, Gly and Leu, becomes equal. This is compatible with the idea that the inequality in uptake at pH 7 reflects an element of hydrolysis followed by entry of amino acids in addition to peptide uptake. We shall all have to think much more about the effects of pH.

In work done with Professor Sleisenger (Sleisenger *et al.* 1977), we infused into the duodenum of guinea-pigs either a mixture of amino acids simulating casein or a partial hydrolysate of casein containing a large proportion of small peptides, and then estimated amino acids in portal plasma during and after the infusion, which lasted 20 min. With these two preparations there were similar increments for many amino acids in the portal plasma.

However, in several cases there were discrepancies, the curve following the partial hydrolysate being much lower than that following the amino acid mixture. The discrepancies were greatest with glycine, glutamine, proline and serine. There are many possible explanations for such discrepancies but we weren't able to carry on the work for long enough to investigate all possibilities. One possible explanation is that peptides were entering the portal blood. We know that with our particular analytical technique we would not be able to pick up the peptides. This is additional evidence suggesting that the whole question of entry of peptides into the portal blood should be more thoroughly explored.

Meister: Are you using dipeptides here?

Matthews: We were using a mixture of peptides of mean chain length about three residues, and free amino acids. More than half the nitrogen was in peptides. Apart from this, we don't know the composition. The mixture was produced by hydrolysis of casein by papain followed by renal peptidases.

Meister: Is the uptake of glutamine, for example, determined by the difference method?

Matthews: It is determined by the difference between the resting value in the portal blood and the value in the portal blood during and after intraduodenal infusion of the amino acid mixture of enzymic hydrolysate.

Meister: What may be forgotten here is that the enterocyte has a tremendous capacity for metabolizing amino acids, especially glutamine.

Matthews: I thought that most experiments *in vitro* (and I must admit that our work was *in vivo*) had shown that most amino acids were only metabolized to a small extent—a few per cent.

Parsons: We have found that in perfused rat intestine, glutamine is rapidly removed from the vascular circulation (P. J. Hanson & D. S. Parsons, unpublished work).

Matthews: But why should there be extensive metabolism of glutamine when it is presented as peptides but not such a high proportion metabolized when it is presented as an amino acid?

Parsons: Substances going in as peptides might enter a different pool. Certainly there seem to be different pools within the epithelium, the ease of access to which depends on whether the substrate is presented via the lumen or via the blood (see Gangl & Ockner 1975; Hanson & Parsons 1976).

Adibi: Dr Gardner (1975) did a very nice piece of work which suggested that some amino acids appearing on the serosal surface are peptide-bound when a protein digest is put on the mucosal surface.

EPITHELIAL TRANSPORT OF PEPTIDES

Parsons: We are now moving from membrane transport to the epithelial transport of peptides. During the absorption of dietary protein, the output from the epithelium into the portal blood side seems likely to be a mixture of amino acids and perhaps of peptides. The old arguments that under physiological conditions peptides did not enter the portal blood and general circulation because they were toxic, and/or were excreted in the urine and therefore wasted, have clearly been put aside by the experiments of Dr Adibi.

Gardner: My results and Professor Matthews' results may have been prejudiced in favour of finding transport of intact peptides by the fact that we used casein hydrolysates which are not necessarily typical. Professor Mellander's comments about stability to hydrolysis of casein peptides may be relevant. We will obviously have to do these absorption experiments with proteins other than casein and with varying degrees of hydrolysis. Most of us use partial hydrolysates in which the mixtures contain 30–50% peptide nitrogen. Dr Silk had 80%, which was exceptional, and we need to do more experiments with mixtures containing a high peptide content like that. A high total peptide content may also represent a more physiological state of affairs. Each individual peptide may be present in very low concentrations but tremendous numbers of different peptides are probably produced *in vivo* during pancreatic digestion of a protein meal. This may be a key difference between *in vitro* experiments and the physiological state.

Parsons: That is, peptides may accumulate inside the cell under *in vitro* conditions so that the intracellular peptidases would have a higher substrate concentration on which to operate. So in your *in vitro* preparation you might be less likely to see peptides appearing on the serosal side than might appear in the portal blood *in vivo*.

Gardner: Yes, that is one point which I discussed earlier. The sort of experiments I have been doing should be repeated with vascular perfusion, with which it should be possible to mimic the *in vivo* situation and to prevent the accumulation of peptides in the tissue which produces high local concentrations. Under these circumstances we may find more peptides crossing the mucosa and going to where the portal blood would be.

My other point was that in the form in which the digest is presented to the mucosa the concentration of each individual peptide may be very low indeed, but there is a colossal number of different peptides and so the physiological mixture is not the same as some people have studied.

Meister: Has it been established whether peptides are normally present in the portal blood?

Matthews: The presence of peptides in the portal blood, on a fairly substantial scale, during absorption of a protein meal, has never been completely satisfactorily excluded. There have been dozens of reports since the early 1900s of the rise in amino acids in portal plasma after a protein meal but very few investigations by modern techniques of this question of the presence of peptides. In one study in dogs (Levenson *et al.* 1959) the authors deliberately looked for peptides in portal venous plasma after feeding protein. They expected to detect peptides, if they were present, by ion-exchange chromatography of deproteinized plasma filtrates. They didn't find them and they reported negatively. But both Dr Gardner and I have noticed that in some circumstances one can put solutions (enzymic hydrolysates) which are extremely rich in peptides on the amino acid analyser and see absolutely nothing. So the question does need reinvestigation. The work of Dent & Schilling (1949) and Christensen (1949) on gastrectomized dogs given heavy protein meals has always been taken as the first satisfactory demonstration, using good techniques (paper chromatography and also estimation of α-amino nitrogen), that only amino acids enter the portal blood. The chromatograms of portal plasma showed an increase in peptides. However, there was a suspicious rise in bound α-amino nitrogen in portal plasma. Christensen did not interpret this rise as due to peptides, but nowadays one might suggest that it was. Incidentally, Dent & Schilling did report peptiduria in their dogs after protein feeding.

Meister: What did Christensen say the increase in bound amino acids in portal plasma was due to?

Matthews: He did not rule out the possibility that it was due to peptides, but he thought it was more probable that it was due to non-peptide amino acid conjugates. When certain amino acids, for example glycine, are introduced into the circulation there is a rise in conjugates that are not peptides.

Adibi: Another piece of evidence is that six dipeptides put into the gut of the rat were all found in portal blood.

Matthews: But they were introduced at very high concentrations (100 mM).

Meister: The usual amino acid analyser is generally adapted to the determination of free amino acids. A modified system may be desirable for the detection of peptides.

Adibi: It is possible to identify peptides by amino acid analysis. Some peptides have a very flat peak but some dipeptides could well be identified on the amino acid analyser.

Matthews: Can single ones be seen with your system?

Adibi: Yes.

Fottrell: Pyrrolidone carboxylic acid peptides may escape detection by ordinary techniques, so one has to be careful.

Josefsson: Dr Adibi, with physiological amounts of dipeptides or peptides put into the gut, do you really expect, from the clearance data, to find any dipeptides in the portal blood?

Adibi: I was hoping that Professor Matthews and Professor Sleisenger, who have studied this, would answer that question. Our knowledge is quite deficient in terms of portal blood composition after meals.

Parsons: Is it a question of the balance between the rate of transport of peptides into the cell and the rate of hydrolysis?

Adibi: My guess is that we might find these dipeptides in physiological conditions in the portal blood. Professor Matthews' data suggest this.

Josefsson: The doses of dipeptides found in the portal blood were very large.

Allen: With a slower input into plasma, such as oral administration, the chances of the natural peptides building up to a detectable concentration are very low. If you could find peptides which have a longer half-life when they are administered intravenously, then you would have a greater chance of significant levels being achieved after oral dosing.

Silk: Professor Matthews used a unique experimental model, in which he was able to measure portal vein concentrations directly. We mustn't muddle up, in the discussion, peripheral venous levels of peptides and portal vein levels. It is the portal vein concentrations that interest us when we are talking about mechanisms of absorption, because Dr Adibi has very elegantly demonstrated that removal of peptides from the systemic circulation is very rapid.

Parsons: There have been very few critical studies in which peptides have been looked for in the portal (not systemic) circulation during protein absorption.

Silk: We are now studying patients with cirrhosis of the liver who have had portocaval shunts. In this situation the portal blood is shunted directly into the systemic circulation and it may be possible to detect peptide-bound amino acids by this type of approach.

Meister: The problem could probably be solved using the existing techniques. In the early 1950s, Stein & Moore (1954) analysed peripheral blood plasma and specifically looked for peptides by doing analyses before and after acid hydrolysis. They made corrections for compounds such as glutamine and asparagine and found that there was not an appreciable amount of peptide in peripheral human blood plasma. I believe that the same technique can be applied to portal blood plasma.

Matthews: There is a difficulty here. We and Dr Gardner are using a system which is basically similar to the original one of Stein and Moore. We have no difficulty in detecting and estimating individual small peptides, which usually give beautifully clear peaks. If there was a multitude of different peptides at very low concentrations such as might enter the portal blood, those peptides might escape detection.

Meister: If there are appreciable quantities of peptides in portal blood, the amino acid analyser with the hydrolysis technique should indicate their presence. One could compare portal blood with blood from the systemic circulation to see whether there is actually a difference.

Matthews: But there is a real analytical difficulty. If we put an aqueous solution to which we have added a large quantity of a partial enzymic hydrolysate of casein on an amino acid analyser, we see nothing but amino acids. In the portal plasma of the guinea-pig the fasting values for amino acids released after acid hydrolysis are unexpectedly high. This makes it more difficult to detect an increase in them during absorption.

Rhodes: What precipitating method did Stein and Moore use?

Meister: Picric acid.

Gardner: Several people have looked quantitatively at amino acids in the portal blood. Dawson was probably the first to obtain a reliable estimate for recovery of absorbed amino acid in portal blood. But this could only account for about 41% of absorbed amino nitrogen appearing as free amino acids in plasma during digestion and absorption of ^{14}C-labelled algal protein (Dawson & Porter 1962). There appears to be an analytical discrepancy which has been swept under the carpet.

Adibi: The problem is very complex. Stein and Moore did very elegant work but at that time they didn't realize the complexity of the very low reacting (ninhydrin) agent.

Meister: I believe that if you do amino acid analyses before and after hydrolysis with HCl, you should be able to settle this question with the usual ninhydrin reagent. If as much as 60% of the absorbed amino acid nitrogen is in the form of peptides it should be quite simple to determine this.

Milne: There are a lot of natural peptides in low concentration in plasma. They hydrolyse with hydrochloric acid but because of their low amount the differences in amino acid content between the original and hydrolysed specimens are very small. In blood a difference method has a lot of experimental inaccuracies. This to me is the difficulty.

Meister: But a 60% difference is practically a twofold increase. There should be no problem in seeing this.

Radhakrishnan: There are new methods for separating peptides as a class from amino acids. Also, some rather sensitive monitoring procedures are now available (Gruber *et al.* 1976).

Adibi: We would like to hear about those. Are they quantitative? The problem is not new. Quantitating peptides has eluded protein chemists for many years.

Rhodes: It has been shown (Prockop *et al.* 1962) that 25 g gelatin in man will result in a rise in serum hydroxyproline peptides and 7–8% of the hydroproline appears in the urine in peptide form. Casein and gelatin could be compared in these systems.

Matthews: Gelatin was studied as early as 1922 (Folin & Berglund 1922). A very large rise in non-protein, non-urea, non-amino acid nitrogen, which would include peptide nitrogen, was found in the peripheral plasma of man after ingestion of 135 g of gelatin.

Nützenadel: Younoszai & Cynek (1975) measured the amount of hydroxyproline in the peptide-bound form and the free form in blood. Only one-tenth of the total amount of hydroxyproline is in the peptide-bound form.

Radhakrishnan: That is not right. Free hydroxyproline is very easily metabolized. What is excreted in the urine is about 10% of the total calculated collagen breakdown. Almost all of the urinary hydroxyproline is in the bound form.

Milne: In blood there is at least three times as much free hydroxyproline as bound hydroxyproline. In urine it is 95% bound because the clearance of the peptide is very much higher than clearance of the free hydroxyproline. In blood the technical difficulties are much greater than in urine, because a very large basal level of the amino acid is being taken from a smaller increase after hydrolysis. The errors in urine are bad enough by the difference method but they are much greater in blood. But I agree that the question should be explored in portal blood.

Adibi: A difficulty in this area of research, besides the technique, is the supply of these peptides. They are excessively expensive and very hard to come by. We would urge friends and colleagues to help us with research in this area.

RATE-LIMITING STEPS: HYDROLYSIS OR TRANSPORT?

Parsons: The next question is whether the rate-limiting step in peptide uptake is transfer or hydrolysis. The capacity of brush border membranes for hydrolysing disaccharides greatly exceeds the capacity for transporting the relevant monomers. I have the impression that this is probably true for some of the peptidases.

Gardner: The rate of transport and the rate of hydrolysis are certainly interdependent. For instance, I can show that the rate of peptide absorption is proportional to the rate of hydrolysis, but it is equally true to say that the rate of hydrolysis is proportional to the absorption rate (Gardner 1975). I think that the rate-limiting step in hydrolysis is absorption. I base this partly on the observation that if one works at low peptide concentrations the rate of peptide absorption is directly proportional to the peptide concentration. The chances of fortuitously obtaining such a close relation for 16 or so different amino acids, if hydrolysis were the rate-limiting step, are statistically very slim. I think that, although peptide absorption rate or transport rate depends on the intraluminal concentration, in turn the rate of peptide hydrolysis depends on the rate of absorption. It was interesting that John Payne mentioned that in his cells the peptidase hydrolysis activity was a thousand times greater than the transport activity.

Payne: No, I said that the peptidase activity was perhaps a thousand times greater than that needed to supply the amino acids from a peptide so as to maintain the maximum rate of protein synthesis. But that doesn't really relate to the rate of transport.

Gardner: So those are two slightly different but not unconnected issues This brings us back to the question of concentration gradients across the luminal membrane and to the energetic question. Although high intracellular peptidase activity inevitably produces steep downhill concentration gradients inside cells, I doubt whether this is energetically important. I feel that downhill transport of many molecules, probably including peptides, is an energy-requiring process. There are two provocative issues here. I suggest that it is transport which is rate-limiting.

Silk: To a large extent, workers in the field of disaccharide transport have been fortunate as there seem to be no intracellular disaccharidases, which means that the dimer sugar is exclusively hydrolysed at the cell surface. The problem that has been perplexing me for some time is that there may be more than one mode of uptake of dipetides. On the one hand we are now convinced that dipeptides can be transported into the mucosal cell intact. On the other hand we have also heard that, at least in certain instances, superficial hydrolysis of dipeptides precedes the transport step. The kinetic constants for luminal disappearance of dipeptides *in vivo* are not only high (Adibi & Soleimanpour 1974), but seemingly greater than the kinetic constants derived for the superficially oriented brush border enzymes (Kim & Brophy 1976). Would it confuse the field if I propose a hypothesis that intact uptake of peptides occurs only when the superficial brush border enzymes are saturated? Thus if these superficial enzymes have a high affinity for a particular peptide, perhaps the role of those enzymes is as important as that of brush border disaccharidases during disaccharide transport.

Josefsson: All those problems could be sorted out by using Gly-Pro or Gly-Gly-Pro peptides. We know that there is only one enzyme so far in the intestinal tract which can split amino acyl-proline peptide bonds, and that is the proline dipeptidase in the cytosol.

Silk: Unfortunately I don't believe that those are necessarily representative peptides. Furthermore, we have all been studying peptide transport using concentrations of individual peptide that vary from 2 to 100 mM. These are likely to be unphysiological concentrations and it is probably wrong to draw firm conclusions from such experiments.

Josefsson: You could do a transport experiment at normal physiological concentrations with a peptide which you know is split by a superficial membrane-located peptidase and compare it with one that is not split by these enzymes, but only by an enzyme located in the cytosol.

Adibi: The concentration required to saturate the transport system *in vivo* is much greater than the concentration required to saturate the membrane enzyme. I do not know whether the K_t determined *in vitro* would have any meaning for something that continues to work dynamically *in vivo*.

Radhakrishnan: We do not have enough data on the membrane enzymes to discuss the problem of rate-limiting steps. If you accept the kind of activity that we find for peptide hydrolysis in the cytosol, I don't think there is any rate-limiting step to discuss. The cytosol enzymes would chop off any peptide that entered the cell before we knew what was happening.

References

ADIBI, S. A. & SOLEIMANPOUR, M. R. (1974) Functional characterization of dipeptide transport systems in human jejunum. *J. Clin. Invest. 53*, 1368-1374

CHRISTENSEN, H. N. (1949) Conjugated amino acids in portal plasma of dogs after protein feeding. *Biochem. J. 44*, 333-335

DAWSON, R. & PORTER, J. W. G. (1962) An investigation into protein digestion with ^{14}C-labelled protein. 2. The transport of ^{14}C-labelled nitrogenous compounds in the rat and cat. *Br. J. Nutr. 16*, 27-38

DENT, C. E. & SCHILLING, J. A. (1949) Studies on the absorption of proteins: the amino acid pattern in the portal blood. *Biochem. J. 44*, 318-333

FOLIN, O. & BERGLUND, H. (1922) The retention and distribution of amino acids with especial reference to the urea formation. *J. Biol. Chem. 51*, 395-418

GANGL, A. & OCKNER, R. K. (1975) Intestinal metabolism of plasma free fatty acids. Intracellular compartmentation and mechanisms of control. *J. Clin. Invest. 55*, 803-813

GARDNER, M. L. G. (1975) Absorption of amino acids and peptides from a complex mixture in the isolated small intestine of the rat. *J. Physiol. (Lond.) 253*, 233-256

GRUBER, K. A., STEIN, S., BRINK, L., RADHAKRISHNAN, A. N. & UDENFRIEND, S. (1976) Fluorometric assay of vasopressin and oxytocin: A general approach to the assay of peptides in tissues. *Proc. Natl. Acad. Sci. U.S.A. 73*, 1314-1318

HANSON, P. J. & PARSONS, D. S. (1976) The utilization of glucose and production of lactate by *in vitro* preparations of rat small intestine: effects of vascular perfusion. *J. Physiol. (Lond.) 255*, 775-795

KIM, Y. S. & BROPHY, E. J. (1976) Rat intestinal brush border membrane peptidases 1. Solubilization, purification and physicochemical properties of two different forms of the enzyme. *J. Biol. Chem. 251*, 3199-3205

LEVENSON, S. M., ROSEN, H. & UPJOHN, H. L. (1959) Nature and appearance of protein digestion products in upper mesenteric blood. *Proc. Soc. Exp. Biol. Med. 101*, 178-180

MATTHEWS, D. M. (1975) Intestinal absorption of peptides. *Physiol. Rev. 55*, 537-608

PARSONS, D. S. (1972) in *Transport Across the Intestine* (Burland, W. L. & Samuel, P. D., eds.), Williams & Wilkins, Baltimore

PROCKOP, D. J., KEISER, H. R. & SJOERDSMA, A. (1962) Gastrointestinal absorption and renal excretion of hydroxyproline peptides. *Lancet 2*, 527-528

SLEISENGER, M. H., PELLING, D., BURSTON, D. & MATTHEWS, D. M. (1977) Amino acid concentrations in portal venous plasma during absorption from the small intestine of the guinea-pig of an amino acid mixture simulating casein and a partial enzymic hydrolysate of casein. *Clin. Sci. Mol. Med. 52*, 259-267

SMYTH, D. H. (1963) Intestinal absorption, in *Recent Advances in Physiology* (Creese, R., ed.), Churchill, London

STARLING, E. H. (1906) *Recent Advances in the Physiology of Digestion*, Constable, London

STEIN, W. H. & MOORE, S. (1954) The free amino acids of human blood plasma. *J. Biol. Chem. 211*, 915

WACKER, H., LEHKY, P., VANDERHAEGHE, F. & STEIN, E. A. (1976) On the subunit structure of particulate aminopeptidase from pig kidney. *Biochim. Biophys. Acta 429*, 546-554

YOUNOSZAI, M. K. & CYNEK, A. (1975) *In vivo* D-glucose absorption in developing rat small intestine. *Pediatr. Res. 9*, 130-133

Closing remarks

D. S. PARSONS

Department of Biochemistry, University of Oxford

I think we can all agree that in 1976, the beginning of the last quarter of the twentieth century, the concept of the transport of peptides across membranes is alive and well and the transport itself is flourishing in a wide variety of living cells. The process has been observed in a variety of sites which range from the scutellum of germinating barley to the mammary gland of the blue whale and which, in addition to the intestine, include kidney, liver, brain, muscle, erythrocytes, ascites, tumour cells, fibroblasts in tissue culture and possibly the yolk sac.

As far as the intestine is concerned, we have heard that the absorption and digestion of peptides has been studied in the pig, the monkey, rat and guinea-pig, in fetal, newborn and suckling rabbits and in adult Europeans and Indians.

The nomenclature of the peptidases has, I believe, been greatly clarified during this meeting. It is now clear that the answer to the question 'When is a dipeptidase not a dipeptidase?' must be 'When it also hydrolyses tri- and higher oligopeptides', for the true dipeptidase exhibits no activity towards the higher oligopeptides. Another point about the aminooligopeptidases, as they should properly be called, is that specific enzymes are named after the substrates towards which they exhibit high activity. Thus we have heard of glycylleucine dipeptidase and of proline dipeptidase. Exciting advances have been made in elucidating the composition and molecular architecture of some of the peptidases present in brush border membranes. These are evidently glycoprotein molecules that span the membrane, the two hydrophilic ends being separated by the hydrophobic centre that is embedded in the lipid phase of the membrane. It is rather remarkable that the kidney brush border peptidases, which are endopeptidases, account for about 10% of the protein in the membranes, so that there must be 400–500 peptidase molecules in each microvillus.

A very interesting question is 'What exactly is the relationship between the molecules that constitute the "peptidases" that are present in the brush border, and the molecular mechanisms underlying the propulsion of intact peptides across the plasma membrane?' However, this point and the nature of the mechanisms underlying this propulsion and the sources of energy available await elucidation.

A point of fundamental importance is that glycylglycine and glycylleucine disappear at prodigiously high rates after they have been introduced into the blood-stream; this disappearance is not due to excretion by the kidney. The peptides are taken up by the tissues; those that are not hydrolysed, such as glycylsarcosine, actually accumulate within the tissue cells. It seems possible that the transport of amino-N in the blood, to and fro between various tissues including the intestinal tract, occurs not only as amino acids but also to a significant extent as peptides. The widespread distribution and nature of systems for the uptake of peptides in animal tissues will, I am sure, be an important topic at future symposia. It should not be forgotten that such uptake processes were first described in microorganisms.

I would now like to exercise what might perhaps be called my 'droit de Président' and say something about the nomenclature of peptide uptake and digestion.

Peptide uptake (peptide absorption)

It seems that peptide uptake (or absorption) can occur in one of two ways, or by a mixture of both. In the first way the peptide is hydrolysed outside the cell, or on its surface (see below), and the uptake occurs in the form of the free amino acids. In the second way, the peptide is taken up into the membrane, before hydrolysis, by systems that are distinct from those for the uptake of the constituent amino acids. In this case, after uptake, the peptide may be subjected to hydrolysis, possibly within the membrane (but see below) or after expulsion from it into the cell interior. Such a definition of peptide uptake is really an operational definition because there is a quite straightforward experimental test of whether it is working in a particular case. The test is to expose the cells to a concentration of the constituent amino acids that is sufficiently high to saturate the uptake in the form of free amino acids. If the peptide is then added and there is a further uptake of amino-N, then there is a peptide uptake system.

Peptide hydrolysis (peptide digestion)

In spite of some of the powerful arguments elegantly adduced earlier today, I still think that the key to any classification of peptide digestion lies in the nature of the answer to the question 'Where is the active site of the molecule that is the peptidase?' In the old days one would have used the classification that was based upon the location of the enzymes, but we now know that the molecules of the peptidases span the membranes so that in each case one end is outside the cell, another is inside and the rest—the bit in the middle—is embedded in the membrane. I believe that it is this that makes a classification based on the location of the active site necessary.

Bearing all this in mind, the classification will then go like this:

Extracellular digestion. Here the enzymes, and therefore the active sites, are present in solution in the extracellular fluid.

Cellular digestion. Here the enzyme, and therefore the active site, forms a part of or is attached to the cell. By 'part of', I mean that the enzyme is synthesized in the cell and forms part of its structure, for example as a component of a membrane, etc. By 'attached', I mean that the enzyme is adsorbed to the cell surface, the molecule having been synthesized elsewhere. For example, it might be formed in the pancreas and become attached (adsorbed) to jejunal mucosal cells, as Professor Ugolev has suggested.

Cellular digestion can be divided into membrane digestion and intracellular digestion. Dealing first with intracellular digestion, this can be subdivided in various ways that I do not propose to go into. The important feature of intracellular digestion, according to the classification that I am developing, is that the enzymes are not an integral part of the plasma membrane (brush border) and that the active sites are therefore not associated with that membrane.

Membrane digestion. I suggest that if the active site is located on the outside of the membrane, then the digestion can be called surface digestion. This will be true whether the enzyme to which the active site belongs is an integral part of the membrane, or is of exogenous origin and merely adsorbed to it. If the active site is deep within the membrane, then the process can be called intramembrane digestion. If the active site is at the inner face of the membrane, I suppose the system could then be called inner surface digestion. Where the active site is embedded within the membrane, a nice point arises if it is found that the co-substrate for hydrolysis (water) is derived primarily from the outside or from the inside of the cell.

You will appreciate that in deriving this classification, I am attempting to be forward-looking to the time when the molecular architecture of all the hydrolases and, in the appropriate instances, the mode of insertion into the membrane are fully understood. I also hope that the suggestions are sufficiently provocative to make people go off and do experiments to show the unsatisfactory nature of the classification.

Finally, I am sure that as a result of what we have heard about peptide transport and hydrolysis, the members of this symposium will all leave bursting with new ideas for more experiments.

Index of contributors

Entries in **bold** *type indicate papers; other entries refer to discussion contributions*

Indexes compiled by William Hill

Subject index